ELEMENTS OF RISK

ELEMENTS OF RISK

THE CHEMICAL INDUSTRY AND ITS THREAT TO AMERICA

Cathy Trost

Times
BOOKS

Published by TIMES BOOKS,
The New York Times Book Co., Inc.
130 Fifth Avenue, New York, N.Y. 10011

Published simultaneously in Canada by
Fitzhenry & Whiteside, Ltd., Toronto

Copyright © 1984 by Cathy Trost

All rights reserved. No part of this book may be
reproduced in any form or by any electronic
or mechanical means, including information storage
and retrieval systems, without permission in writing
from the publisher, except by a reviewer who may
quote brief passages in a review.

Library of Congress Cataloging in Publication Data
Trost, Cathy.
  Elements of risk.

  Includes index.
  1. Chemical industry—Environmental aspects—United
States.   I. Title.
TD195.C45T76   1984   363.7′384′0973   83-45927
ISBN 0-8129-1114-8

Designed by Giorgetta Bell McRee/Early Birds

Manufactured in the United States of America

84 85 86 87 88   5 4 3 2 1

For their advice and support, many thanks to: Paul Magnusson, Raphael Sagalyn, Jonathan Segal, Ruth Fecych, Elizabeth Haskell, Robert Haskell, Dolly Katz, Carol Morello, Ellen Grant, Bill Grant, Deborah Blum, Susan Sward, Frank Greve, Bill Hart, Kate McCarry, John Hyde, Louise Swartzwalder, Cecilia Trost, Lori Magnusson, Betty Magnusson, Warren Magnusson, Merrilee Trost, and Cecil Trost; and special thanks to the Alicia Patterson Foundation.

In completing one discovery we never fail to get an imperfect knowledge of others of which we could have no idea before, so that we cannot solve one doubt without creating several new ones.

—JOSEPH PRIESTLEY
*Experiments and Observations
on Different Kinds of Air* (1775–86)

ELEMENTS
OF RISK

chapter 1

When the timber harvest ended in Michigan, and the world's largest stand of white pine was tapped clean, and the lumberjacks who once walked seventy miles from Edenville to Saginaw on a river paved with logs had moved farther west, twenty houses burned down in Midland in one week. Fires were set for the insurance money, and when the insurance companies began to leave town, the villagers were gripped by rumor and hysteria. People said, "If you're going to burn out, better do it now." The town that had beaten back a wilderness at the junction of the Tittabawassee and Chippewa rivers was sunk in despair. "Midland had eaten the cake of its natural resources," an observer wrote. "It was a town all but sacked by its own population."

Michigan's Green Rush had been $1 billion more valuable than California's Gold Rush, and for a time Midland was invested with the spirited defiance that comes with money. One hundred sawmills lined the Saginaw River, and fourteen saloons lined Midland's Main Street. The town lit up on Saturday nights, when the Shanty Boys and the River Hogs rode in from the outlying lumber camps to drink and brawl. Eye gouging and ear chewing were popular recreations. If a lumberjack said he'd put the boots to someone, it meant that he'd jumped up and down on the fellow with his feet.

The pine board sidewalks, dented by their heavy boots, stood empty in the silence of the lumberjacks' retreat. From Great Lake to Great Lake, in abandoned towns pockmarked by pine stumps, in a region tormented by long and bitter winters, those too weak or conniving to follow the lumbermen west stayed behind to fight fire and ice. Midland was no different, but it was there that a young schemer by the name of Herbert Henry Dow plotted among the dreamless.

Dow arrived in Midland on August 14, 1890, with $100 in cash and $275 borrowed from a family friend. Electricity had come to Detroit, 127 miles to the south, only four years before, and Henry Ford would not build his first car for another three years. Dow was a twenty-four-year-old college chemistry teacher from Cleveland whose mustache and beard put people in mind of Union General Ulysses S. Grant. On August 16 he deposited $275 in the bank and began boarding at the Antler House. When the locals learned that he wanted to invest his money in the ground, they took to calling him Crazy Dow.

But Dow knew that Midland was sitting on an 1,800-foot-thick bed of pure rock salt. Men had been pumping brine from the prehistoric underground sea and evaporating it by heat from their waste lumber to make salt for a dozen years, but the take was small (46 cents a barrel), and fuel was getting costly. The villagers were bred in lumber and salt and had limited vision. Herbert Henry Dow was bred in chemical research, and he saw endless possibilities in a ribbon of brine. He knew that Midland's brine was rich in bromine, and he knew that bromine was ten times more valuable than salt for the money it would bring from sales to patent medicine and photographic film manufacturers. Dow had devised a way to separate bromine from the brine deposits, not by the slow and wasteful method of evaporation but by the more efficient use of electricity. "The populace frankly called this man a fool," wrote Michigan historian Laurence A. Conrad. "He was, in fact, suspected, distrusted, and genuinely despised. The town's darkest years were his also."

Dow bought an old brine well at the west end of Main Street and began extracting bromine from the brine. He struggled to make his schemes work. He was obstreperous and tight with a dollar, and he could be diverted from his calculations and inventions only by the manipulations of chess. After a time he also became susceptible to the attentions of Grace Anna Ball, a teacher at the Post Street School, whom he married in 1892. In the same year, with the assistance of his family friend, he founded the Midland Chemical Company. When he discovered that the brine contained more chlorine than bromine, Dow expanded. An hour after he had thrown the switch in his new chlorine plant, it blew up. There were no injuries, but his machinery was destroyed. When his board of directors voted to close the chlorine plant, Dow moved back to Ohio for six months to refine his plans. An explosion in his makeshift chlorine plant there knocked him unconscious, and he was dragged to safety by a workman. Undeterred, Dow returned to Midland and, with $200,000 put up by fifty-seven investors, founded the Dow Chemical Company in the spring of 1897. The stock was divided into 20,000 shares at $10 a share. Dow was general manager and chief stockholder. One of the earliest investors was Hetty Green, a millionaire so miserly with her fortune that she lived in Hoboken in a rented $15-a-month flat.

The Dow Chemical Company was not an overnight success. At one point it was losing $163 a day. But the bromides and bleach fashioned from the bromine and chlorine in the brine eventually began paying the bills. Bleaching powder was much in demand by textile and cotton mills. But explosions continued to rock the plant, and chlorine gas so permeated the building at times that workers tried to repel the fumes by pressing bottles of grain alcohol to their noses. Such problems were not Dow's alone; they could be found almost anywhere business was being conducted in a country that was moving forward so purposefully it did not know the meaning of "reverse." Across the Great Lakes in Niagara Falls, New York, workers in the bleach chamber at the Hooker Electrochemical Company wore hand-whittled wooden frame goggles and wrapped wet flannel around their faces to filter out the chlorine gas. They worked twelve-hour shifts, for 15 cents an hour, and it is said that new employees "often quit before the timekeeper had made out their cards."

With the passing of the lumber kings and the coming of the chemical and mineral barons, the air and rivers in factory towns turned corrosive with progress. Dow Chemical dumped its wastes out the back door and into the conveniently located Tittabawassee River. The Chippewa Indians had called it the shining river, but it soon lost its luster. "We have thousands of dollars, good dollars, running into the river every year," lamented Herbert Henry Dow, who didn't like to see anything go to waste. When he expanded into the chloroform business, fumes from the plant caused people living downwind to vomit, their vegetables and shrubs to die, and their kitchen utensils to rust. Two lawsuits were brought against Dow by neighboring residents beginning in 1900. "It was charged that his chemical plant had seriously depreciated property values in the town and that the stinks he made were a menace to the public health," observed the *Michigan Manufacturer* in an article some years later describing how Midland came to be "a center of scientific fabrications." The article stated: "Individual citizens attached themselves to the suits, claiming damages for all sorts of absurd reasons. Either one of the lawsuits would have ruined his business utterly; but neither one was successful."

Dow and his company prevailed, fighting both the battles of science and their neighbors, and Dow kept at the brine until he mastered it completely, extracting from it not only bromine, chlorine, and sodium but magnesium and calcium, too. Dow liked to say that his success in life was partly the result of his being right 52 times out of 100, and wrong only 48. But scientific research with no immediate commercial application was considered a waste of time by many boards of directors in the early 1900's, and Dow's penchant for experimentation did not sit well with some of his own stockholders. "Mr. Dow no doubt has a gift of discovery," one disgruntled investor wrote. "He is an enthusiast along that line, but not a safe man to expend the money of poor widows and orphans." But his business continually grew, branching out into synthetic dyes, chemical fertilizers, food preservatives, solvents, and workhorse chemicals like caustic soda—most of them supplied in bulk to other chemical companies, which were springing up across the country in the new century.

Du Pont had built a gunpowder factory on a Delaware riverbank in 1802, and from there it grew into the country's largest chemical company. Later in the century Schoellkopf Aniline & Chemical Works—predecessor to the Allied Chemical Company—was founded in Buffalo, New York. One year after the century turned, John F. Queeny started a saccharin business in St. Louis with $1,500 of his own money and $3,500 in borrowed funds and called it Monsanto, after his wife's maiden name. In 1905 a politically connected civil engineer named Elon Huntington Hooker, who later served as treasurer of Theodore Roosevelt's Bull Moose party, founded the Hooker Electrochemical Company in a pear orchard in Niagara Falls, New York. In 1907 another civil engineer bought the American rights to a German process for making cyanamid, a fertilizer component, and started American Cyanamid. Shell Chemical and Occidental Chemical sprang from oil companies of the same names, which were founded in 1912 and 1920 respectively. War did not create the chemical industry, but the industry used war as a feedstock to help it grow. From a handful of companies turning out explosives and artificial dyes, the industry became a set of giant corporations that supplied raw materials to power every other major industry in the nation and made finished goods found in some form or another in every American home.

In the early days the industrial pioneers were preoccupied with breaking the German domination of the world chemical market and, in some cases, shaking the eccentric image handed down by their ancestors. The man who protected a courtful of astrologers, artists, and alchemists in sixteenth-century Prague and who believed that base metals could be turned into gold and a drug to confer immortality could be invented was known as the mad Holy Roman Emperor Rudolf II. The people of Midland, confronted three centuries later by a man who believed that salt could be turned into photographic film by way of electricity, could not be faulted for thinking that Mad Rudolf and Crazy Dow were bowling in the same league.

Once the chemical industry had disengaged itself from another superstition—the widely held nineteenth-century theory that organic compounds found in animals and vegetables were governed by special "vital forces" outside the realm of chemistry, which made them impossible to duplicate in the laboratory—the real business of chemical production began. "I must tell you that I can make urea without the need of kidneys or of any animal whatever," the German Friedrich Wöhler wrote another chemist in 1828, after he had transformed salt and ammonium cyanate into urea, a compound found in animal and human urine. It was the first production of an organic compound from inorganic materials, but the vital force theory was not entirely abandoned until midcentury, when a German scientist synthesized acetic acid identical to the acid that occurs as one of the products of distillation of wood. The age of organic chemistry burst open as scientists began isolating new compounds from plant, animal, and mineral sources and manipulating them into substances never before seen on earth. Coal tar became the industry's principal raw material as scientists learned how to turn it into synthetic dyes (aniline), perfumes (heliotrope and coumarin), flavors (vanilla), sugar substitutes (saccharin), drugs (sulfa), explosives (picric acid and TNT), and plastics. Some of the excitement of the time was expressed by a German chemist who said:

> Go out on a starry night and you realize how infinite is the number of stars that are visible. Suppose each star is an organic compound. Then suppose you were to go to any one of those stars and imagine the infinite number of new stars that would then become visible. This in a way illustrates our conception of the possibilities of the multiplication of organic compounds.

The Germans were distracted by World War I and lost their coveted supremacy in coal-tar chemicals to the Americans, who poured research and development money into the growing industry. So did the government, which contracted with Dow for 15 tons of picric acid for explosives and 10,000 pounds of deadly mustard gas a day. The Dow staff developed much of the military's information on the manufacture of mustard gas. But the company was also supplying bromine for sedatives and acetylsalicylic acid for aspirin. Dow had become, by 1918, a "virtual arsenal," with nearly its entire chemical output going to the government. Midland was so crowded with Dow workers during the peak war production years that the company subsidized the Stag Hotel, a boardinghouse that offered 120 "clean and comfortable beds" for $6.50 a week, and helped finance the building of houses for employees.

One of Herbert Henry Dow's many preoccupations was with phenol, one of the new coal-tar chemicals, which was used as an antiseptic for a time and which became an ingredient of the pioneer plastics. Dow designed a new process to make phenol, and when his factory was pressed into wartime service to supply phenol-based picric acid for explosives, he soon found himself running the largest synthetic phenol plant in the world. It was the heyday of organic chemistry, and whole

Elements of Risk / 7

classes of chemical products were being created from the mixing and matching of compounds with one another. Dow was competitive, and he pitted his researchers against one another as if they were thoroughbreds in a horse race, sometimes putting several people to work on the same problem with the admonition that cost was not as important as quick results. That was a policy partly generated by Dow's philosophy that "the first firm to make a success of any given line of manufacture has a distinct advantage." But he was not a risk taker unless convinced that the results would show a profit, and he kept a tight rein on finances. Executives and factory workers alike were required to punch a time clock.

By the 1930's the chemical industry had overcome its postwar slump to rank among the fastest-growing industries in America. From a few struggling private companies at the turn of the century it had become, by 1935, twenty-six separate concerns with total assets of $1.7 billion. Many had merged during the postwar years to form large, publicly held corporations. The industry's new base was petrochemicals, the hydrocarbons found in natural gas and oil, which form the building blocks of a staggering number of chemical products. The chemical industry was hyped as breathlessly then as the computer industry is today. BIG FIELD AWAITS YOUNG CHEMISTS was a headline over a typical newspaper account of the "important part now played by chemicals in many phases of daily life, including food, housing, lighting, heating, transportation, wearing apparel, drugs and perfume." According to the *Detroit Free Press,* "Attractive prospects" awaited the "young man or woman looking for a niche in America's complex and highly competitive industrial set up."

Herbert Henry Dow died of cirrhosis of the liver in the first year of the Great Depression, 1930. Just a few months before his death Dow had been awarded the industry's prestigious Perkin Medal for the part he played in creating a national chemical industry out of brine and electricity. When he died, Dow left to his thirty-three-year-old son, Willard, a company the sales of which had multiplied nearly four times during the previous decade and the stock of which split four for one and then peaked at $500 a share. To his workers, Dow left an unshakable belief in the certainties of science and the redemptive powers of technology and a share of the good life in a company town that was without rival. It was a philosophy molded by the wonders he had witnessed in his own life, which he attributed to industrial progress, by which he meant mass production. He once marveled to a childhood friend that whereas men had once been forced to work from sunrise to sunset to provide for their families, now a man needed to work only "44 or 48 hours a week" to obtain "luxuries that were not dreamed of when we were boys."

Success and his family (he had seven children) softened Dow's rough edges over the years, and he took time off for a few of his devotions. Inevitably his devotions benefited either his company or his adopted town. He was an avid fruit tree grower and planted an orchard of 5,000 trees, with exotic varieties of apples, pears, and plums. Dow spent long hours experimenting in his orchard, so much experimenting, some said, that the orchard never lived up to his expectation that it would produce enough to pay its own way. After he had met horticulturist Luther Burbank in California in 1905, Dow put his chemists to work concocting insecticide sprays and soil nutrients. Two years later Dow marketed its first farm product—liquid lime sulfur for the control of orchard pests—and the company subsequently became a leading manufacturer of agricultural chemicals.

Dow's affection for gardens infected the whole town, and for a time he served as Midland's superintendent of parks. He hired a famous Tokyo landscape artist to duplicate Japanese gardens at his house in Midland and then offered the artist's services free of charge to any Midland resident who wanted them. He paid to have a new fifty-six room modified Tudor-style county courthouse designed, helped finance its construction, and commissioned artists from New York and Detroit to paint across its flanks brilliantly colored murals (fashioned from plastic magnesia stucco which Dow produced from brine). Dow founded a country club and a community center and kept the town's churches painted at his own expense. At his company Dow established one of industry's first profit-sharing plans. Midland could not help flourishing under Dow's benevolent eye. It grew from a tumbledown lumber town into a thriving city of 10,000 residents, most of them employed by Dow and all of them blessed with such a generous benefactor that during the Depression, which swamped Detroit in poverty and fatigue, Midland was famous as "the town that doesn't know there's a Depression going on." The official Midland County seal acknowledged the roots of its good fortune: Inlaid in its design was a pine tree, a brine well, and a horn of plenty.

A few years after Dow's death, while the rest of the country was climbing out of its slough, Midland residents were building 500 new houses a year. A staggering 72 percent of them owned their own homes. More than sixty were designed by Dow's second son, Alden, an architect of some renown who had studied under Frank Lloyd Wright and who favored flat roofs, modernistic angles, and great expanses of glass and plastic. Alden Dow built his own home on a pond in his father's gardens (which were donated to the city) with a studio that sat underwater, 100-foot glass walls, and a gabled living room ceiling covered with basket-woven Ivory Ethocel made by Dow and chemically treated so that it glowed like moonlight at night. The Dow family's generosity spread to Midland. Over the years, through direct gifts from individual members of the Dow family and a $325,000 trust fund established in the 1930's by the founder's wife, Grace Dow, Midland acquired a new hospital built on a forty-acre wooded tract donated by Mrs. Dow and designed by her son Alden; a $1.6 million Presbyterian church with thirteen-ton pillars cut from Indiana limestone and a bell tower painted in gold leaf financed by the Dow vice-president who invented Saran (as in Saran Wrap) in honor of his sister; a $1.5 million Community Center, designed by Alden Dow, where a bride could rent a mirrored ballroom for her wedding reception for $2 an hour; the Alden Dow-designed $1.25 million Grace A. Dow Memorial Library; an $8.6 million Center for the Arts, partially financed by Dow and designed by Alden Dow with an auditorium, a theater, and a "Hall of Ideas"; the Dow Symphony Orchestra; a riverside parkway and a county jail; a residential subdivision, carved out of the Dow family orchards, complete with a man-made lagoon for canoes and powerboats, the lots of which were sold for $5,000 and up, with proceeds going to the city; a softball stadium on Dow-donated riverfront land where the Dow Chemicals played ball; an eighteen-hole golf course where anyone could play for a 60-cent greens fee and an Olympic-size pool where anyone could swim for 10 cents; and a Lutheran church extending outward in eight sections from the altar, which architect Dow described as symbolic of Midland's growth. "Growth is unlimited, as long as it radiates from the principles of God," said Alden Dow, whose influence was visible even at the Dow plant, where steel supporting lines were painted red and yellow as a balance to the blue sky, and the chemical holding tanks were painted in shades of green, yellow, and blue-green to offset the brown soil. Alden Dow also designed Discovery Square, which was the confluence of the library, the arts center, and the gardens. To Dow's way of thinking, the square represented the "important relationship between science and the arts: fact and feeling, both necessary ingredients for quality in life." Every Christmas Grace Dow had the courthouse strung with 6,000 colored lights in honor of her husband, and tourists came from all over Michigan to see the display. "Like that aunt for whom it is so hard to buy a Christmas present—because she has almost everything," said one writer of Midland.

The Dow family's politics (conservative) and religion (Presbyterian) became the town's politics and religion. Partly because of the Dow influence and backed by popular vote, Midland was a town where no hard liquor was sold by the glass after 1908.

But neither Dow nor his successors believed in free rides, and almost everything Midland got, Midland had to help pay for. Dow gave the city a new library on condition that the city pay for the annual upkeep.

And the city fathers—declaring that Midland had a moral obligation to make its own way—rejected federal aid for a sewage disposal plant and barred school cafeterias from participating in state and federally subsidized lunch and milk programs.

The city was special in another way. It was filled with young, upper-middle-class, highly educated people who were drawn to work in Dow's laboratories and to live in Dow's marvelous town. In the late 1930's a newspaper writer described Midland as "Michigan's different city":

> Erudite doctors of philosophy—there are estimated to be 100 of them in this city of 10,000—go to work in blue work shirts.
>
> The country club has no bar.
>
> About 1,000 residents are college graduates.
>
> At 4 years of age, children can pronounce Tittabawassee River and at 6 such words as phytohormones and polystyrene are a part of their vocabulary.
>
> When you enter a person's home you sniff and the expert can tell in what department his host works—indigo, carbon tetrachloride, wintergreen or whatnot.
>
> The city has six residents listed in Who's Who. The average city of 75,000 isn't at all ashamed if it has two.
>
> The Dow Chemical Company employs about 500 scientists and technicians . . . many look like college boys and nowhere will you see more adept presentations of the Big Apple and the Susy-Q than in Midland.
>
> With most of the city's residents under 35, the city is strong for sports, games and entertainments.
>
> And the city's residents actually like those chemical smells which float over the town—they mean prosperity.

By the end of the 1930's Dow was the country's fastest-growing chemical company, and fifth largest, behind Du Pont, Allied, Union Carbide, and American Cyanamid. Like his father, Willard Dow believed that research was the key to the company's progress, and he set a course that was as arrogant as it was forcefully forward. He urged his scientists not to preoccupy themselves with what the competition was doing because it would hamper "our own original thinking." He had no patience for "always having to find out what the literature says before we attempt to carry out a development of our own." His synthesis and research chemists were pushed hard to produce new compounds and to refine those compounds into new products. Like traffic cops with quotas to fill, the worth of a Dow scientist was often measured in the quantity of compounds he created and the number of patents he obtained. In the 1950's a Dow laboratory technician wrote a poem dedicated to the fecundity of a colleague. Its spirit was playful, but the verse speaks to the heart of the Dow philosophy of science and progress.

> *There was a chemist named Ray*
> *Who never went in for horseplay*
> *He turned the crank,*
> *With never a prank,*
> *And out came two compounds a day.*
> *"You'll find there's plenty of hay,"*
> *Says this chemist named Ray,*
> *"If you show your mettle,*
> *Push hard on the pedal,*
> *And turn out two compounds a day."*

Raymond "Ray" Rigterink created more than 2,000 new compounds that led to major new products for Dow's agricultural business, and he was listed as sole inventor on forty-seven U.S. patents. His work lent credence to Dow's view that the tradition of research, fostered by scientific freedom and encouraged by the profit incentive, was the strength of America. Like many conservatives, Willard Dow began to feel that too much power was being concentrated in Washington during President Franklin Roosevelt's New Deal years. He was one of the first industrialists to warn of "big government," and he began actively to promote a theory of business self-reliance as counterpoint to the encroachment of the regulatory state. It was a time when the chemical industry operated with little long-range concern for the effects of its products on humans and the environment and with virtually no laws forcing it to test its products for safety or to pass on warnings to workers who made the products and customers who used them. But Dow was as sensitive to the threat of regulation as to the reality. His view was refined by his successors, but it essentially contained three major points: An ambitious, profit-minded industry was the engine that drove the development of America into a great nation; technology was providing Americans with a better life and a better future; and the surest way to destroy that better life was to bite the hand that fed it by overregulation.

Dow believed the best regulatory policeman was an in-house patrol, and so in 1933 he hired a biochemist who eventually set up a toxicology laboratory. But when Dr. Don Irish first came to work at Dow, his job parameters were rather vague. "To hear Don tell it," recalled his colleague Dr. Theodore Torkelson, "he said, 'I came to work and Willard Dow told me to find something to do.' " The toxicology lab itself was organized after workers had been fatally exposed to a chemical. "Ultimately there was an accident, and a couple of people became sick and I guess ultimately died from an exposure . . . that should not have been a problem," Dr. Torkelson remembered. "And Irish went around and did toxicological experiments on skin absorption with rabbits and was able to identify that this stream had gotten phenol in it, which was known to cause problems but which hadn't been suspected in this case. That is when toxicology got started at Dow. Because the instruction from Willard Dow was something to the effect of 'Don't let it happen again.' "

Dr. Irish's first laboratory was in the corner of a converted garage. The rabbits he purchased from neighboring farmers for experiments were housed in a construction shanty next door because they smelled bad. His job was to study the effects of the company's chemical products on animals and try to predict their effect on humans. It was a shoestring operation—he conducted his first vapor studies by pouring chemicals into a barrel and lowering caged animals inside it—but it nevertheless predated similar efforts by other chemical companies by as many as three decades. If a company did toxicological studies at all in the 1930's and 1940's, the work was usually hired out to universities or consultant laboratories. Dow, Du Pont, and Eastman Kodak were among the few companies with in-house labs. "Everybody was at the bottom end of the learning curve—nobody had any experience—and if somebody developed a test for studying skin absorption, they'd share it with everybody else. Research was pretty freewheeling, very undirected. It was kind of the glory days of scientific research when you could do almost anything," Dr. Torkelson recalled. "And so the science of toxicology evolved from essentially nothing over the years."

As Dow's operation grew more sophisticated, toxicological information was used to compose warning labels for the products and to safeguard workers who were exposed to the chemicals during production. That was the theory; in practice, it often failed. Because toxicological testing was as simple as the times themselves, a chemical had to be quickly and noticeably poisonous to warrant a second look. "If it didn't kill you immediately, then it was okay," recalled Dr. Wendell Mullison, a research chemist who joined Dow in 1946.

"The label at that point in time had a different purpose," Dr. Torkelson said. "The industrial chemical—like a solvent—was labeled with very little information. It said virtually nothing. This is the very early stages. Then later on we started labeling 'Vapors hazardous, Avoid skin contact.' But these were primarily designed for a single exposure. And that was primarily for a shipping accident. Because it was assumed that the fellow using it routinely would have other [safety] literature. Now, the agricultural label was a little different because it required a detailed description of how to use the chemical. Very early it had descriptions of toxicity and handling precautions. But again it was based almost entirely on single exposure and acute effects. Just up until a few years ago the label did not even discuss or consider chronic toxicity. That was handled in other literature, like product brochures and formulators' manuals."

The toxicologist's job was to protect human health, but he sometimes butted heads with people in his marketing department who thought severe hazard warnings might deter customers. Dr. Irish once recalled that the marketing people looked at a new, less toxic solvent in the 1940's and said, "Who will pay more because it is safer?" Dr. Torkelson insisted, "Once in a while our people with solvents would complain that we were trying to be too restrictive and would scare away the customer, but we never yielded to them because sooner or later we'd get into trouble if we did."

Willard Dow's aggressive leadership carried the company through the Depression and World War II and positioned it in a front-row seat for the astonishing era of discovery and growth that followed the war. He was an amiable and confident executive who, like most Dow men, wore glasses and dark suits and constantly educated himself, obtaining a science degree from the Michigan College of Mining and Technology and engineering degrees from the University of Michigan and the Illinois Institute of Technology, all of them after he had become president of the Dow Chemical Company. Like his father, Willard Dow liked to grow things. Herbert Henry Dow had his orchard. Willard Dow had a greenhouse on the second floor of his Midland home. Under Willard Dow's tutelage, Dow Chemical expanded forcefully into the area of agricultural chemicals.

Chemical control of insects was not new or particularly modern. Homer talked about using sulfur against crop pests in 1000 B.C., and the Chinese were using arsenic to control garden insects before the tenth century. But the widespread application to farming occurred only in the late nineteenth century, when growers used arsenical compounds, the nicotine in tobacco leaves, the ground flowers of chrysanthemum, paint pigments, and boric acid to combat pests. Once the fledgling chemical industry had teamed up with agricultural experiment stations to promote the use of chemical insecticides, their use soared, and growers increasingly became addicted to the primitive poisons that killed quickly and with a minimum of fuss. Because of the convenience of the new chemicals, the old laborious methods of crop rotation, mechanical removal, and the use of natural predators or resistant varieties of plants naturally fell into disfavor.

The development of synthetic organic insecticides and herbicides, or weed killers, which could be manufactured cheaply from abundant raw materials like the coal-tar phenols and petrochemicals, gave the chemical industry an awesome profit margin and a big incentive to sell them to growers. By 1948—one year before Willard Dow and his wife were killed in an airplane crash near London, Ontario—Dow Chemical was marketing more than fifty insecticides and herbicides.

Scientists at the Dow labs deciphered the code to a vast group of fungus-fighting chemicals, wood preservatives, and weed killers when they wedded together chlorine and phenol to form a new group of compounds called chlorophenols. The firstborn of this marriage were the Dowicides, introduced by Dow in the 1930's to keep a variety of materials, including lumber, book glues, and other textile, leather, and paint products, free of fungus and mold and so to preserve them.

The Dowicides that served as wood preservatives were essentially comprised of a grouping of chlorine and phenol molecules called pentachlorophenol, or simply penta. Dow promoted penta as "life insurance" for lumber. It was used to treat everything from the rides at Chicago's Riverview Amusement Park ("Thrills Without Spills") to the roof of the Minneapolis Courthouse to fence posts, utility poles, highway guard posts, railroad ties, and a sizable quantity of lumber that went into homes, carports, playhouses, toolsheds, and barns built across the country. In the months before Dow first described its new Dowicides in a company prospectus in 1936, several hundred lumber workers in Mississippi began developing severe skin rashes. By 1937 Dow's own workers in Midland, Michigan, were afflicted with the same problem.

Dr. Karl O. Stingily, a physician in Meridian, Mississippi, had seen the first of what would become "three or four hundred" cases of this baffling "new industrial chemical dermatitis" in the spring of 1936. Writing in the *Southern Medical Journal* in 1940, the doctor described the "peculiar type of pustular and ulcerative lesions" that afflicted lumber workers, most of whom were black men because "negro help is chiefly employed in this procedure."

An Alabama physician reported in the same journal that a man sent to him by a lumber company had acne and blackheads covering his face, back, shoulders, arms, and thighs. He was accompanied by his two children, a five-year-old girl and a three-year-old boy, both of whom had blackheads "all over their faces." The millworker told the doctor "that when he came home with his overalls on, the children would grab him around the legs and hug him and he would take them up on his lap."

Dr. Stingily found few clues in the medical literature to explain the disease, which typically progressed from ulcers to hard cystic lesions. The lesions might persist for months or even years, accompanied by leg cramps, thrombosis, and thickening of the skin.

Photographs showed the scarred arms and buttocks of a worker who had "backed up" against lumber wet with the antifungus solution.

Because it was "almost impossible to remove the worker from his source of infection, for economic reasons," Dr. Stingily tried to treat the lesions by scraping them and mopping them with acetic acid or vinegar baths. He concluded that the chlorophenol compounds used in the lumber industry were "toxic to persons who come in contact with the material for any length of time."

In 1937 twenty-one men who had spent time "sacking" powdered chlorophenol products at Dow's Midland plant came down with "acne-like" eruptions. Seventeen of the men, only one of whom was older than twenty-three, had outbreaks severe enough to require treatment. Dr. Milton G. Butler, a medical consultant for Dow, had the men removed from the plant and placed in other jobs while he investigated the disease. In a medical report published in the *Archives of Dermatology and Syphilology* in 1937, Dr. Butler described the "enormous number of comedones [blackheads] and numerous sebaceous cysts" that covered the men's faces, so numerous in some cases as to produce "a black discoloration." In addition, "secondary pustular lesions from the size of a pea to that of a chestnut were observed." He said the eruptions typically began forming behind the ears and then spread to the forehead, cheeks, chin, and back of the neck. Some men had lesions on the arms, buttocks, abdomen, thighs, penis, and scrotum.

Fifteen months after the workers had been removed from the plant, none had completely recovered. Besides "severe scarrings of the pitted variety," a number of men "lost weight and complain of being easily fatigued."

Dr. Butler became convinced that the chlorophenol compounds were the cause of the disease, which he thought fitted the description of something he'd seen described in the medical literature as chloracne, a term first used by a German physician to describe severe acne among workers in a chlorine plant in 1899. "These drugs should not be used as fungicidal agents until further investigation reveals the mechanism whereby they produce this dermatosis," he cautioned. Years later Dr. Butler told Harvard biochemist Robert Baughman that he had lobbied Dow for permission to do biopsies of the workers and animal experiments on the chemical in order to determine how it worked and whether it had caused damage to the internal organs. But he said Dow refused his requests for further studies, and after a time the plant reopened. Dow has no record of Dr. Butler's investigation or request for studies. Dow's own records show that twenty-one men who developed chloracne after working with the Dowicide P wood preservative were removed from exposure and offered biopsies, which they rejected.

The next expansion of the chlorophenols into commerce came with their application as weed and brush killers, or herbicides. Dandelions and crabgrass were a source of irritation to homeowners, especially as

credit: Paul Magnusson

Cathy Trost, author of
ELEMENTS OF RISK: The Chemical Industry and Its Threat to America

Times Books/130 Fifth Avenue/New York, NY 10011
212/620-5900

people began to move out of the cities and into suburbs, where lawns took on symbolic meaning to the "keeping up with the Joneses" generation. But weeds in general were an enormous problem for American industry. Farmers were losing an estimated $3 billion a year after World War II from weed damage to crops. European farmers had battled weeds with highly caustic chemicals like sulfuric acid since the late 1800's. American growers had little interest in the corrosive chemicals, which worked by burning or poisoning the portions of plant tissue they touched, because they were costly and did not give a consistent kill. Instead, the farmers tried to yank weeds out by hand or machine, burn them, or kill them with less caustic chemicals. Weeds were troublesome for railroad and utility companies and cattlemen, too. Brushy, brambly weeds along railroad tracks interfered with trains and had to be hand-pulled, set on fire, or doused with steam; the same methods were used against weeds that grew under power and telephone lines, interfering with power transmission and telephone impulses. The Northwestern Bell Telephone Company, for instance, was paying as much as $1 million a year during the 1950's for brush control in Iowa alone.

In the Southwest prairie fires were commonly started on rangelands to kill oak brush, which sucked the land dry of moisture and shaded the grasses that provided feed for livestock. In the Northwest timbermen hacked, bulldozed, burned, and sprayed scrub oak and other broad-leaved weeds and brush, which covered and deprived pine seedlings of sun and ate holes in valuable timber stands.

Few chemicals were capable of killing weeds selectively—which is to say, without injuring crops and grasses. Growers tried the caustic compounds like sulfuric acid; they tried sodium nitrate, iron sulfate, copper salts, even common table salt. They used flamethrowers "in crops sturdy enough to survive treatment." But nothing worked well.

It was not until researchers began synthesizing chemicals that regulated the growth of plants that science learned how to kill plants efficiently. About the same time that Herbert Henry Dow was investigating Midland as a site for his new chemical plant, the English botanist Charles Darwin was reporting his observation that plants always bend toward their source of light. Darwin believed that the plant tip sent to the lower part of the plant some substance that caused this phototropism. Scientists eventually found a light-sensitive "growth" hormone in the tips of seedlings that controlled plant growth by moving downward in the stems, increasing cell elongation. In the 1930's, nearly half a century after Darwin's breakthrough discovery, the growth hormones were synthesized in the laboratory and found to affect growth in the same way as the plant's own hormones.

Some of the most potent of the newly synthesized growth-regulating compounds were a group called the phenoxyacetic acids. One of these was known as 2,4-dichlorophenoxyacetic acid, or 2,4-D. Researchers reported in 1942 that 2,4-D was 300 times more powerful than another widely used synthetic plant hormone.

Some researchers found that the new chemicals would injure plant tissues if applied in too large a dose and would kill plants that were overdosed. But most botanists were thinking in terms of growth, not death, and only one of them looked at the discarded pile of overdosed plants with any interest. His name was Dr. E. J. Kraus, and he was head of the botany department at the University of Chicago. Dr. Kraus was the first to suggest that the growth regulators might work as herbicides if purposefully applied to weeds in toxic doses.

In the early 1940's Dr. Kraus was supervising the doctoral programs of John Mitchell and Charles Hamner, who were working at the United States Department of Agriculture's Plant Industry Station in Beltsville, Maryland. Dr. Kraus outlined his novel theory of using growth regulators to kill plants in a letter to Mitchell and Hamner in the summer of 1941. Under the cover of secrecy because of the war, the two researchers quickly began running tests on the synthetic compounds, including the phenoxy chemicals 2,4-D and 2,4,5-trichlorophenoxyacetic acid, or 2,4,5-T. The two were close chemical relatives, differing only in the number of chlorine atoms each possessed.

At about the same time in 1941, with the war entering its third year, Secretary of War Henry L. Stimson was persuaded by some of America's most prominent scientists to have the National Academy of Sciences and the National Research Council set up committees to monitor the potential dangers of biological warfare. The following year the civilian War Research Service took charge of all aspects of biological warfare.

"What could a botanist do to help the war effort?" said Dow's Wendell Mullison, who studied under Dr. Kraus at the University of Chicago. Dr. Kraus suspected that the biological warfare committees might be interested in "the toxic properties of growth regulating substances for the destruction of crops or the limitation of crop production." In late 1942 he suggested that testing should be carried out on field crops. As a result, the University of Chicago was given a $3,500 contract to do herbicide research for the army.

The military had been looking for a new weapon to fight the Japanese, a weapon reminiscent of the naval blockade in its ability to starve an enemy, but less elaborate to deploy. What was needed was an agent that would destroy Japanese food crops and thus weaken and eventually cripple the country. Dr. Kraus's final report on his herbicide work at the University of Chicago was forwarded to the War Research Service. As a result, the military decided in early 1944 to include herbicide research in its testing program at Camp Detrick, a National Guard airfield near Frederick, Maryland, which was the country's leading center of biological warfare research.

From February 1944 through September 1945 Dr. Kraus and an assistant directed the synthesis and testing of nearly 1,100 substances as part of a "special projects division" at Camp Detrick. They had already shown that a strong dose of the phenoxy compounds 2,4-D and 2,4,5-T were effective in killing rice grown in indoor tanks, and now they began spraying the chemicals on field-grown crops.

The British, meanwhile, were working along similar lines and in fact had already isolated several compounds that had powerful growth-retarding properties. Completely by serendipity, as Dow's Dr. Mullison remembered it, the British sprayed their chemicals across some cereal crops to see if they could speed growth. Instead, "they suddenly noticed there were no weeds in this field," Mullison said. "That caused them to think that maybe this was a herbicide." The British briefly considered spraying the chemicals across German farms but decided their available aircraft were insufficient to do the job properly and instead turned their technology over to the Americans, whose anticrop research was already well advanced.

From a panoply of candidates, three compounds were subsequently developed as anticrop agents at Camp Detrick. Two of them were the phenoxy compounds, 2,4-D, which was code named LN8, and 2,4,5-T, code named LN14. In sufficient quantities each could wipe out an important Japanese crop. The Americans calculated that 20,000 tons of 2,4-D, for instance, could destroy the entire Japanese rice crop.

But the war ended before any of the anticrop agents could be used. According to a report prepared for the U.S. Joint Chiefs of Staff Technical Warfare Committee on Crop Destruction, which was long kept secret, the Americans had "built up a stock of material and were planning an attack on the main islands of Japan early in 1946, calculated to destroy some 30 percent of the total rice crop." In May 1946 War Research Service director George W. Merck disclosed that "only the rapid ending of the war prevented field trials in an active theatre of synthetic agents that would, without injury to human or animal life, affect the growing crops and make them useless." A few years later the British became the first to deploy the anticrop chemicals when they sprayed 2,4,5-T on Chinese guerrilla crops in Malaysia.

The chemical industry had been sidetracked by the war, which demanded the full attention of America's technology in producing weapons to fight the war and drugs to heal the warriors. The federal government traditionally turned to industry for assistance during wartime, beginning in 1804, when Thomas Jefferson's military buyers were first in line for Du Pont's black gunpowder, and continuing through the

Elements of Risk / 19

Mexican, Civil, First and Second World, and Vietnam wars. During the Second World War the government asked for—and got—an entirely new synthetic rubber industry to replace cut-off supplies of natural rubber from Asia. Monsanto helped supply the wartime synthetic rubber demands. In its annual reports to shareholders in 1943, Monsanto said, "It is with regret that we abandon our past practice of transmitting to our shareholders an informative and interpretive annual report. The necessity of secrecy imposed by our national interest surrounds much of the activities upon which the company has been engaged." The Hooker Chemical Company supplied the government's Chemical Warfare Service with the equipment to turn chlorine into poison gases and assisted in meeting the demand for magnesium and picric acid. (Picric acid is a yellow dye. It turned Hooker workers' skin so alarmingly sallow that baseball fans fled their seats when a jaundiced-looking Hooker plant manager sat down next to them at a World Series game.)

Dow built a 72-million-pound-capacity magnesium plant in Texas to supply the lightweight metal needed for building aircraft and airborne equipment. As the fighting wore on, the bulk of Dow's production was directed once again to the war effort. Dow's slogan was "Chemicals Indispensable to Industry and Victory." Dow plants operated on a twenty-four-hour, seven-day-a-week schedule. The army-navy flag flew over the plants, and employees wore E pins awarded by the government for excellence in the production of war equipment. Dow was cited for its production of strategic chemicals and magnesium for the war effort, although, as the company solemnly noted in a promotional brochure, "except as the needs of our nation's security have become manifest, the Dow Chemical Company has never included in its own program the production of materials designed solely for a destiny of destruction."

The industry underwent phenomenal forced growth during the war years. Chemical industry sales more than tripled from prewar levels of $4.3 billion to $13.8 billion in 1949. Dow's sales volume increased 381 percent, and during one three-year period alone, sales were hiked from $39 million to $124 million. After the war overexpanded plants and inventories coupled with government cancellation of $35 billion worth of war contracts caused a "reconversion depression" in the industry. But companies were soon taking full advantage of new plants and new processes to satisfy pent-up consumer demand. Americans had worn out everything they owned during the war years and had saved $140 billion along the way, so the postwar years witnessed an unprecedented demand for industrial goods. Massive expansion programs were launched. The rate of growth was spectacular. From 1923 to 1950, while the nation's industrial activity doubled, chemical production increased almost fivefold. "In the last decade alone," chemical research and engineering and chemical business management "have together created an industry now possibly the chief generating force in the United States social and economic pattern," industrial analysts observed in 1950.

Most of the industry's expansion could be directly linked to wartime research, which paid off handsome postwar dividends as bug killers and weed killers discovered by chemical and biological warfare scientists were put to civilian use, and a chemical called DDT, which had been used to protect the boys in uniform from foreign diseases, came home to a thunderous reception in America.

DDT was the most famous of the chlorinated hydrocarbon insecticides that emerged in postwar America as the farmer's savior and the homeowner's friend. In 1939 the Swiss chemist Paul Müller was searching for a more effective moth repellent. Instead, he stumbled across a chemical called dichlorodiphenyltrichloroethane, or DDT, which had been synthesized some years before and ignored. Müller found DDT to be devastating to flies. Both the Allies and the Germans quickly began producing the chemical to combat what one newspaper writer called "an enemy which fighters in the Pacific respect far more than they do the Japs—disease bearing mosquitoes and flies." What was needed was an insecticide to check typhus-carrying cooties and malaria-spreading mosquitoes. DDT was used with enormous success in combating a typhus epidemic in Naples in 1943. The army surgeon general's report on the trial use of DDT on the beaches at Saipan was released to nearly hysterical acclaim in the United States in 1944. "DDT will exceed even penicillin in its ultimate usefulness," Army Brigadier General J. S. Simmons predicted in a report in the *Detroit Free Press* on November 26, 1944. "DDT will be to preventive medicine what Lister's discovery of antiseptics was to surgery," added Lieutenant Colonel A. L. Ahnfeldt.

Newspaper reports guaranteed DDT's popularity when it was finally marketed in America in 1945. They stated:

> Insects need only to walk on it and they keel over soon afterwards—victims of a deadly "hot foot."—*Detroit Free Press,* November 26, 1944.
>
> The stuff kills both by contact and when eaten and is harmless to humans and their domestic animals and pets. Walls and ceilings covered with a spray remain deadly to flies for three months.—Associated Press, June 1, 1944.
>
> On dogs and cats the powder kills fleas. Beds properly sprayed just once have killed and kept away bedbugs for 300 days. Dairy cattle made nervous by flies have been quieted by spraying.—Associated Press, June 1, 1944.

> The unique thing about DDT, compared with other insecticides, is its lasting effect.—*Detroit Free Press,* November 26, 1944.

WE HAVE DDT! banners hung from store windows. Want ads in daily newspapers appealed for DDT salesmen, and chemical jobbers announced that they had the chemical in stock. "DDT received a free advertising buildup in general that rivals the campaigns of America's leading sellers of cigarettes and soap," Dow reported in its agricultural trade journal, *Down to Earth,* in November 1945.

It was the age of the "wonder" chemical, and Americans wanted one of everything, no questions asked. DDT would give them a "bugless utopia." Plastics—which would eventually account for more than a third of Dow's sales—would give them "swear-proof" doors and window frames that would not stick in damp weather and refrigerators so light in weight that a "95-pound woman could move it about with ease." Chemists were called "modern Merlins," and their factories "Houses of Wonder."

Sales of agricultural chemicals soared after the war. Dow announced at war's end that it would devote its peacetime resources "to the pursuit of better living for the American people" and singled out its growing line of agricultural chemicals as among the company's "greatest gifts to human welfare." One of Dow's first postwar "gifts" was its version of the herbicide 2,4-D.

War's end had reduced the American military's drive to develop anticrop agents, but American industry caught the scent. When they finished their war research, Dr. Kraus and his colleague John Mitchell concentrated on the great possibilities they believed the phenoxy compounds 2,4-D and 2,4,5-T held for agricultural use. Sworn to secrecy about the chemical warfare aspects of their research, the botanists had known for some time, but had been unable to circulate the news, that the compounds were ideal selective systemic herbicides. In other words, they killed some plants without harming others, and they worked by spreading throughout the plant's system. An extremely small amount of the chemical placed on a single leaf would translocate or spread throughout the plant, eventually killing even the roots.

One of the first public revelations that the growth-regulating compounds could be used as herbicides had come in an August 1944 article in *Science* magazine by Charles Hamner and H. B. Tukey. Their report, "The Herbicidal Action of 2,4-D and 2,4,5-T on Bindweed," was a summary of experiments they had recently conducted at the New York State Agricultural Experiment Station in Geneva, New York. They had sprayed two 100-foot rows of apple nursery stock that were heavily infested with the feisty weed. Within a few hours after application the weeds "appeared wilted." There was a "slight upward folding of the leaves," which were "somewhat stiff to the touch." Within twenty-four hours the weeds had taken on a dull green color and were lying flat on the ground. Their petals failed to open, and there was no growth of young shoots. The belowground parts became enlarged and split. By the fifth day the leaves were yellow. By the tenth day they were dry and dead.

"That article in *Science* really caused a tremendous impact on the development of herbicides in this country," recalled Dow's Dr. Mullison. "This was partly due to the economy of the times—labor was getting scarcer and these were labor-saving chemicals—and partly because they played an important part in increasing crop yields, because without weeds you could grow crops closer together."

Excitement about the new herbicides rapidly built with the publication later that year of the results of John Mitchell's successful deployment of 2,4-D against weeds at Beltsville. On August 11 Mitchell had sprayed plots in a lawn dotted with dandelions. By August 29 there was a "complete kill." Dr. Kraus dug up the dandelion plants in disbelief and was startled to see that their roots were dead as well.

More than two years later, Kraus and Mitchell summarized their years of secret research in the magazine *The Botanical Gazette*. "The use of growth-regulating substances in the control of weeds and other vegetation," they wrote, "has now become widely publicized and exploited."

Some of their experiments had been carried out in collaboration with a young and aptly named botanist, Fanny Fern Davis, who was in charge of the U.S. Golf Association's greens section. During the war years she had been searching for a chemical to stimulate the growth of grass and turf and had tried some of the chemical growth regulators without success. John Mitchell knew of Mrs. Davis's work and suggested that she try 2,4-D not as a growth stimulator but as a dandelion killer. "Not being one to let grass grow under my feet, I made a date . . . the very next day to try out this hormone on golf course turf of various types," Mrs. Davis recalled years later. She treated parts of the practice green, bluegrass, and fairways at the Chevy Chase Club in suburban Maryland with 2,4-D. During the following weeks the weeds curled and disintegrated, leaving pencil-sized holes in the ground where dandelion roots once had been. The grasses were untouched.

With the chemical industry watching her every move and mixing up different formulations of the chemical for Mrs. Davis to test, she moved on to the weedy turf on the Mall between the Capitol and the Lincoln Memorial in Washington, D.C. She planted four-by-four and ten-by-ten demonstration plots in front of the Natural History Building, using various formulations. Later she applied the weed killer to large sections of the White House lawn. The weeds shriveled and died, leaving the grasses unharmed. Thus was reinforced the concept of a selective weed killer—something that would kill weeds without harming grass and crops.

The chemical industry was delighted by the stunning effectiveness of the phenoxy herbicides. Several companies laid plans to market 2,4-D commercially in 1945. In May 1945 Franklin D. Jones of the American Chemical Paint Company applied for a patent on 2,4-D, which the U.S. Patent Office granted in December under the title "Methods and Compositions for Killing Weeds." His company quickly began selling Weedone and thus became the first of many to market systemic herbicides on a commercial scale. The herbicides were not only effective but easy to use and inexpensive (less than $3 a pound, a price that declined to 50 cents by 1950).

The Dow Chemical Company entered the herbicide sweepstakes that year with its version of 2,4-D, which was sold in powdered or liquid form, or in twelve-tablet packs for $1, to home gardeners to kill broadleaved weeds like dandelions, poison ivy, and poison oak without damaging lawns; and in bulk quantities to farmers for control of weeds in rice, sorghum, and sugarcane fields, to cattlemen for rangelands, and to railroad and utility companies for clearing rights-of-way around tracks and poles. "Like any chemical it has its limitations," Dow observed, "but the results on the whole are nothing short of miraculous."

Death to Weeds, a color "moving picture" which enjoyed "extremely heavy bookings" among grower groups across the country, extolled the benefits of Dow's 2,4-D to grains in Michigan, rice and sorghum fields in California, range and farmlands in Texas, rights-of-way in Pennsylvania, and sugarcane fields in Louisiana. Fancy promotions were quickly superseded by word of mouth. Dow's Wendell Mullison described a typical transaction. "A guy in Kansas would have a wheat field infested with yellow mustard weed, and the county [agricultural] agent would say, 'I heard about this new stuff called 2,4-D.' And when the farmer used it, he got a harvestable crop. It made all the difference between nothing and reasonable."

Everyone knew that the herbicide caused plant stems to twist and bend, the roots to develop abnormal swellings, and the leaves to turn yellow and die. But no one knew the exact mechanism of how it worked. "Manufacturers rushed to market 2,4-D for the demanding public," one observer pointed out, but "the public's enthusiasm for the newly unveiled herbicide was equalled by universal ignorance of the chemical and how it worked." The first extensive field tests of 2,4-D and government-ordered toxicological tests on the chemical's safety weren't even begun until after the chemical had been put on the market in 1945. When the annual North Central State Weed Control Conference was held that year, the "weedmen" were reassured that 2,4-D applied in twice the normal dosage to pastures produced no toxic effects to grazing sheep and cows. As a sort of icing on the cake, the dynamic Dr. Kraus stood up to make an announcement. Dow's Wendell Mullison remembered what he said. "Gentlemen, I have taken five hundred milligrams of 2,4-D daily for twenty-one days," Dr. Kraus declared. "Look at me and judge for yourself if I am sick."

But even as 2,4-D was enjoying stardom, researchers were recognizing its limitations and were investigating its chemical relatives to see if they could better control the toughest and most resistant woody species of plants. John Davidson and his research group at Dow's South Haven, Michigan, field research station suggested that Dow market 2,4,5-T as a selective herbicide. They found that 2,4,5-T—one of the anticrop agents developed during World War II along with 2,4-D—worked well against wild blackberry and raspberry. An even more efficient kill was provided if the two chemicals were mixed together. A fifty-fifty mixture of 2,4-D and 2,4,5-T was successfully used against brushy vegetation clogging utility rights-of-way in a mountainous part of Pennsylvania, and in the spring of 1948 Dow began a limited sales program of 2,4,5-T alone and in combination with 2,4-D. Dow's Dr. Mullison, who patented some of the earliest versions of the herbicides, devised products that were less volatile than those marketed by other companies and so penetrated the plant better. "The results look exceptionally promising," Dow said in a promotional piece about its new herbicides in 1949. The promise was fulfilled when up to 85 percent of the brush along 742 miles of the Delaware, Lackawanna & Western Railroad Company tracks in Pennsylvania was destroyed by a mixture of 2,4-D and 2,4,5-T. "Chemical weed killing," Dow said, "is here to stay."

Dow subsequently added another phenoxy herbicide to its growing product line. The latest addition was known as silvex (which Dow sold under the trademark name Kuron), and it outperformed its chemical cousin 2,4,5-T in several ways. Silvex was said to be even more effective against woody and broad-leaved weeds, particularly against certain species of oak. As with the other herbicides, silvex was taken up by the leaves and stems and then translocated, or spread, to other parts of the plant, causing the familiar sped-up growth-and-death cycle. The major difference was that silvex was slower-acting, and it took a few days longer to kill its prey. Dow originally sold silvex to prevent fruit drop and to improve the color of fruit. In higher doses, though, it killed the trees. Dow's John Davidson suggested that its potency might be harnessed as a herbicide against the tough oak. "That was one of my biggest mistakes," Dr. Mullison, who screened silvex when it came through Dow's biochemical research laboratories in Midland, said with a grimace. "It was very active. I called the chemist and said that it was about the same as 2,4-D. He said it cost twice as much as 2,4-D. So I dropped it. In thinking it was too expensive, I never thought about selectivity. John's group at South Haven picked it up subsequently."

Dow toxicological studies on silvex showed it to be relatively nontoxic to animals in acute or brief exposures. A 550-pound steer fed silvex through a stomach tube every day for fifteen days remained healthy, and its internal organs were undamaged on autopsy. The acute oral LD50 was similar to that of the other phenoxy herbicides. (LD stands for "lethal dose"; 50 for "half the animals tested.") It took about 500 milligrams of silvex for every kilogram, or 2.2 pounds, of body weight to kill half a population of rats. To accomplish the same thing with an acutely toxic pesticide like parathion required as little as 1.7 milligrams per kilogram of body weight. Scientists warned only that skin irritation might result from repeated contact with concentrated doses of the chemical and that drift from the herbicide might damage other field crops. "Silvex possesses some unique properties which warrant further investigation," Dow scientists observed in a January 1954 technical bulletin distributed to agriculture schools and growers. "... It should be emphasized that the practical field hazard from drift to cotton and other crops has not been fully evaluated." In the same year—while those further investigations were being made—Dow put silvex on the market as a herbicide.

chapter 2

When the American version of the plagues of Egypt—first the Colorado potato beetle, then the grasshopper epidemics, the chinch bugs, and farther south, the boll weevil—swept across the Great Plains in the 1800's, farmers tried to beat them back with everything from patent bug-killing machines to days of prayer and fasting. Chemical poisons were the only solution, so, when a new plague overtook the farmers and this one was resistant to poison and to fire, it was as horrific as if God had visited on them rivers turned to blood. The farmers dropped to their knees in defeat before an enemy they could not even see because it was smaller than a speck of dust. There was no comfort in knowing that the invisible pest had brought down Genghis Khan, who invaded Europe because the worms destroyed grain crops in Asia, and the Inca Empire, where corn crops were ravaged. By the time the Dow Chemical Company and its chief competitor, Shell Chemical, devised a way to destroy them the diminutive worms called nematodes were the despair of farmers on four continents.

Nematodes are no ordinary earthworms or nightcrawlers used as fish bait but are worms so small as to be hidden beneath the head of a pin, worms so numerous they exceed the world's insect population. The smallest are but one-hundredth of an inch long; the biggest and brawniest attain lengths of one-quarter inch (although one rare worm that lives under human skin can measure four feet). If everything on earth except these worms were removed, one could still see the outline of cities, rivers, forests, mountains, and even animals, such is the magnitude of their kingdom. Most of them—the group known as free-living—feed on decaying organic matter. Others are parasitic on animals, fish, themselves, and humans. The most notorious group is that known as plant parasitic nematodes, for these are the worms that destroy farmers' crops and these are the worms that humans have tried most ardently to conquer. John Turberville Needham first found them crawling around in crushed wheat kernels under his eighteenth-century microscope and called them eelworms; later they were called roundworms. In nineteenth-century Germany they were the scourge of the sugar beet farmer, who tried to get rid of them by rotating crops, setting out trap plants to seduce them away from beets, and injecting flammable liquid fumigants into the soil. By the first decade of the twentieth century the United States Department of Agriculture (USDA) was seriously concerned about the nearly invisible invaders, which, once looked for, could be found on nearly every American fruit, vegetable, and grain crop. The first shipment of cherry trees from Japan to Washington, D.C., in 1910, was infected with nematodes and had to be destroyed; a second shipment, now blooming annually around the Jefferson Memorial, was judged safe. Fifty years later nematodes were blamed for annually destroying one-tenth of American crop production, and farmers in California were so defeated they stopped planting some of the nematode's favorite crops.

Of the plant parasitic nematodes there are several hundred types—cabbage cyst nematodes, smooth-headed nematodes, sting nematodes, pin nematodes, stubby root nematodes, dagger nematodes, burrowing nematodes, kidney-shaped nematodes, and true spiral nematodes—but the one that does the widest and most expensive damage to crops is the root-knot nematode, which is found in more than 1,700 host plants covering most of the vegetables, nursery crops, field crops, nuts, fruits, and vines grown in this country.

Root-knot nematode larvae invade plant roots through the tip, or cap, where cell division is most active. Then they migrate through the cells and position themselves in the vascular tissue, or central cylinder of the root. Damage from the invasion generally prevents root elongation.

The nematode injects a substance into the vascular tissue that causes certain cells to enlarge into what are known as giant cells. The nematode feeds on these cells, disrupting the plant's intake of food and water. The root system shrinks, the plant's top growth deteriorates, and the farmer is left with reduced crop yields.

The nematode, which lives no longer than three months, molts like an insect. It molts five times, and each time it sheds a skin it gets bigger and takes on a different shape. Despite its tiny size, the nematode has complete digestive, reproductive, nervous, and excretory systems. The female nematode resembles a tiny pearl. She lays her eggs—as many as 3,000—in a gelatinous mass, which protects the eggs from drying. During the hottest part of the summer, they grow into larvae and adults.

It would seem a simple matter to find a pesticide that could be injected into the soil around the plants to kill the nematodes. But in the

1930's and 1940's chemicals used against nematodes were so flammable they could set a whole field on fire, so poisonous they took down the plants with the worms, and so expensive they weren't worth the trouble.

Clyde W. McBeth spent nearly half a century trying to understand and destroy the enemy nematode, tracking worms from Africa to Hawaii, first for academia, then for government, and finally for industry. He remembered the primitive chemicals used in the prewar years—some, like the tear gas chloropicrin, developed as chemical warfare agents by the Germans during the First World War, and others, like carbon disulfide, tricky to use and so volatile that one spark from a plow against a rock could turn a field into an inferno. They were fumigants, injected into the soil to kill pests with their poisonous vapors.

The first real breakthrough came in the early 1940's, when scientists at the Pineapple Research Institute in Hawaii, trying to save the island chain's pest-ridden pineapple crop, applied a noxious, black, apparently useless waste product from the Shell Chemical Company's glycerin production to a test plot called the graveyard. "Everything they could get a hold of, they put in there," recalled a graveyard habitué. The chemical not only made the plants grow better but also—to the researchers' utter surprise—worked against nematodes. The chemical was called dichloropropane-dichloropropene. Shell named it D-D.

Clyde McBeth was working for the United States Agriculture Department's nematology division, one of fewer than a dozen men in the country professionally devoted to the tiny worms. "Shell got excited about D-D, and they came to our experimental station in Georgia. They wanted us to test D-D there," McBeth recalled. "We found that it was very good. It was volatile, but less volatile than chloropicrin, and didn't escape all over the place like chloropicrin did. It was also much easier to handle."

Shell put D-D on the market in 1943. It was relatively expensive to use and sufficiently deadly to plants that it had to be injected into the soil weeks before crops were planted. But it killed nematodes better than anything on the market, and Shell set about selling D-D to nematode-plagued farmers. The problem was, most of them didn't know they *had* nematodes. They knew only that something was terribly wrong with their crops. "Although a few scientific workers knew the importance of these pests," McBeth said, "there was not much that could be done to control them. With the advent of D-D, it became possible to demonstrate the damage and correct the problem on a field scale."

Shell spread the word through agricultural journals, farm supply stores, county agricultural extension agents, company-sponsored workshops, and even swing-style advertising jingles sung by Andrews Sisters types, which were played on radio stations in farming areas. Driving through the country in the mid-1940's, you were likely to hear on your car radio the "D-D Hoedown," the "Nematode Lament," or, as follows, the "D-D Hop":

> Kill the worm, kill the worm, kill the worm
> Kill the nematode worm
> He's the worm at the root of it all
>
> If your crops don't grow like they ought to grow
> There's just one cure that you ought to know
> Before you plant just treat your fields
> Get bigger crops, get better yields.
>
> The nematode's a tiny mite
> But against your fields he's dynamite
> Just use D-D and do it right.
>
> Yes the nematode he's as mean as sin
> though he's no bigger round than the head of a pin
> You can kill these rascals easily
> Just fumigate and use D-D
> They cause root rot that spoil your plants
> But against D-D they've got no chance

The message was clear and seductive: "Bigger crops for bigger dough." As the happy narrator of "D-D Hoedown" put it: "I'm getting back $4 now for every $1 I spend."

Shell's success sent other chemical companies racing to find their versions of D-D. Chemical companies were as competitive as football teams, and Dow's scientists at the company's Seal Beach, California, agricultural research laboratory had been searching for their version of D-D during the war years. They hit paydirt in 1942 with the discovery that ethylene dibromide, or EDB, worked well against both nematodes and a distant relative, the wireworm. "As a matter of fact," Dow boasted in an agricultural journal circulated to growers, "results of ethylene dibromide fumigation have been so successful that reduction in planting rate has often been a NECESSITY to prevent overcrowding." The experimental fumigant was applied to 500 acres of cultivated land throughout the western states in 1945. In one test, crop yields jumped to 13,850 pounds of sweet potatoes per acre on land formerly capable of eking out only 816 pounds per acre. The chemical debuted on the market under the name Dowfume W-10 in 1945 in response to what Dow called "spontaneous grower demand," and it quickly put a dent in D-D sales.

The new nematocide, as this class of pesticides was called, was still potent enough to kill everything it touched and had to be spread through the soil weeks before crops were planted. What was needed was a nematocide that could be applied to living plants—perennial crops like vineyards and citrus trees—without killing them.

It was the scientists at Hawaii's Pineapple Research Institute who found the solution, again by mistake. Taking a lead from the composition of Dow's ethylene dibromide, they began experimenting with other bromine compounds and happened on one, called 1,2-dibromo-3-chloropropane, or DBCP, which was tested against nematodes. The results were nearly as good as they had been with D-D, although the growth was not as outstanding, according to Clyde McBeth, who had left the government in 1945 to work for the Shell Chemical Company.

One of Shell's fieldmen had seen how DBCP performed in Hawaii, and he reported the results to Shell's research facility in Modesto, California. McBeth was working in Modesto as Shell's chief nematologist in charge of screening and developing chemicals for nematode control. In 1951 the Modesto chemists included a sample of DBCP in their screening program. Shell gave DBCP a code name—Experimental Nematocide OS 1897. "We got hold of some DBCP and found it was nearly twenty times as effective as D-D," McBeth recalled. It was more expensive, but it produced higher yields for the dollar than did D-D, and best of all, farmers could apply DBCP safely to the roots of living plants either by injecting it into the soil around the plots or by spiking their irrigation water with it. "We had an awful time getting a low enough dosage," McBeth said. "We cut it in half, and it would still be 100 percent effective. So we cut it in half again."

McBeth spent part of the summer of 1952 studying nematode problems in pineapples at the Pineapple Research Institute. DBCP was applied to field plots before pineapples were planted there. The researchers noted that it took up to six weeks for the nematodes to disappear from the soil, which was about the length of time it would take them to starve to death. "DBCP would knock 'em down to a very low population in the field," McBeth said. "It took 'em four years to build up again."

Thirty years later, retired from Shell, McBeth still did not know exactly how DBCP works. "It doesn't actually kill nematodes," he said. "We don't know exactly what it does."

He did not agree with Dr. Earl Anderson, a colleague from the Pineapple Research Institute, who said, flatly, that DBCP works by inducing sterility in nematodes.

"Some people say it prevents them from reproducing. Our work says it prevents them from feeding," McBeth explained. "Maybe it affects the sensory organs so they can't find the root. Maybe it disorients them. Our plant physiologist called it a narcotic—he agreed they probably starved to death."

The chemical's mechanism of action—how it works—was not a high-priority question in the 1950's, when the two companies that had pioneered nematocide research—Shell and Dow—began running laboratory tests on DBCP with an eye toward putting it on the market. Toxicological testing at that time was simple and straightforward. "Very few people had even been trained in toxicology at that time," said Dr. Ted Torkelson, who came to Dow as a twenty-two-year-old biochemist in 1953. "Fact is, when I was hired, I didn't know what toxicology was."

In the fifties small groups of laboratory animals were fed the chemical or forced to breathe it in special chambers or had it applied to their skin. They were observed for a period of hours or days to monitor acute effects and a period of months to monitor chronic effects. "When I came to Dow, six months was a long study," Dr. Torkelson recalled. "The reason they ran six months is that in the old lab the air conditioning wouldn't allow them to run in the summertime, so they would start in the fall and end in the spring." Animals were autopsied when they died or when the experiment ended. Researchers were concerned about changes in weight, blood chemistry, and organ tissues—usually the lung, liver, and kidney. These were thought to be target organs—which is to say, chemical agents would migrate to these body parts, almost as if they were missiles programmed to hit a target. The lungs inhale air and feed its oxygen to the blood, the liver breaks down chemical contaminants in the blood, and the kidneys filter waste products out of the body in the urine. Explained Dr. Torkelson: "The liver is the chief metabolic organ, and anything you inhale or eat goes through the liver. So if it's going to be metabolized, it probably will be done in the liver. The kidneys are thought to be target organs because much of what you get rid of goes through the kidneys and is concentrated there. The lung is a target organ because it is often the first one the chemical comes in contact with."

Chronic tests were rarely followed for the life of an animal (two years for a rat). Instead, animals were given a variety of doses of a particular chemical. Generally researchers found a particular dose-response curve —which is to say, exaggerated effects at the highest doses became less severe as the doses waned. The final step was to find a no-effect level, or the amount that researchers calculated could be absorbed by an animal with no harmful effects. Once this threshold amount was found, it was reduced further to give what was supposed to be a wide safety margin for human exposure.

Once these cursory examinations had been finished, companies often put the chemical product on the market while continuing to investigate its effects. Such was the case with DBCP, which was put on sale in the midst of ongoing experiments, and such was the case with the herbicide

2,4,5-T and its relatives, which were marketed with only the sparsest of information on their long-term effects.

This hurry-up-and-go attitude was partially a phenomenon of its time. There was little room in the 1950's for advocates of the slow, thoughtful approach in any portion of life—business, science, or politics. The country was so firmly in control of itself and had tied technology so tightly to patriotism that to be skeptical, to be a Robert Oppenheimer working to "retard" the hydrogen bomb program or an "alarmist" scientist warning of potential dangers of radioactive fallout, was to be a traitor. Nationwide publicity linking cigarettes to heart disease for the first time in 1954 was countered by advertisements that pointed out reassuringly that "More Doctors Smoke Camels Than Any Other Cigarette." "The deadliest sin was to be controversial," observed William Manchester in describing a generation that wanted "the good, sensible life" and that was "proud to be conservative, prosperous, conformist and vigilant defenders of the American way of life." The largest group of college undergraduates were business majors, and industry leaders were lionized (General Motors president Harlow Curtice was *Time*'s Man of the Year in 1956). A free market, left to its own devices, was thought to be the most efficient path to productivity. In 1957 the Soviets simultaneously launched *Sputnik 1* and the space race by taunting Americans with the specter of Russian superiority. Obeisance to technocracy took on patriotic as well as religious overtones.

The chemical industry was in its florescence, and its leaders spared no hyperbole in promoting it—and private enterprise in general—as the last line of defense against moral and economic collapse, world famine, and Communist takeover. "It must be crystal clear to every American that our steadily expanding industrial system, with its emphasis on private competition, enterprise, and individual self reliance, is the major source of our strength in these critical times," the Dow Chemical Company editorialized during the peak of the cold war years. Dow employees were sent to school on company time to learn "How Our Business Operates" and took classes in "The Importance of Competition," "Individual Freedoms," and "Dow and Progress." In speeches Dow president Leland I. Doan quoted the historian Edward Gibbon ("All that is human must retrograde if it does not advance") and biologist Thomas Huxley ("The rung of the ladder was never meant to rest upon, but only to hold a man's foot long enough to enable him to put the other somewhat higher").

Leland "Lee" Doan had assiduously adhered to the teachings of Huxley in his own corporate ladder climb. Doan started out as a twenty-two-year-old plant helper in 1917 and rose steadily through the ranks, first to director of sales and then to vice-president. Along the way he also married the boss's sister. When Willard Dow was killed in the

1949 airplane accident, the fifty-four-year-old Doan was the obvious heir to the throne. Doan was a product of his times—a committed capitalist and a fervent cold warrior. In one of his first public speeches —to the Midland Chamber of Commerce in 1950—Doan exhorted the men in his audience to go out and sell "this economic system of ours, the best product the world has ever known" to anyone they could corner at parties, schools, social clubs, even the street corner, where, surprisingly, "much of the anti-profit, anti-capital, anti-big business talk circulates." They were to show the disbeliever around the "good streets" and "fine schools" of Midland itself and, then, at day's end, stand him in sight of the Dow plant "with the sun lighting up a river bank studded with red brick, and green and red and yellow tanks and pipes.

"And you can say to him in all sincerity:

" 'That, Mister, is Main Street . . . U.S.A.' "

Four years later scientists at Doan's sprawling chemical plant just off "Main Street, U.S.A." began running toxicological tests on DBCP, the chemical so full of promise as a nematocide for living crops.

Shell Chemical, Dow's competitor and copioneer in nematocide research, was equally interested in marketing DBCP. Shell did not have an in-house biochemical research laboratory, as did Dow, but relied instead on a group of toxicologists at the University of California Medical School in San Francisco who had been conducting directed research for the chemical company since the petrochemical boom of the 1930's, when a wave of new synthetic compounds began pouring from chemical laboratories across the country. The story is told that a worker made dizzy by one of the new chemicals dropped his wrench one day and narrowly missed beaning a Shell president who was walking through the plant. The president suggested more testing might be in order for certain of the new petrochemical compounds, and he struck an arrangement with the chairman of the University of California Medical School's pharmacology department. "Directed research" is just another way of saying that Shell asked the university toxicologists to answer specific questions about specific chemical compounds and then financed the research with annual grants. In the 1950's these grants were in the neighborhood of $3,000 a year.

The university toxicologists had their first look at DBCP in the spring of 1954, when Shell asked them to gather information—with an eye toward establishing safe handling procedures—on DBCP and several other chemicals that were "approaching advanced stages of development." Shell also wanted to know if DBCP really was a Class B poison, the rating attached to it by the Interstate Commerce Commission for purposes of handling during transportation. Chemicals were ranked in five categories, according to how toxic, or poisonous, they were (with Class A and B chemicals carrying more severe label warnings than the rest), under the guidelines of the Federal Insecticide, Fungicide and Rodenticide Act, or FIFRA. So many synthetic chemical pesticides and insecticides had been unleashed during the industry's postwar expansion that the government responded in 1947 with FIFRA, which essentially was a labeling and registration law to protect farmers and chemical applicators from unsafe products.

The toxicologist who directed the DBCP research was Dr. Charles Hine, a powerfully built man in his late thirties who had degrees in chemistry, medicine, pharmacology, and toxicology. He had joined the University of California Medical School teaching staff in 1947 after stints with the National Naval Medical Center, where he studied everything from the toxicity of materials liberated by overheated submarine engines to dangerous materials libated by drunken sailors. He also served as liaison with the government's chemical warfare program at the Edgewood Arsenal and with the San Francisco Radiation Defense Laboratory, where he conducted partially classified work on the ill effects of radiation. Dr. Hine found himself increasingly drawn to the field of occupational health, which was in its infancy. He wanted to study the adverse effects of chemicals in the workplace as well as in the laboratory, and his interests coincided with those of Shell Chemical, which was looking for a toxicologist to monitor the health of its research and development workers. Dr. Hine learned about Shell's search on a chance visit to the university. The chairman of the pharmacology department offered him a deal. If Dr. Hine wanted to work with Shell, a teaching appointment to the university medical school with time to do toxicological research as well as outside time for a clinical practice in order to accommodate Shell's needs could be arranged. Dr. Hine thought it sounded like a "fine deal," and Shell obliged by giving him a consultant's contract and an office in its research facility across the bay from San Francisco in Emeryville. In a convoluted arrangement not unusual at a time when industry contracted out as much research as the government does today, Hine was working for the University of California, where his research included Shell-funded studies, and part time as a paid consultant for Shell itself.

The university reserved the right to accept or pass on a Shell research request, but Dr. Hine was eager to work on DBCP because he had just finished testing a compound with a similar chemical composition, and it seemed a natural extension. Shell sent a quart of the straw-colored DBCP liquid over to Dr. Hine and his research assistants, who set to work. The first tests indicated that the nematocide was not sufficiently toxic to be classified as a Class B poison, a judgment that kept the skull and crossbones warning off its label when it was subsequently put on the market. The second, third, and fourth tests confirmed the relative safety of the chemical. Rats fed DBCP-spiked meals for three months developed no significant problems. DBCP brushed across the shaved backs of rabbits provoked only "minimal irritating effect" and didn't sear the corneas when dropped into the rabbits' eyes. Rats that breathed single and repeated doses of the chemical in exposure chambers began to show respiratory distress and decreased weight gain only when overwhelmed with a large single dose or after inhaling smaller doses over a period of five to ten seven-hour days.

Dow had been running similar tests at its laboratories in Michigan under the supervision of Dr. Torkelson, a sandy-haired scientist who in size was Jeff to Dr. Hine's Mutt. Dr. Torkelson supervised Dow's new inhalation laboratory, which had been expanded along with the entire biochemical research department in 1955. His lab had eleven inhalation chambers, including five big walk-in models where the long-term DBCP studies were run. The chambers held two dozen rats and a dozen guinea pigs, half a dozen rabbits, and several larger animals, either monkeys or dogs. DBCP was metered into an airstream and pumped into the chambers at different dose levels.

Dr. Torkelson knew that the volatile DBCP was more toxic than some of the other chemicals he was studying simply because to work as a fumigant, it had to be highly active at low concentrations. "If you spill DDT in a corner, so what? Go get a vacuum and clean it up," he said by way of example. "But boy, if you spill a little DBCP over there, it's going to fill the whole room." Dr. Torkelson's tests were almost identical to Dr. Hine's in methodology, but they were producing different results. "We had exposed animals at the same concentration that Charlie Hine's group had said was without effect," he said. "And we killed them! It wasn't a matter of a little difference. Ours were dead; theirs were alive."

The two labs communicated about the problem. Dr. Hine thought that the organ changes noted by the Dow group were "surprising," but he agreed to review and rerun some of his tests. The new tests did indeed turn up results similar to those occurring in the Dow labs.

Both labs continued to test at different dose levels. The lowest dose tested by Dr. Hine's group was five parts per million, which is the equivalent of five parts DBCP in a million parts of air or like putting five people with black hair in a group of one million people all who had blonde hair. Even at the lowest dose, the rats had retarded growth, organ damage, and undersized testes. At double the dose, or ten parts per million, all but one of the surviving rats had testes half the normal size. Those rats that survived the twenty parts per million exposure were completely azoospermic, or sterile. At forty parts per million, all rats were dead.

In his written "Confidential Report UC No. 278" dated April 21, 1958, Dr. Hine observed that "among the rats that died, the gross lesions were especially prominent in lungs, kidneys and testes. Testes were usually extremely atrophied."

The Dow group's report—which was circulated to a wide group of people inside, but not outside, the company—followed on July 23, 1958. Repeated skin and inhalation exposure tests on rats, rabbits, guinea pigs, and monkeys showed that DBCP was "readily absorbed through the skin" and "high in toxicity by inhalation." Breathing twelve parts per million DBCP over a course of fifty to sixty-six seven-hour days—what a worker might breathe in three months' time in a factory if air concentrations were that high—killed nearly half the animals. Their abnormally small testicles had reduced numbers of sperm cells, some of them malformed, or none at all. The rats also showed liver and kidney damage. The two monkeys in the experiment were the most seriously affected by chronic exposure to DBCP. They grew ill, weak and listless, and susceptible to infections.

"These data also show that liver, lung and kidney effects might be expected," the Dow group concluded. "Testicular atrophy may result from prolonged, repeated exposure."

Neither of the two research groups had tested down to the no-effect level—which is to say, they did not know the highest dose an animal could assimilate without harm. The "no-effect level" is a synonym for "safety" and the much disputed cornerstone of traditional toxicology, which believes there is a safe level for virtually every substance on earth, below which no injury will occur. The Dow group found damage at its lowest dose tested—twelve parts per million—as did Dr. Hine at five parts per million. Without knowing the no-effect level, both research groups made judgment calls, suggesting in their separate reports that workers should not be exposed to more than one part per million DBCP. "In view of the severity of the effects observed [at the twelve parts per million level], a tentative hygiene standard of 1 part per million is suggested," the Dow group wrote in its report. "Further experimental data must be obtained before such a standard can be set with confidence, however."

Except for the effect on testicles, the Shell-Dow findings fitted almost exactly into the types of target organ effects that toxicologists were trained to hunt. "Something was wrong with the liver and kidney," Dr. Torkelson said. "But the testes appeared most likely to be more of a secondary, nonspecific effect." The scientists believed that the testicular problems were caused by the animal's general ill health, not by the direct action of the chemical. As for finding an exposure level below which a chemical was thought to be safe, the chemical industry tended to counsel the "need to be realistic," as one of Dow's scientists wrote in a 1959 report. It observed:

> Unfortunately, there is not enough manpower, time or money to allow such exhaustive studies to be made upon the thousands of chemicals or compositions facing the public today, nor is it generally necessary or desirable that such be done. From the standpoint of the worker and the public, it is far more desirable to expend a generous portion of the resources available in determining the physiological properties of many chemicals, rather than exhaustively studying a very few.

Still, the DBCP researchers believed more experiments were necessary before a safe level of worker exposure could be authoritatively set. But at the time the two reports were completed in the spring and summer of 1958, both Dow and Shell were already manufacturing DBCP, and had been doing so for some time. Shell started production of DBCP in 1956 at an insecticide plant it had taken over from the Julius Hyman Company four years earlier on an isolated plateau outside Denver, Colorado. Dr. Hine had not yet completed his tests on DBCP when Shell began making it, but the company nevertheless distributed a data sheet that confidently described the chemical as "moderately toxic" if inhaled or eaten, "slightly toxic" if absorbed by the skin, and not toxic enough to be rated as a Class B poison.

Casting around for a commercial name for its new nematocide, Shell came up with Nemagon, which sounded like a child's pronunciation of "nematodes all gone." Shell had been eager to start making Nemagon ever since the good news began filtering back from its Modesto researchers about the new economical nematocide that could be used on living plants. After the chemical company, through a series of seminars and Tupperware-like demonstrations on actual crops, had proved the existence of nematodes to farmers, such a clamor for Nemagon developed that Clyde McBeth, Shell's nematode man, was getting six calls a day from farmers anxious to buy it. "The demonstration plots were the best way to get the information out, and we had them all through the West," McBeth recalled. Shell would persuade a farmer to apply Nemagon to his field before planting cotton, for instance. Once the crop had come up, fine and hearty, university agricultural extension agents took all the cotton growers in the area on a field-day excursion to the demonstration plot, where McBeth and other Shell advisers showed them living proof of Nemagon's effectiveness. Afterward Shell might treat the growers to dinner and a slide show about nematodes or even a movie starring the tiny worms. The movie showed magnified pictures of the nematodes at work and horrified the growers with the astounding news that a nematode colony could increase from 250 to 1.5 million worms in three months' time. "Scientists continue to understand the minute but mightily important plant parasitic nematode," the film's narrator intoned as two men, barefaced and bare-handed, poured a five-gallon bucket of DBCP into a funnel in preparation for field application.

Dow began making its own DBCP-based nematocide in 1957, one year after Shell commenced production. Dow workers manufactured the product, which was baptized Fumazone, in Midland, Michigan. "Fumazone" was thought up by Dow's "electronic computer," which searched for new chemical trademark names by combining sets of letters and syllables into all possible combinations—i.e., Dowadine, Dowadyne, Dowafine, Dowaline, Dowamine, and so forth. Just like Shell, Dow had a nematode movie—*Thief in the Soil*—a ten-minute, full-color drama describing the damage inflicted by nematodes. (Movies were common advertising gimmicks employed by the chemical industry to display new products. *Uncle Henry Saves the Play*, for instance, was an ode to the dry-cleaning industry—a big market for Dow solvents.)

Dow spread the word about Fumazone through advertisements in trade journals and through special pamphlets and informative brochures distributed to growers. "Farmers no longer have to watch their prize acreage sapped of its profit by nematodes and other soil-borne pests. They can fight back—and win astonishing profits—with the new soil fumigants," Dow promised in a 1958 promotional booklet called *Plunder Underground,* which was decorated with drawings of worms and crammed with tantalizing facts about Fumazone and other worm combatants. The same year little worms wearing military helmets and carrying bayonets slithered across the top of a Dow promotion for Fumazone called *The Underground Battlefield.* "Most soils are teeming with crop enemies" was the message. Dow's entry into the DBCP market coincided with the discovery by its own and Shell's toxicologists that the chemical had more serious side effects than previously thought, and its information sheets reflected concern, albeit mildly, about the hazards. Dow advised applicators to wear gloves, shoes, and garments made of resistant materials and to avoid breathing the vapor or fumes.

There had been a dispute early on over who would get the various patents governing the manufacture, formulation, and sale of DBCP products. After some initial assignment, bickering, and shuffling of rights, Dow and Shell eventually entered into a cross-licensing arrangement, which essentially gave the two companies exclusive control of all the patents necessary to make and sell DBCP as a nematocide. Shell began shipping DBCP to Best Fertilizers in Lathrop, California, where workers blended it into various formulas and repackaged it for sale.

Shell's marketing and medical staff, meanwhile, had been monitoring Dr. Hine's research as it progressed over four years. His final report in 1958 on the lung, kidney, and testicular effects of repeated vapor exposures caused some concern. A copy of the report was sent to Shell's technical service staff—the people responsible for getting a chemical registered with the government—at marketing headquarters in New York City in May 1958. Edmund Feichtmeir, Shell's products application manager, typed up a message and attached it to the report: "I know from discussions with Dr. Hine that data of this type has been secured by Dow Chemical Co. Dow in particular has been very upset by the effect noted on the testes."

A week later Louis Lykken, Shell's technical services, or registration, man, passed the report to Dr. Mitchell Zavon, Shell's medical director for agricultural chemicals, with the parroted addendum: "We understand that Dow have [sic] similar data and are [sic] very upset by the effect noted on the testes."

Little more than a week after that Dr. Zavon sent a letter back to Shell-New York outlining his concern about workers engaged in the manufacture of Nemagon in the Denver plant. "It appears to be more urgent that we pay additional attention to Nemagon and its effect, if any, on those people exposed to the material," he said.

Dr. Torkelson said Dow was not upset about the effect on the testes but was concerned that Dr. Hine's original results were so divergent from its own. " 'Upset' is probably not a good word, but yeah, we were wondering what was going wrong," he said.

The annual Gordon Toxicology Conference was coming up toward the end of that summer of 1958 on the East Coast. The conferences were held at college campuses and limited to the number of toxicologists who could fit in the dormitories. There were speeches and discussions about the latest developments in toxicology. "I remember times coming back after an evening session, and we'd go down to the bar and sit there for hours just talking about technical matters over a beer, with the discouraging of taking notes," Dr. Torkelson said. "Every time somebody keeps records, you have to be more careful."

Shell consultant Dr. Hine met there with his colleagues from Dow. Each faction brought its separate research data on DBCP; the plan was to discuss the data during breaks at the conference with an eye toward merging the two sets of information into one for publication in a scientific journal. Dow toxicology chief Verald K. Rowe took the data home with him to Michigan, and Dr. Torkelson was appointed the task of weaving the reports together. The combined report was not published in the scientific journal *Toxicology and Applied Pharmacology* for another three years. In the interim Shell's medical director asked Dow to update him on the toxicological effects of DBCP, and the company sent along the essence of the two reports, along with an admonition from Dr. Rowe. "I believe the material deserves a tough label," he said.

"Part of our goal here obviously was to protect the worker," Dr. Torkelson said. "In fact, that was the major goal. At that time we were probably one of the leading companies in industrial hygiene. Industrial hygiene and toxicology are just two parts of the same thing. It doesn't do any good at all to study the animals unless you apply it to the people.

"Based on our animal data, we would recommend [worker exposure] guides," he continued. "A bunch of us would sit down and say, 'This looks like exposures ought to be limited to a hundred parts per million because this stuff has no toxicity.' Or in the case of DBCP, we said, 'Hey, we got a bear cat here; we've got to limit exposures to less than one part per million.' There are very few chemicals where I have had to recommend what we recommended on DBCP. We just said, 'Don't get it in your eyes; don't eat it.' This was an extremely toxic material, and we just said, 'This is a son of a bitch! Handle it carefully.'"

Dr. Zavon told his Shell colleagues about Dow's concerns. Dow's team had pointed out that levels as low as five parts per million in the air "were a hazard to exposed persons" and suggested that Shell's labels for Nemagon be toughened to reflect this hazard more accurately. The first thing Dr. Zavon wanted to do was measure the concentration of Nemagon in the air at Shell's Denver plant. Two Shell officials discussing this plan in the autumn of 1960 observed, "Dr. Zavon has no intention of alarming plant personnel but he would like to discuss the subject in detail with [plant manager] Mr. McGilvray and others who you feel should be alerted to this potential hazard." Dr. Zavon subsequently visited the plant and found that worker exposure was "under reasonably good control." But the data showing harmful effects at exposure levels as low as five parts per million nagged at him. "It would appear to be wise to tighten up the operation even further," he wrote in an internal memo, suggesting that air concentrations be reduced "as close to a maximum of 1 part per million as possible."

At long last, in the spring of the following year the finishing touches were applied to the combined Dow-Shell report on DBCP. "I doubt if you can believe this," Dow's Dr. Rowe wrote to Shell's Dr. Hine shortly before the study was submitted to *Toxicology and Applied Pharmacology* in March 1961, "but enclosed are six copies of a paper entitled 'Toxicologic Investigations of 1,2-Dibromo-3-Chloropropane.' ... You can see that we have made an effort to put the data from the two labs together and I hope it makes sense." The report was published in the magazine's September 1961 edition. It summarized the results of toxicologic studies on laboratory animals, studies that the authors said were conducted in order to allow the assessment of toxic hazards associated with the chemical's manufacture, handling, and use. The major findings included damage to the liver, kidneys, and various tissues, including sperm cells. In summing up the three-month vapor exposure tests at the twelve parts per million level, which killed 40 to 50 percent of forty rats (deaths that were largely attributed to lung infections), the authors wrote, "The most striking observation at autopsy was severe atrophy and degeneration of the testes of all species [rats, guinea pigs, rabbits, and monkeys]. In the rats this was characterized by degenerative changes in the seminiferous tubules [where sperm is produced], an increase in Sertoli cells [which nourish growing sperm], reduction in the number of sperm cells, and development of abnormal forms of sperm cells." Similar but less widespread damage to the testes occurred after repeated exposures to five parts per million, the lowest level tested.

The report went so far as to attempt to counter the "severe effect in the testes of animals" with hormone injections (they failed) and to put two human volunteers into a 130-cubic-foot room to test the atmosphere for odors of DBCP. They reported smelling a "definite, not unpleasant" odor at a level of 1.7 parts per million. This level became what was known as the concentration that had "warning properties" —in the case of DBCP, if you could smell it, there was too much in the air.

The authors suggested, as they had in their separate reports, that the concentration of DBCP in the workplace be kept below one part per million. "The warning odor MUST NOT be ignored," they wrote. Further safety precautions were proposed: adequate ventilation systems under normal working conditions; full-face gas masks equipped with vapor canisters or some type of breathing device in the event of spills. In addition, workers should wear standard protective clothing impermeable to the material, preferably Compar rubber or polyethylene as opposed to standard rubber or neoprene gloves, which did not afford adequate protection. Clothing and shoes, naturally, should not be allowed to become contaminated with the chemical.

"The '61 report said it all," explained Dr. Torkelson. "It was designed to take everything we knew so that it would be public and would be followed by anybody who was diligent enough to do even the most limited search." The report was circulated to the magazine's 600 or so subscribers and to chemical companies that subscribed to the *Chemical Abstracts* and *Biological Abstracts* indexing service. "In grad school back in the fifties every chemistry department had *Chemical Abstracts*," Dr. Torkelson said. "It came every two weeks. It was available in every university, in technical libraries; every chemical company would have had it. It was real easy to search by chemical name. If one had looked for 'testes,' it would have been harder to find." Some 1,300 reprints of the article were distributed by Shell and Dow. The report was also submitted to the federal government in support of a joint petition by the two companies for the establishment of tolerances, or maximum allowable amounts, of DBCP residues on forty-four food crops. The Food, Drug, and Cosmetic Act, passed back in 1938, and subsequent amendments had given the public statutory protection from pesticides by investing in the Food and Drug Administration (FDA) authority to set tolerances on any pesticide that left residues on raw agricultural products. Nematocides were thought not to leave residues and thus were exempt from routine federal pesticides registration laws until 1959, when the law was amended to include them. Once the tolerance had been set by the FDA, the USDA took over, approving pesticide labels if data showed that the chemical was safe and effective when used as directed and registering only those uses for a chemical that left residues lower than the tolerances. The process was designed to provide adequate warnings on pesticide labels so that the worker and field or home applicator would be protected from harm. At the same time the consumer would be protected from eating food that contained dangerous levels of pesticide residues. But the law often failed to live up to its promises. Scientists were primarily looking at the quick, readily observable effects of a chemical on an animal, not at the long-term damage. Analytical equipment was incapable of detecting the minute amounts of residue left behind by some chemicals. Tests required for registration of a pesticide in the 1950's generally were cheap to run ($10,000 tops) and quickly finished (three months). Tests to determine if a chemical was carcinogenic, or capable of causing cancer, were in their infancy. There was virtually no monitoring of reproductive organ damage, cell mutations, or birth defects. The most popular test by far was the LD50, which measured death, the most toxic effect of all. A chemical's LD50 was the dose that killed half the animals tested over a specific period of time.

Because testing was limited and because its residues evaded detection, a pesticide like DBCP could, and did, slip through the regulatory holes. If a chemical exhibited alien effects that did not fit into any of the regulatory boxes, the scientist and the company he was working for had two choices: press on with more studies and alert the government to a potential problem; or ignore the effects, get the chemical registered, and sell it. Even alerting the government was no guarantee of safety because its technical knowledge often lagged behind industry's. The scientist was under none but a moral obligation in choosing his path, and in the 1950's morality was often synonymous with progress, although Dr. Torkelson insisted, "I've never felt the pressure to short-change. I've always said, 'If it's low toxicity, you want to tell people; if it's bad news, you've *got* to tell 'em.'" Dr. Etcyl Blair, a research chemist who was hired by Dow to synthesize new agricultural chemicals in 1951, killed a herbicide project to which the company had devoted years of work and millions of dollars because it produced some troublesome health effects in animals and did not break down without sunlight. Monsanto subsequently developed a herbicide to fit that segment of the market, and Dow did not. "Monsanto has five hundred million dollars in that product," Blair said. "So it's big business. That decision probably set our whole program back light-years."

In 1960 Shell asked its consultant, Dr. Hine, to write a summary on DBCP for use in a technical data sheet, which was given to customers who wanted information about particular chemicals. Shell also planned to use Dr. Hine's data in support of the government registration. The agreed-upon price for twenty hours' work was $500, a cost to be split by Shell and Dow because the results would be used to support the joint petition. The draft report was completed in May 1961, but Louis Lykken, Shell's technical services man in New York, had some problems with it. He thought that the paper "read with considerable difficulty" and needed to be "tightened up." Lykken flew to San Francisco to meet with Dr. Hine and to explain the changes he had proposed in comments scrawled in the margins of the report. Dr. Hine had written that DBCP could be classified as "moderately to highly toxic" following one respiratory exposure and "highly toxic" on repeated exposures, even at five parts per million levels. Excessive exposures to the vapors could cause "cumulative toxic effects" on the liver, kidney, respiratory tract, and certain surface tissues. "The most striking effect at autopsy," he wrote, is "severe atrophy and degeneration of the testes in all species studied."

Hine repeated the caution to maintain workplace concentrations below one part per million, adding that protective clothing impermeable to the material should be worn if skin contact was likely. Lykken scratched out the part about impermeable clothing and wrote next to it, "Impractical." In the margin adjacent to Dr. Hine's observation that repeated exposure to the chemical could have an effect on the human reproductive system, Lykken scrawled, "Do not extrapolate." He added, "Leave out speculation about possible harmful conditions to man. This is not a treatise on safe use." He further advised Hine to "delete" all references to testing that the doctor thought advisable—including metabolism, or how the chemical was broken down in the body, and effects on reproduction and lactation, to name several—but that had not been done.

In a follow-up letter, Lykken admonished Dr. Hine that "human use experience must take precedence over animal experience and the report ought to show this clearly. In this regard, the recommendations for use of impermeable clothing is [sic] impractical. Millions of pounds of DBCP have been used and manufactured and, to our knowledge, no one has ever worn impermeable clothing."

Lykken enclosed, as a model, a copy of a report written by Shell's medical director, Dr. Zavon, on another Shell pesticide called endrin. He further advised that demand for Hine's summary report had now lessened because the Food and Drug Administration had extended the date by which tolerances for DBCP had to be established. Dr. Hine says he became preoccupied with other work and never rewrote the report to Lykken's specifications.

Shell and Dow were, by this time in 1961, working together to complete the registration process and obtain approved labels for their DBCP products. Trouble developed that summer when a Shell man called on the USDA in Washington with a proposed label for Nemagon. Dr. John Leary and his colleagues in the USDA's pesticides regulation division told the Shell representative the label was being rejected because data received from independent labs on the West Coast showed that Nemagon applied to the skin of five rabbits caused all of them to deteriorate physically and killed two of them.

Shell officials were not pleased with the implications of the USDA ruling, even though the regulators promised to reconsider if Shell and Dow submitted new data to "prove or disprove" the chemical's safety. One marketing man worried that "elaborate precautionary statements such as 'discard contaminated shoes and clothing' " would have to be applied to Nemagon labels in order for it to be used around the house and garden. The USDA subsequently gave Shell, which was acting as the front man, four demands (Dow's Washington representative sent a Western Union econogram to the company's Midland headquarters about an upcoming meeting with USDA officials regarding the demands, but Dow has no record of ever receiving the demands themselves).

As the first condition, the USDA wanted this information on the label:

> WARNING: May be fatal if swallowed, inhaled or absorbed through skin. Irritating to eyes, nose and throat. Do not breathe vapor. Use only in well ventilated area. Do not get in eyes, on skin or on clothing. Wear polyethylene gloves and goggles when handling material. In case of contact immediately remove contaminated shoes and clothing and wash skin with soap and water; flush eyes with plenty of water for at least 15 minutes and get medical attention. Wash thoroughly before eating or smoking. Wash clothing and air shoes thoroughly before re-use. Do not contaminate feed and foodstuffs. To protect fish and wildlife, do not contaminate streams, lakes or ponds with this material. In case of accidental spillage indoors, have available a self-contained breathing

apparatus, or airline respirator, or a full face gas mask with canister for protection against organic vapors and gases.

The USDA also wanted to see more skin toxicity data and Shell's handbook describing the use of Nemagon in controlling nematodes. Thirdly, the label would have to include the warning "For Pest Control Operators Only" to discourage use by casual home gardeners.

"In addition," the USDA request said, "in view of the testicular atrophy demonstrated to occur in experimental animals, we would like to have information regarding health records of those individuals who have been employed for an extended period in the manufacture or formulation of products containing 1,2-dibromo-3-chloropropane."

Shell thought the four demands were out of bounds. A flurry of memos ensued as management worked to create a counterattack strategy.

August 21 We have discussed with Dr. Zavon USDA's views on precautionary labeling and the hazards associated with this pesticide chemical. He shares our opinion that USDA is being overcautious in their views on labeling products containing this pesticide chemical. It is the consensus that Dr. Zavon and a representative of Dow's toxicology group should meet with USDA Toxicology Section representatives to settle this issue.

August 29 We have just received and reviewed the subject Technical Bulletin [an information brochure on Nemagon] and have some reservations with regard to the adequacy of the statements under Safety Precautions. In light of the fact that the threshold of odor detection has been reported at 1.7 parts per million and the lowest level studied [5 parts per million] has demonstrated damage after repeated exposures, it appears the statement "there is a good margin of safety in handling" would be difficult to justify and might be prosecuted as negligent.

November 9 The Pesticides Regulations Branch of the U.S. Department of Agriculture has expressed concern over the hazards associated with the use of Nemagon soil fumigant and has proposed stringent labeling for the various formulations now being marketed. It is the consensus in the Division Office that the USDA is being overly cautious and the precautionary statements proposed could have an adverse effect on the sale of this product. This matter has been discussed with the USDA representatives and they are willing to relax their labeling requirements if we can provide them with a history of safe use experience in the field and in the manufacturing plant.

Toward that end a health survey of Shell's Nemagon workers was begun, and the workplace air tested for chemical concentration. The plant physician at the Denver facility began administering to workers routine physical examinations, including chest, X-ray, blood, and urine tests and brief visual inspection of the testes. The physician was not told to be on the lookout for testicular effects, and he made no special effort to find them. It was the same battery of tests he routinely administered once a year to workers over the age of forty and once every eighteen months to younger employees. (For a time each new employee in the insecticide plant received an electroencephalogram—EEG—which was thought to detect changes in brain wave patterns caused by chlorinated hydrocarbon insecticides. EEGs given periodically to the same worker could be compared to his initial results in order to track changes.)

The physician found no ill effects in workers, save for minor eye and nose irritations "when Nemagon vapor concentrations reached a level of 10 parts per million." Measurements of Nemagon in the factory air revealed that concentrations were "normally maintained" below the recommended one part per million level but "occasionally" rose to levels as high as fifty parts per million for thirty minutes "or more." The recent reports by Dow and Dr. Hine's group at the University of California had shown that Nemagon was more toxic to animals than had originally been thought, so "during the later runs of the unit, more emphasis has been placed on keeping Nemagon atmospheric contamination as low as possible." Workers wore rubber shoes, coveralls, and plastic gloves. Personal respirators, army types of gas masks, rubber suits, airline respirators, and demand types of oxygen masks were available if needed.

With the results of its manufacturing survey in hand in 1962, Shell concluded that Nemagon had caused no adverse effects in workers, for it had been "used extensively in the United States since 1956" with no cases of intoxication or injury reported. A twenty-four-hour rabbit test was included to demonstrate the chemical's harmlessness to skin, and the company sent along the whole report to the USDA with the hopeful message that Nemagon could be used without "undue hazard," even by those who were nonprofessional or home gardeners. The government was persuaded, and in 1963 proposed tolerances for DBCP on food crops were announced in the *Federal Register*. The following year Nemagon and Fumazone were formally registered—without labels containing stronger safety warnings. Although Shell played the front role,

Dow had cooperated with Shell in the joint toxicological study published in 1961, was made privy to the USDA's concerns about the chemical, may have joined Shell in meetings with the USDA in Washington, D.C., and thereby benefited from the joint registration process.

So it was that people using DBCP around the home, or applying it to crops, or mixing and canning it in pesticide factories were confronted by a label the harshest admonitions of which were "Do not breathe vapors" (or, in Shell's case, "Avoid prolonged breathing" of vapors) and "Use only in well ventilated area" (absent from Shell's label). As for workers, the recommendation to keep air concentrations below one part per million was just that: a recommendation. The American Conference of Government and Industrial Hygienists met annually to discuss threshold limit values, or guidelines for exposure to industrial chemicals, but no such guideline had been established for DBCP. Neither the states nor the federal government had ever set a workplace exposure standard for the chemical. In other words, it was left up to the individual plant to set limits and enforce them.

Dow's and Shell's farmer education and promotion program—and DBCP's unquestionable effectiveness—had begun to pay off handily. The nematocide was so popular by now that something akin to a price war was being waged between the two major manufacturers. "Dow came in with Fumazone and undercut Nemagon prices—they said they had a better method," Clyde McBeth, Shell's nematode man, recalled. "Dow and Shell men crisscrossed California in competition. We had demonstration plots side by side to show that Nemagon was just as good as Fumazone."

Shell had been shipping technical-grade DBCP to Best Fertilizers for formulation into Nemagon since 1958, but that relationship came to an abrupt halt in 1960. Jack Horner, who ran Best's nematocide division, told his colleague Clyde McBeth, "I hope we can still be friends, but we're not going to be working together anymore." After severing ties with Shell, Best tried and rejected several plans and finally bought its own manufacturing reactor so it could make DBCP from scratch. Then, in 1964—one year after the Occidental Petroleum Corporation had bought out Best Fertilizers and installed its chemical division in the plant—management struck a deal with Dow to make Fumazone. Dow furnished the raw materials, Occidental reacted them to make DBCP and added solvents to dilute the mixture to various concentrations, then bagged and shipped the finished Fumazone to Dow customers. Under terms of a contract signed that year, Occidental pledged the annual production of half a million pounds.

Back at Dow headquarters in Midland, Dow entered the chemical in its medical files for the first time in 1964. Company toxicologists listed effects of all of Dow's chemicals, and these were typed up and entered on medical cards for use by company doctors in case of accidents or information requests. The medical cards were primarily designed for cases of single exposure to a chemical, as when a worker breathed fumes from an accidental spill. Fumazone's medical card noted "moderately irritating" eye effects, "slightly irritating" skin effects (although small amounts could be absorbed through the skin and could produce systemic injury, even death), and testicular degeneration, which "may result from chronic exposure to active material." Liver and kidney injury was a possible consequence of excessive exposure from ingestion or inhalation of vapors. The vapors were capable of producing anesthesia.

After 1964 the medical cards on Fumazone were slightly modified. "The cards addressed single exposures, and you don't see testicular changes on single exposures," said Dr. Torkelson. "It was put on some of the cards because whoever wrote it knew about it and thought it was important enough to put on there along with liver and kidney effects. When it was retested, the testes didn't show an effect from a single exposure, and a new card was made out."

On the new cards, all the possible effects were essentially the same as before 1964, save for one: Warnings about testicular effects were removed.

chapter 3

In September 1954 the Dow Chemical Company became the corporate sponsor of NBC's television series *Medic,* which dramatized authentic medical case histories. Dow used its commercial time on *Medic* to advertise Saran Wrap, the first product the company had ever marketed directly to the public. The transparent plastic food wrap had been test-marketed the year before in four Ohio cities and promoted on three of NBC's most popular television programs—the *Today* show, the *Kate Smith Hour,* and Sid Caesar and Imogene Coca's *Your Show of Shows* —and in the Sunday newspaper supplement *This Week.* The promotional blitz for Saran Wrap was calculated to make Dow a household word. The company had too long been buried under the anonymity of its industrial chemical products, and it was aiming for a higher public profile.

Dow's promotions came at a time when the entire chemical industry was engaged in an image-building program. The Manufacturing Chemists Association, the industry's lobbying arm, was savvy enough to know that the industry was dependent on public opinion for its continued success, and it had put together Chemical Progress Week earlier in the year to draw attention to the fact that 40 percent of all chemical sales were of products that had not existed ten years before, that each of the seventy-two major types of manufacturing depended on chemical products, that the industry spent $275 million a year on research, and that its safety record was twice as good as industry's record as a whole. "The chemical industry rapidly has become a major contributor to a fantastic industrial society that has produced a living standard undreamed of a generation ago," the trade group said.

Through its wartime contributions, its remarkable discoveries, and its self-promotions, the chemical industry had gained the kind of respectability that evaded it during the early years, a respectability partially built on two oft-repeated themes: Progress was impossible without chemistry, and safety would not be traded off for profit or advancement. The industry issued soothing platitudes about the depth and width of its research to guarantee a product's safety before it was marketed. "With safety woven so deeply into every phase of Dow's research and development programs," went one of the company's promotions, "the public can be assured of complete safety of Dow products under the recommended conditions of use."

It was a time of great prosperity and confidence, for both the chemical industry and the country. Dow's annual growth rate was exceeding 14 percent a year, and it was in the middle of a $431 million capital expansion program. The propped-up economy of the war years had been replaced by a highly competitive market, which taught that self-reliant commercial initiative was the way and that progress was the truth.

There was a wide-eyed appreciation of all things industrial. Dow called its petrochemical stacks "20th Century minarets" growing "slim and tall against the sky" and defined progress in numbers: the number of new plants capitalized that year; the number of dollars spent over a number of years on industrial expansion; the number of new patents granted for new chemical compounds; the number of products on the market.

There was an undisguised loathing for all things that hampered industry. In the summer of 1952 Dow asked the question "Where do we go from here?" in one of its many trade magazines. A signpost offered two paths. One veered off toward government spending, inflation, higher taxes, and controls. The other led to opportunity, individual freedom, incentive, and stability. The path to opportunity was open only to those who "resist" the idea that Americans must be "regulated and shepherded" and "milked" like "a herd of goats" and who throw off any convictions about the "necessity or rightness of big government."

Science and scientists were preserved in a solution of innocence and purity. Dow said its scientific research was "born in the Yankee ingenuity which gave our forefathers the Colt Revolver and the Winchester Rifle." Scientists everywhere were accorded something akin to papal infallibility in their pronouncements. The sentence "Scientists have determined that it is not hazardous to man and domestic animals when used in accordance with directions" became as common and as trusted as the benediction at mass. And as much as Dow fought unnecessary government regulation, it didn't hurt to remind its customers that "federal and state agencies have been doing an excellent job in carrying out their functions of protecting the consumer."

What Americans came to believe unquestioningly, then, was that science equaled new chemical discoveries equaled progress, and little harm could come from those new chemicals because science kept a check on them. With government promising to regulate the side effects of chemicals, the feeling of security was reinforced.

The stupendous rise in the use of agricultural chemicals during the 1940's and 1950's goes to the heart of the matter. In 1948, 38 million pounds of DDT were sold and the Swiss chemist Paul Müller was awarded the Nobel Prize in medicine for his discovery of the powerful pesticide. By 1952 Americans were buying 115 million pounds of DDT a year. Dow told its stockholders that year that "agriculture in recent years has been undergoing vast changes and our company has both contributed to and profited by its modernization in considerable measure." Six years later, with American growers losing $13 million a year to crop pests, Dow reported that its agricultural chemical division was the only area of sales that showed a gain over the previous year, with herbicides accounting for better than half the total.

The laborsaving herbicides were a surefire growth industry in a country with a declining farm population. As Dow pointed out in its promotions, "with one pound of 2,4-D a farmer today can accomplish in an hour what otherwise would require 100 hours of manual labor." By the end of the 1950's chemical weed killers were part of every grower's artillery. The estimated annual loss caused by weeds, and the cost of their control, totaled nearly $5 billion. And industry was selling some 100 herbicides in 6,000 different formulations.

The productive American farm became the symbol that Dow and the rest of the industry thrust forth like the Olympic flame to hammer home the benefits of capitalism and chemicals. "Four out of five people go to bed hungry every night," Dow declared in the summer of 1960. "In this country, we are fortunate to have a sufficient supply of food, largely made possible through the use of chemical products." The industry was backed up by its allies in the agricultural institutions. "Abandoning the use of chemicals on farms and in the food industry would result in an immediate decline in the quantity and overall quality of our food supply," warned Ezra Taft Benson, secretary of agriculture under President Eisenhower, "and cause a rapid rise in food prices paid by the consumer."

Whether the chemicals were safe—to the workers who made them, the farmers who used them, and the people who ate the sprayed food—was an issue debated only in the tightly closed circles of industry and government. There did exist a group of people who opposed the prevailing view that if a chemical did not harm you quickly, it was safe, but this dissent was not organized, and it was buried in the scientific literature. As early as 1942 a book by one of the country's pioneers in occupational and environmental cancer research linked toxic chemicals to cancer in man, but coming as it did several days after the Japanese had bombed Pearl Harbor, its message was undernoticed. *Occupational Tumors and Allied Diseases* was written by Dr. Wilhelm Hueper, a pathologist who left Du Pont's Haskell Laboratory of Industrial Toxicology in the late 1930's.

Dr. Hueper, who later directed the Environmental Cancer Section of the National Cancer Institute, wrote:

> The gigantic growth of modern industry occurring in its main portion within the lifetime of men now living . . . has introduced numerous artificial, heretofore unknown, exogenous factors in constantly increasing number and variety. The creators and beneficiaries of the industrial development are thereby made potential victims of health hazards which cause numerous and diverse acute as well as chronic and insidious diseases never observed before.

Dr. Hueper believed that industrial carcinogens first and most forcefully affected workers who were in daily contact with them and then spread from the plant into the general environment either as part of the product itself or through air and water pollution. His theory was not especially new: The first doctor to suspect that environmental agents played a role in human disease was Percivall Pott, a British surgeon who published his *Chirurgical Observations of Cancer of the Scrotum of Chimney Sweeps* in 1775. Pott deduced that the accumulation of soot in the genitals of London chimney sweeps was responsible for the high incidence of cancer of the scrotum. In 1822 another British physician identified the connection between arsenic and skin cancer. By the end of the nineteenth century half a dozen sources of industrial carcinogens were known. But the affirmations of science often trailed by centuries the first suspicions that chemicals in the workplace and the environment were causing health problems. A classic case in point was the rapid turnover in husbands among women in a German-Czech mining region in the early 1500's. More than four centuries later scientists discovered that radioactive pitchblende in the mines was causing the men who worked there to die young.

Dr. Hueper's work in the 1940's, and similar research by Dr. Malcolm Hargraves of the Mayo Clinic and others in the 1950's, were discounted as anecdotal and inconclusive by the mainstream medical community, although even as conservative a body as the American Medical Association worried out loud in 1951 about the growing concentration of pesticides in human body fat. "And a new principle of toxicology has, it seems, become firmly entrenched in the literature," wrote Dr. Morton Biskind in the *American Journal of Digestive Diseases* in 1953. "No matter how lethal a poison may be for all other forms of animal life, if it doesn't kill human beings INSTANTLY, it is safe. When nevertheless it unmistakeably does kill a human, this was the victim's own fault—either he was 'allergic' to it (the uncompensable sin!) or he didn't use it properly."

Warnings about pesticide dangers occasionally erupted in the popular press in the 1950's. "No nation in the world has so high a record of polio and of degenerative diseases at middle-age as this one," observed *The Cleveland Plain Dealer* in 1951. "No nation uses strange poisons so freely and indiscriminately without investigating or knowing anything concerning the effects of these poisons." But the bulk of popular reporting leaned in the direction of openmouthed awe, no questions asked. "The fabulous Dow Chemical Company opened its doors to the working press for the first time and revealed itself as a hitherto hidden house of wonders," the *Detroit Free Press* reported in 1950. "The clothes you are wearing, the ice cream you had for lunch, your wife's permanent wave, the pharmaceuticals in your medicine chest, your children's toys and your automobile all most likely have ingredients in them which came from Dow."

Industry and the bulk of science were so united and so convincing in their defense of the safety of chemicals that the public could not help feeling protected. That security was bolstered by a group of federal and state laws that, by the end of the 1950's, purported to protect food from harmful chemical residues, require stringent label warnings for the most toxic of chemicals, and provide a mechanism whereby chemicals could be banned if proved unsafe. None of that was true, but only a few people recognized it. The chemical industry tranquilized lingering doubts by promising that it had not only legal but "moral" responsibilities to its employees and customers. "The regulatory is not the only factor involved in determining what we do," a Shell man told a congressional committee in the early 1960's. "We think moral and legal obligations are as important, or maybe more so."

When irksome problems began to develop in the postwar years with chemicals in general, and the herbicides in particular, they were easy to ignore. This was the generation that was hitching its star to the wagon of technology, and there was something very close to hero worship going on between Americans and science. It was a relationship that precluded doubts.

So it was that two mass outbreaks of skin disorders among workers at chemical factories in West Virginia and West Germany in 1949—one year after the herbicide 2,4,5-T had been put on the market—were virtually ignored by the American herbicide industry, as had been earlier exposures that disfigured German chlorine workers in 1899, the Mississippi lumber workers treating wood with chlorophenols in the mid-1930's, and the Dow workers making chlorophenol compounds in 1937.

The West German workers were making the wood preservative pentachlorophenol. They also were doing some experimental work with 2,4,5-trichlorophenol, the raw material from which the herbicides 2,4,5-T and silvex are made. They had been working with the chemicals for fewer than five months when the foreman of the plant reported that one of his workers had severe skin eruptions. Nine other workers developed skin problems during the next few months. All but one were still troubled with the skin disease a year later, when they were examined by doctors. In addition to the skin problems, most of the workers had pain, weakness, and a sensation of pricking and tingling in their lower backs, buttocks, and thighs. Some had heart palpitations and shortness of breath and diminished interest in sex. Most of these symptoms developed weeks, or even months, after exposure.

In 1951 German doctors writing in the American medical journal *Industrial Medicine and Surgery* clearly established a connection between the workers' exposure to pentachlorophenol and their chloracne and neuralgic conditions. They reported that rats injected with pentachlorophenol became listless, defecated frequently, showed slight motor weakness, and fell into convulsive seizures prior to death. Autopsy showed extensive damage to the cardiovascular system; swollen thymus (the gland that regulates the body's immunological responses); dilation or enlargement of the heart; congestion of the lungs, trachea, and bronchi; an enlarged and sometimes swollen, fluid-filled liver; and congested kidneys. The doctors argued convincingly that chloracne was just the most visible manifestation of a general "intoxication," which affected the internal organs as well as the skin. They singled out the possible toxic effects on the liver for particular attention.

Eventually seventeen workers from the German plant were treated by doctors. All of them had chloracne, eleven had bronchitis, five suffered damage to the muscular layer of the heart wall, two had liver cirrhosis (one fatally), and nine had symptoms of neuritis, most of them involving severe pains in the lower limbs. Seven workers had physical complaints ranging from continuous fatigue to depression, lack of vitality, nervousness, slight headaches, disturbed sleep, and decrease in libido and potency.

In America, meanwhile, Monsanto began making 2,4,5-T in 1948 at its factory in a sleepy town in West Virginia's Kanawha River valley called Nitro, which took its name from the nitrocellulose produced there for gunpowder in World War I. Workers earned an extra 4 cents an hour—they called it the T rate—for operating the manufacturing vessels and attending to other chores in the herbicide production unit.

But an accident in 1949 in the area where the raw ingredient 2,4,5-trichlorophenol was being made left 228 workers, laboratory and medical personnel, and even the company's safety director and several workers' wives who had never been in the plant with chloracne. Like the West German workers, some developed even more severe problems, including nausea, headaches, muscular aches and pain, fatigue, shortness of breath, swollen livers, loss of sensation in the fingers and toes, emotional instability, and intolerance to cold. United States Public Health Service doctors who visited Nitro after the explosion indicated in a report that there was a connection between the chloracne and the chemicals produced in the plant.

As before, the workers were examined and treated, the diseases reported in the scientific press, and the plant continued production, as did dozens of other plants making 2,4,5-trichlorophenol and 2,4,5-T around the world. And the reports of industrial problems continued. There were four more incidents in West Germany from 1949 to 1954; three more in the United States and one in France during the 1950's; two in the United States and one each in Italy, the Netherlands, France, the United Kingdom, Czechoslovakia, and the Soviet Union in the 1960's.

But there were no answers—and precious few investigations—until German scientists ran some animal experiments in the aftermath of major chloracne outbreaks among workers in three West German factories in the mid-1950's. The first involved some sixty workers who had been engaged in the manufacture of 2,4,5-trichlorophenol for several years at a Badischer Anilin & Soda-Fabrik (BASF) factory. The chamber where trichlorophenol was manufactured—called the autoclave—had been overheating for some months, and all the plant workers and many of their wives, children, and even house pets developed chloracne. But no one tried to find out what was causing the chloracne until the overheated reactor exploded on November 17, 1953, spreading vapors throughout the plant and causing a wide range of illnesses, including severe and sometimes fatal liver disorders. A room next to the autoclave was so contaminated that even one year after the accident both humans and animals were affected after being in the room for only a short time. A scientist writing about the Hamburg toxin called it "the dreaded substance X."

At about the same time workers in two C. H. Boehringer Sohn trichlorophenol plants in Ingelheim and Hamburg developed chloracne. Experimental production of trichlorophenol had begun at the Ingelheim plant in the summer of 1951. In the spring of the following year the Hamburg workers began processing trichlorophenol into 2,4,5-T. The Hamburg workers spent most of their time shoveling trichlorophenol flakes from open barrels into the reactor. The shoveling operation stirred up clouds of fine dust, which spread throughout the work area. Between 1952 and 1954 nineteen workers at Ingelheim and eighteen at Hamburg developed chloracne.

When the workers fell ill, the plants were closed, and company officials consulted physicians from the University of Hamburg's skin clinic. While the plants were being cleaned out with ammonia and caustic soda, two rabbits were placed on each of the floors with the doors and windows closed. All the rabbits died within five days. Autopsies showed pronounced liver damage.

Dr. Karl Schulz, a young German physician, and Dr. Josef Kimmig, director of the Hamburg skin clinic, collaborated with Georg Sorge, the former manager of the Hamburg plant, on a search for the cause of the chloracne outbreak. They knew from reading reports about the BASF reactor explosion in 1953 that the "chloracne inciter" was not only extremely toxic but extraordinarily stubborn. The Hamburg researchers started by brushing trichlorophenol on rabbits' ears and feeding it to rabbits and cats. The rabbits developed skin irritations identical to chloracne and marked liver damage. But damage occurred only with the technical grade of trichlorophenol used in the factory, not with pure trichlorophenol. That led the researchers to believe that a mystery toxic compound was formed during the manufacture of trichlorophenol. They set to work synthesizing compounds that were likely formed as impurities in the manufacture of trichlorophenol. Most failed to produce chloracne when painted on a rabbit's ear. One compound, dibenzofuran, had a strong effect, and it was the prime suspect until, completely by coincidence, Dr. Schulz learned about a laboratory assistant at the University of Hamburg's wood research institute who had never been inside a trichlorophenol factory but who had developed severe chloracne. The man was referred to Dr. Schulz after seeking treatment at hospitals all over Hamburg for a severe case of chloracne on his face, neck, and other parts of his body. As it turned out, the wood research lab workers had been trying to come up with a better wood preservative than pentachlorophenol. In the process of chlorinating a chemical, this lab worker had produced a white crystalline material, which he placed in a drying box. When he opened the box to check the material, the vapors struck him "in such a way that he could not go to work the next day because of severe facial dermatitis." During the next two weeks all the persons working in the lab developed chloracne. The research project was immediately canceled.

Dr. Schulz obtained a few hundred milligrams of the white material from the box and was able to identify it as tetrachlorodibenzo-p-dioxin. It was part of a vast family of compounds called polychlorinated dibenzo-p-dioxins, of which there were seventy-five members. Dr. Schulz had stumbled on the "tetra" branch of the dioxin family—a branch that in itself had twenty-two members with the same molecular weight and the same number (four) of chlorine atoms. Each was distinguished from the others only by the positioning of the chlorine atoms on the main dioxin structure. The most toxic member by far of the dioxin family was the one with its four chlorine atoms attached at the two, three, seven, and eight positions, or 2,3,7,8-tetrachlorodibenzo-p-dioxin. The most famous and most deadly member, it came to be known in common parlance simply as dioxin.

The German doctors surmised that dioxin was formed as an unexpected and unwanted contaminant during the manufacture of 2,4,5-trichlorophenol before it was reacted with other chemicals to form the herbicides 2,4,5-T and silvex, some of the wood preservatives like pentachlorophenol, and disinfectants like hexachlorophene. In swift order, they synthesized dioxin in their own laboratory. They also isolated it from the waste residues of the factory's trichlorophenol. Dioxin brushed on rabbits' ears in impossibly small concentrations—as low as 0.001 percent of the total solution—produced severe skin reactions. Only when the dioxin was diluted to 0.01 percent and painted on rabbits' ears at two-day intervals was there any chance that the animals would even remain alive. Severe chloracne developed on the ears "uniformly and regularly" a week after the skin had been painted. Equally tiny doses of dioxin fed to animals led to severe liver damage. A single feeding of 0.1 milligram of dioxin per kilogram of the rabbit's weight killed it. Cause of death was liver damage.

The researchers had been puzzled at the beginning of their experiments when control animals that had no contact with trichlorophenol but that lived in cages next to animals treated with the chemical developed liver damage. Even control animals put in empty cages that had once housed treated animals had liver problems. The doctors had thought a virus was responsible, until the enormous toxicity of dioxin was revealed.

As the final act of this intricate piece of research, Dr. Schulz brushed a 0.01 percent solution of the chemical across his forearm to prove that dioxin was indeed the cause of human skin chloracne. Within several days the typical symptoms of chloracne covered his skin.

The findings were published in four German medical journals in 1957. The startling effects the researchers described were caused by 2,3,7,8-tetrachlorodibenzo-p-dioxin, but they had not specifically isolated that family member, so they referred variously to a close relative or to the tetrachlorodibenzo-p-dioxin group as a whole. Even though they had misidentified it, the researchers understood dioxin's extraordinary potency in causing chloracne and noted in particular its "high general toxicity" and its "unexpectedly high" action on the liver. Research on dioxin became widely available in America in late 1958, when one of Dr. Schulz's articles appeared in the December 10 issue of *Chemical Abstracts,* an indexing service that is read as heavily in the chemical industry as is *The Wall Street Journal* in business circles. The article clearly pointed to the presence of liver damage in several Boehringer workers exposed to 2,4,5-trichlorophenol.

After repeated cleanings, including sandblasting of the walls, and after using a new production method devised by one of its chemists to reduce dioxin levels, the Boehringer plant in Hamburg had gone back into the business of manufacturing trichlorophenol in early 1957. Georg Sorge, the former plant supervisor who collaborated with the University of Hamburg dermatologists in their medical investigation, believed that dioxin could be controlled by manipulating the temperature and pressure of the production process. He developed a low-temperature trichlorophenol process, which was more costly but which produced a less dangerous product. When trichlorophenol was manufactured Sorge's way, it showed negative results in Dr. Schulz's animal tests.

After the chloracne outbreak at the Boehringer plant in 1955, company officials had contacted the Swiss chemical firm Givaudan in search of safety information about the trichlorophenol production process. Givaudan referred the letter to the Dow Chemical Company, which responded by sending Boehringer a description of "the hazards and precautions for safe handling of 2,4,5-trichlorophenol."

It was only natural, then, that the German company spread the news of its new and safer trichlorophenol process to other chemical firms. In February 1957 Boehringer wrote a three-page letter to all manufacturers of trichlorophenol and 2,4,5-T worldwide, warning them of problems in production and alerting them to its safer low-temperature production process to avoid "the formation of chloracne exciters." The letter described the danger points in the process and the limits that had to be observed in order to avoid producing the acne exciter. Among other things, Boehringer suggested that a temperature limit of 155 degrees centigrade be imposed during the production of trichlorophenol to reduce the irritant. It was willing to market the technology to manufacturers that wanted to avoid production problems.

When the letter arrived at Dow headquarters in Midland, Dow sent its thanks back to Boehringer for "the description of the work you have done on the preparation of Trichlorophenoxyacetic Acid to avoid the formation of Chloracne exciters and consequent dermatitis to your workmen. This work is of very great interest to us. . . ."

The letter was subsequently "filed and forgotten." According to Dow's Garry Hamlin, a public affairs officer who helped piece together the history of 2,4,5-T at Dow, the company was unaware of the document and its significance. "It definitely was filed because it turned up in somebody's files," he said. "The best I can reconstruct it is that we received the letter, we did not have chloracne problems in trichlorophenol production or 2,4,5-T production at the time, and someone filed it. We did have it. That's all I can say." Dow did, however, ask Philadelphia dermatologist Joseph V. Klauder to conduct a series of human experiments to gauge the potential health effects of 2,4,5-T and silvex. In skin tests on fifty-one volunteers, Klauder noted that some of the materials were "highly allergenic" and that a 5 percent solution of silvex caused "primary skin irritation."

Two other American chemical companies should have been interested in the Boehringer news. Monsanto's herbicide workers were still feeling the aftereffects of the 1949 reactor explosion in Nitro, West Virginia. A similar explosion the following year—1950—spread chloracne throughout the sprawling brick Diamond Alkali herbicide plant located on the Passaic River in the heavily industrialized Ironbound section of Newark, New Jersey. Five years later, as health problems continued to plague workers, Diamond Alkali asked the Pittsburgh-based Industrial Hygiene Foundation (now the Industrial Health Foundation) to carry out health studies to determine the cause of the chloracne. The industry-supported foundation had been organized after 470 men had died and 1,500 had been disabled by lethal silica dust during the construction of a tunnel in West Virginia during the Depression. IHF president Dr. Daniel Braun said the foundation was started in 1935 by a group of industries that had a silica problem to fund research in silicosis. "The potential claims caused the organization concern," he said. The foundation had launched a massive public relations campaign to counteract the expected "flood of claims" from silicosis victims and the accompanying "improperly considered proposals for legislation." It was predicated on the idea that, as its membership committee chairman once told his corporate clients, "a survey report from an outside, independent agency carries more weight in court or before a compensation commission than does a report prepared by your own people. . . . Where industry attacks a great social-economic problem voluntarily, there is no necessity for government to step in and regulate."

The foundation study for Diamond Alkali lasted seven years. The study itself no longer exists, but doctors who visited the plant reported in 1963 that Diamond-sponsored studies found "areas of high exposure to chloracnegens." Said IHF president Braun, "They [Diamond Alkali] submitted materials to the [IHF] lab to be tested. The lab reported to them the results. We don't have a copy [of the report]." In 1959, while the new study was ongoing, the foundation sent a bulletin alerting its member companies to the German studies linking dioxin to chloracne.

While the American herbicide industry ignored the mounting evidence that dioxin was a dangerous impurity formed during the high-temperature manufacture of trichlorophenol, the Germans were closely tracking the effects of the chemical in its exposed factory workers and finding that dioxin induced not only physical but mental disturbances. The most carefully watched of all the German workers were those from the Hamburg plant, who were treated by the same university physicians who had originally pinpointed the dioxin problem in their plant. Five years after the initial outbreak of chloracne, nine of thirty-one workers were still receiving medical attention from University of Hamburg doctors for chloracne, chronic neuromuscular weakness of the legs, and, especially, marked psychopathological, or mental, disturbances.

"The course of the dermatological manifestations proved to be extremely obstinate in our cases," Dr. Schulz and his colleagues wrote in a 1961 medical article. Even though the doctors drained the blackheads and applied antibacterial solutions to the skin or prescribed antibiotic pills in severe cases, some men had relapses months after they had been removed from the plant. The most seriously affected workers were afflicted by "closely arranged pitted scars which have a disfiguring effect," especially on the face.

All the affected workers had pronounced fatigue and weakness, which was often accompanied by pain in their legs. Some of the workers had headaches and attacks of giddiness. Many had fine tremors in their hands and sweaty hands and legs. Most had abdominal troubles—a feeling of fullness or pressure in the stomach and liver region and slight pain. Biopsies showed that three men had liver damage.

Most remarkable to the doctors were the distinctive mental and behavioral changes that took place in the workers during the years following their exposure to dioxin. Dr. Schulz wrote:

> With a very large degree of agreement, a subjective syndrome of complaints was reported by the patients under investigation, this syndrome extending from the psychoneuropathic complaints in the region of the extremities, cardiovascular and abdominal symptoms to the mental/spiritual sphere, especially in the modes of behavior associated with the vital forces. Considered in detail, there were reports of disturbances in the vital senses such as general sense of weakness, feeling of fatigue, indisposition, sense of insecurity, inner restlessness and a feeling of illness. The basic mental mood was reported to be deteriorated and lowered towards behavior characterized by dissatisfaction or sullenness and irritation. Not infrequently, a mood component of fear and unease was present. General loss of strength and reduced inner vitality and impulsion were symptoms noted in each of the cases observed. [The workers] described reduction in initiative and

interests, weak willpower, reduced efficiency, and more rapid exhaustion in physical and mental/spiritual matters.

The doctors also reported that the workers suffered from deteriorated food and sexual appetites; there was a sharp reduction in potency, libido, and a lessened interest in food, which caused substantial fluctuations in weight. One worker complained of constantly changing eating habits—a wish for nothing but black bread, milk soup, and three liters of milk on certain days. Most workers had sleep disturbances. Some had reductions in memory and perception. Many were overly sensitive to alcohol, light, and noise.

Standardized psychological tests administered to the workers showed there was a significantly raised percentage of degeneration—"this," the doctors said, "providing a certain indication of an acquired decrease in mental capacity."

Follow-up studies on another group of German workers—those employed in the BASF factory that had been rocked by a reactor explosion in the autumn of 1953—showed remarkable similarities between their medical problems and those exhibited by the Hamburg workers. Chloracne lingered on their faces for years after they had stopped working in the plant. Some suffered additionally from swelling of the skin, excessive hair growth, loss of appetite, loss of weight, gastritis, severe liver damage, pulmonary emphysema, kidney damage, and a wide variety of neurological and psychopathological symptoms, including muscular disturbances, breaks in memory and concentration, decrease in initiative and interests, depressions, disturbances in libido and potency, and weakness in mental capacity.

The Germans would not provide exact figures, but they reported that several workers died as a result of liver damage, and one from intestinal cancer.

In one case, a fifty-seven-year-old mechanic wearing a full protective suit who, five years after the accident, entered one of the chambers where trichlorophenol had formerly been prepared, developed within a matter of days chloracne, headaches, and loss of hearing. One month later he was hospitalized with angina, then acute pancreatitis, and finally a painful tumor in the upper abdomen. When he died, an autopsy showed intestinal ulceration and degeneration of liver and fatty tissue.

In another case a young bricklayer who spent two hours in the manufacturing chamber repairing a wall developed severe chloracne. After a year's time he began running a fever, and a massive opaque area appeared on his left lung. His condition improved, but then, some years later, he began suffering acute psychosis with insomnia, loss of affect, hearing of voices, suicidal tendencies, physical discomfort, and a burning sensation in the back. He consequently hanged himself.

* * *

In 1957, when the Germans let the world know about dioxin and when the Russians launched *Sputnik,* the Dow Chemical Company launched a $2 million program to curb air pollution in Midland. It was a signal, however faint, that times were changing and that an industry was waking up to hard facts about its effects on humans and the environment. But if change was slow in coming in the rest of the country, it was glacial in company towns like Midland, where one businessman sniffed at the malodorous air and said, "The smell tells us Dow is humming."

One newspaper writer observed, "Residents long ago resigned themselves to the fact that there must be some disadvantages to keep a growing, profitable industrial concern here. It is visitors to the city who find the odor most offensive."

In the high-growth post-World War II years, when Midland had a population of 75,000, most residents were under thirty-five, fully 10 percent were college graduates, and 1 in every 100 was a Ph.D. Sheltered by a paternalistic company in an idyllic town, where Dow workers were never laid off, not even during the Depression, the people of Midland believed implicitly in the certainties of industry and science. For a time the Dow philosophy reflected the majority view of America. As times changed—as the rationality of science became suspect and Americans discovered the side effects of technology and the inability of science to control them—Dow and its employees took on a bunker mentality and stood increasingly alone.

One of the first people to challenge the faith in technological superiority espoused by Dow and the country was Rachel Carson, a marine biologist who had worked for the U.S. Bureau of Fisheries in Washington, D.C., and written two popular books, *The Sea Around Us* and *The Edge of the Sea.* Miss Carson followed the postwar research in organic chemicals and was aware of the amazing developments in pesticide and herbicide products. She became particularly interested in DDT, the tough chlorinated hydrocarbon insecticide that America brought home from the war. After its introduction in 1945, DDT became enormously popular because it was cheap, was convenient to use, and had long-lasting effects. DDT was a central nervous poison in high doses, but its acute toxicity to humans in normal exposures was low.

The effects of long-term, low-level exposures to DDT were unknown when the chemical was put on the market. Some scientists knew that DDT accumulated in the body's fatty deposits and could be passed on to children through breast milk. They believed that cows that grazed on DDT-sprayed forage would store the chemical in similar ways and excrete it in their milk. They feared that people drinking DDT-contaminated milk and eating DDT-sprayed fruits and vegetables would accumulate harmful levels of the insecticide in their bodies over time. The American Medical Association warned in 1948 that the chronic toxicity to humans of most of the new insecticides, including DDT, was "entirely unexplored." But the warnings were rarely broadcast outside scientific circles, and the demand for the insecticides was so high that the government allowed them to be sold.

As the evidence about DDT's harmful effects slowly accumulated, the actors and actresses came onstage to play parts in a script that would become a sort of summer stock road show in years to come. A public scandal involving DDT was widely written about in newspapers, government hearings were held, industry fought back, lawsuits were filed, corrective laws were passed, and the chemical was removed from the market. The ban was widely hailed, except by some scientists who said that society was throwing out the baby with the bathwater by completely banning a useful chemical instead of minimizing its risks and finding ways to use it safely. In DDT's case the play took nearly thirty years from opening to final curtain. Except for minor twists and turns, the plot was the same for other chemicals put on trial in ensuing years. Most were judged guilty for good reason; some were convicted for bad reasons. The strength of public opinion, as molded by and reported in the press, increasingly made the difference between conviction and acquittal.

In the case of DDT, as with almost every chemical put on the market during the quarter century that followed World War II's end, very little was known about how it worked or how it might affect humans and the environment over the long haul. As one entomologist wrote in the late 1960's, "It is a striking fact that knowledge of mode of action has rarely preceded the use of any insecticide. Even today we do not know precisely how DDT induces its toxic action."

Early on scientists suspected that DDT's persistence might be a double-edged sword. The chemical was showing up in human and bovine milk as early as the late 1940's and in the fatty tissue of people who did not work near the chemical, so were absorbing it in their diets. The government warned farmers not to use DDT around cows, but the American public didn't really get aroused until the Beech-Nut Packing Company let it be known that it couldn't find enough DDT-free vegetables to use in its baby foods and was reluctantly preparing them with contaminated produce. Congressional hearings were held in 1950 and 1951, not specifically on DDT but on the safety of food additives and other chemicals. For the first time ever chemical industry lobbyists showed up in force, and there was sharp debate over the safety of DDT residues on food and about the laws that sanctioned them.

The debate fractured along partisan lines. University and government scientists declared that DDT had not been proved safe, that more research was needed on its chronic effects, and that the laws governing its use should be revamped. The chemical industry and its agricultural allies—chief among them the United States Department of Agriculture—argued that DDT was safe and that the residues allowed by law were not harmful to the public. "Existing legislation makes possible adequate protection to the public," testified Lea S. Hitchner, president of the National Agricultural Chemicals Association. "I know of no other industry which has to comply with more laws and regulations in order to sell its products."

The industry's essential argument was that even though DDT was widely present in food and in humans, caused liver damage in animal experiments, and had never been tested to a no-effect level below which adverse long-term effects were absent, what was important was that the people who worked around DDT all the time—particularly farm applicators—had no untoward health effects. In other words, if clinical symptoms of poisoning weren't almost immediately visible, then the chemical was safe.

At hearing's end, the House Select Committee to Investigate the Use of Chemicals in Food Products passed the Miller Amendment to the Food, Drug, and Cosmetic Act. The amendment provided for the registration of pesticides before they could be sold, and it placed on the industry the burden of proof of safety. In order to register a chemical for use on food crops, the manufacturer had to present data showing that the expected residues were too low to pose a danger to human health.

Industry found itself under the regulatory eye not just of the United States Department of Agriculture but also of the Food and Drug Administration. It argued that regulatory overkill would not only stifle research but place more emphasis on the opinions of bureaucrats than on the informed consensus of scientists who had spent years testing the product. Dow executive W. W. Sutherland sounded the gong of independence when he told the American Association of Economic Entomologists in December 1951 that the Miller Amendment would give "certain government agencies arbitrary powers over the marketing of products. . . . " Sutherland declared that research required by the government to prove a product's safety "can go on forever without establishing all the factors which practical use of a product may eventually show up. A reasonable approach must be taken if progress is not to be seriously impeded."

DDT continued to be used in large quantities, especially in massive spraying programs to eradicate the gypsy moth and the bark beetle (the carrier of Dutch elm disease), which were often conducted from trucks in suburban neighborhoods. Soon there were complaints about massive bird deaths and fish kills. As the toll mounted, scientists revealed that food chains in the environment could concentrate DDT and that the chemical was interfering with bird reproduction. Their concern was that long-term use of DDT might have a more damaging effect on the environment than on humans. Alarmed by the evidence, several Long Island residents filed suit against the Department of Agriculture in an attempt to stop DDT spraying there. A three-week trial ended with victory for the DDT spray program, but Rachel Carson, who had been observing the trial with the thought of writing a brief book about it, became so intrigued with the issue that she continued her research.

The biologist expanded her research over the next four years to include the effects of all pesticides and herbicides on humans and the environment. She was troubled by the chronic effects of the persistent chemicals on people and by the disturbances caused to the balance of nature. She knew also that the benefits of chemicals were overstated in some cases, especially as they related to insect eradication. She pointed out that annual crop loss to insects had in fact risen from 10 percent in the early 1950's to 25 percent by the end of the decade—evidence, she said, that industry was exaggerating the usefulness of chemicals and that many kinds of insects had developed a resistance to them.

Beginning on June 16, 1962, her research was published under the title "Silent Spring" in three parts in *The New Yorker* magazine prior to its release as a book that fall by Houghton Mifflin. *Silent Spring* was an elegantly written account of the contamination of the environment and the effects on human health of pesticides. In her view, this was, along with the threat of nuclear war, "the central problem of our age."

As chronicled by her biographer Frank Graham, Jr., in *Since Silent Spring,* Rachel Carson became the immediate center of controversy. The chemical and agricultural industries responded forcefully to a book they considered to be long on emotionalism and short on scientific fact. Some companies instructed their scientists to read the articles in *The New Yorker* line by line to probe for weak spots. The National Agricultural Chemicals Association—upset about the "misrepresentation of an industry which has tried to do right"—upgraded its public information program and sent out new booklets reaffirming the benefits of pesticides. Little more than a month after the articles first appeared, *The New York Times* ran a front-page story headlined SILENT SPRING IS NOW NOISY SUMMER. It reported:

> The $300,000,000 pesticides industry has been highly irritated by a quiet woman author whose previous works on science have been praised for the beauty and precision of the writing. The industry feels that she has presented a one-sided case and has chosen to ignore the enormous benefits in increased food

production and decreased incidence of disease that have accrued from the development and use of modern pesticides.

E. M. Adams, assistant director of Dow Chemical's biochemistry research laboratory, told *The Times* that Miss Carson "has indulged in hindsight. In many cases we have to learn from experience and often it is difficult to exercise the proper foresight." Dow's man acknowledged, however, that the possible benefits of some pesticide uses, such as large-scale spraying, had to be balanced against the possible ills. But he pointed to extensive testing programs and federal regulations as proof that "What we have done, we have not done carelessly or without consideration. The industry is not made up of money grubbers."

The president of the Montrose Chemical Corporation, which as an affiliate of the Stauffer Chemical Company was the nation's largest producer of DDT, minced fewer words. He said that Miss Carson wrote "not as a scientist, but rather as a fanatic defender of the cult of the balance of nature." The biggest threat to that balance was not pesticides, he said, but modern medicines and sanitation.

Lamented *Chemical Week,* an industry trade journal: "Industry must again take up the Sisyphean task of repeating—again and again —that its research is aimed at profit through knowledge—not the sale of more and more pesticides whether they kill us nor not."

One company brought pressure on Houghton Mifflin not to publish the book. Louis A. McLean, secretary and general counsel of the Velsicol Chemical Corporation, wrote a five-page letter to the publisher citing Miss Carson's "inaccurate and disparaging statements" about two popular insecticides made exclusively by Velsicol, chlordane and heptachlor. In addition, McLean pointed out that:

> Unfortunately, in addition to the sincere opinions by natural food faddists, Audubon groups and others, members of the chemical industry in this country and in Western Europe must deal with sinister influences, whose attacks on the chemical industry have a dual purpose: 1. to create the false impression that all business is grasping and immoral, and 2. to reduce the use of agricultural chemicals in this country and in the countries of Western Europe so that our supply of food will be reduced to east curtain [sic] parity. Many innocent groups are financed and led into attacks on the chemical industry by these sinister parties.

Houghton Mifflin had the book reviewed by an independent toxicologist and held firm to its decision to publish it. Advance sales for *Silent*

Spring were 40,000 copies, and it was a Book-of-the-Month Club selection when it was published in October. Then came the full force of the industry counterattack. Industry scientists, and those in universities who did industry or agricultural research, attacked the validity of the author's scientific conclusions and, in some cases, the personal integrity and even the sexuality of Miss Carson. ("I thought she was a spinster," one member of the Federal Pest Control Review Board was reported to have said. "What's she so worried about genetics for?")

The book was criticized by newspapers and magazines as well as by industry. The stridently conservative *Time* magazine called *Silent Spring* an "emotional and inaccurate outburst" by an author who had "taken up her pen in alarm and anger." A review in *The New York Times Book Review* faulted the book's biases. "Miss Carson forfeits persuasiveness among those who know she is not telling the whole story," wrote *New York Times* reviewer Walter Sullivan. "She lays herself open to parody. Some unsung hero of the chemical industry has written . . . 'The Desolate Year.' " He was referring to Monsanto's parody of Miss Carson's opening fable about a town struck silent by pesticides. An essay called "The Desolate Year" appeared in Monsanto's monthly magazine describing the horrors of a year without pesticides.

Industry took to the lecture circuit. Robert H. White-Stevens of the American Cyanamid Company, who became the industry's premier spokesman against the book, made twenty-eight speeches about the benefits of pesticides before the end of 1962. Miss Carson was called, among other things, part of "the vociferous, misinformed group of nature balancing, organic gardening, bird loving unreasonable citizenry that has not been convinced of the important place of agricultural chemicals in our economy" and a woman whose "ignorance or bias on some of the considerations throws doubt on her competence to judge policy." Her book was labeled a "hoax" and "science fiction, to be read in the same way that the TV program 'Twilight Zone' is to be watched."

Three months after publication *Silent Spring* had sold more than 100,000 copies and vaulted onto the best-seller charts. The book was ultimately translated into twelve languages. To a world that had long thought all pesticides to be unquestionably safe, Miss Carson's contentions that many were unquestionably dangerous came as a shock. The book made a tremendous impact on public thinking about pesticides and the formation of public policy on their use. By the end of 1962 more than forty bills had been introduced in state legislatures to regulate pesticide use, a panel of President Kennedy's Science Advisory Committee had begun a study of pesticide use in America, and Senator Abraham Ribicoff announced that he would hold a broad congressional review of all federal programs related to problems of environmental hazards.

CBS Reports aired *The Silent Spring of Rachel Carson* in April 1963 despite a deluge of critical mail and the withdrawal of three of five sponsors (only the Kiwi Polish Company and the Brillo Manufacturing Company remained with the show). "It is not my contention that chemical insecticides must never be used," Miss Carson declared. "I do contend that we have put poisonous and biologically potent chemicals indiscriminately into the hands of persons largely or wholly ignorant of their potential for harm. I contend furthermore that we have allowed these chemicals to be used with little or no advance investigation of their effect on soil, water, wildlife, and man himself." Industry spokesman White-Stevens thundered back: "Now, although there are a number of scientific errors, misquotations, and obvious misinterpretations in her book, it must be admitted that much of her material is in part at least scientifically accurate. The area of disagreement between Miss Carson and students of applied agricultural chemicals, however, will lie in her clearly misplaced emphasis." Mr. White-Stevens believed that Miss Carson too sweepingly discounted the safety of pesticides and overbilled their hazards.

One month after the show had aired, President Kennedy's Science Advisory Committee issued a fence-straddling report on pesticide use. The committee believed the dangers of DDT to have been overestimated and underscored the necessity of such chemicals for agriculture and public health, but it agreed there were environmental problems related to the use of pesticides. The committee also pinpointed deficiencies in government regulation of chemicals: The Agriculture Department did not require as much proof of safety as it did efficacy of a chemical, no-effect levels did not exist for several important pesticides, and there were gaping holes in knowledge about the chronic effects of pesticide exposure over a lifetime. The committee recommended that pesticide residues in the air, water, soil, fish, wildlife, and humans should be monitored, that research on toxic effects be increased, that authority for pesticide regulation be transferred from the Agriculture Department to the Food and Drug Administration, and that the accumulation of residues in the environment should be controlled by an orderly reduction in the use of the pesticides themselves. "Elimination of the use of persistent toxic pesticides should be the goal," the committee said.

When the report was released on May 15, television commentator Eric Sevareid told his CBS audience: "Miss Carson had two immediate aims. One was to alert the public; the second to build a fire under the government. She accomplished the first months ago. Tonight's report by the presidential panel is prima facie evidence that she has also accomplished the second."

In the summer of 1963 Miss Carson appeared before Senator Ribicoff's Subcommittee on Reorganization of the Committee on Government Operations, which held hearings for fifteen months on pesticide use. She recommended legislation restricting the sale and use of pesticides. Her supporters included physicians who agreed with the Kennedy science panel conclusions that much more information was needed about long-term effects of pesticides. The opposition argued that pesticides were essential to feeding the world and keeping it healthy. Dr. Mitchell Zavon, the Cincinnati public health official and the medical consultant to the Shell Chemical Company who had worked to get the nematocide DBCP registered, told the committee that Miss Carson was "talking about health effects that will take years to answer. In the meantime, we'd have to cut off food for people around the world. These peddlers of fear are going to feast on the famine of the world—literally." Dr. Charles Hine, another Shell consultant instrumental in the approval of DBCP as a safe nematocide, testified that even though the entire American population now had detectable levels of the more persistent pesticides in its fatty tissues, they were not sufficiently high to present a health threat. It was his opinion that "there are no significant unrecognized effects on human health due to contact with agricultural chemicals in the quantities and amounts used at the present time." But he also believed that vigorous enforcement of pesticide laws should not be relaxed and that research into the toxicology and health effects of pesticides should be intensified.

Industry generally dismissed the science panel's recommendation of two-generation animal reproductive studies as unnecessary and unworkable. "I do not believe that reproduction studies as a routine matter should be conducted on pesticides," asserted Dr. John Frawley, Hercules Inc.'s chief toxicologist. "Actually, I know of no situation, no instance, where reproduction studies have turned up an effect that was not detected in the 2-year chronic feeding study." Dow's Dr. Julius Johnson, who directed the agricultural chemicals division, supported two-generation studies. When Senator Ribicoff asked Dr. Johnson how difficult it was to "sell" his scientific curiosity "to the man on the top floor," he replied: "I would do a disservice to my board and my stockholders if I indicated anything else than that the company is extremely interested in this matter. I have had direct instructions from the executive vice president, to whom I report, to make sure we get the facts on this to every reasonable extent possible. He has children, too."

His peers at Dow contended that Dr. Johnson had anticipated problems with persistent pesticides several years in advance of *Silent Spring* and was responsible for aiming the company's research toward safer products. "Julius Johnson said, 'Aha, I think we've got a new day coming. I think we're going to come to the end of these hard pesticides,'" recalled Herbert "Ted" Doan, president of Dow during the 1960's. "He had a list of things that were bad in terms of long-lasting and bioaccumulative and so forth, and he redirected the research. And so when *Silent Spring* came out, that really confirmed the fact that boy, it was a new world and we had to be on a different tack."

The Ribicoff committee concluded that current pesticide use was not a hazard to humans, but it called for a full study of the effects on humans and the environment. The only substantive change in pesticide regulation generated by the hearings was the elimination of the so-called registration under protest, a method by which industry could keep a chemical on the market after it had been rejected by the Agriculture Department, unless the chemical presented a clear health hazard.

In a summary report, Senator Ribicoff sounded a warning about the need to evaluate pesticide use in an objective and clearheaded manner:

> The reservoir of apprehension in the public mind evolves from three signs of our time: (1) The lack of understanding of science leading to distrust and actual dislike; (2) nostalgia for a simpler life, the good old days, and the "peaceable kingdom"; and (3) a feeling of individual incompetence to avoid the threats of technological side effects (e.g., helplessness against community aerial spraying, unknown source of food stuffs, and total reliance on governmental control and regulation). This anxiety (amounting to fear) is a barrier to facts and presents a bad climate for decisionmaking.

Rachel Carson died of cancer at her Silver Spring, Maryland, home on April 14, 1964. Her funeral was held in the National Cathedral; the largest wreath was sent by Great Britain's Prince Philip. In a Senate floor eulogy Senator Ribicoff paid tribute to "this gentle lady who aroused people everywhere to be concerned with one of the most significant problems of mid-20th Century life—man's contamination of his environment."

Miss Carson's effect on public thinking about pesticides was clear. Her impact on the chemical industry was more murky. Some members of the industry acknowledged that despite its shortcomings, *Silent Spring* forced them to take a hard look at the way their products were affecting the quality of life. "You know, you don't shoot the messenger. She had a good message. *Silent Spring* was a shocking book. It was a good alert," said Dow's Ted Doan. "She truly was a gifted writer and a good biological scientist, and I think we all became very much aware of those things," agreed Dow's Dr. Etcyl Blair, who was directing a group of research chemists in the synthesis of organophosphate pesticides when the book came out. But he, like many scientists, took issue with specific pieces of her research. Dr. Blair thought her criticisms of organophosphate pesticides were too broad-stroked. "It's as gross an error as trying to throw all humanity together," he said. And though he approved of her message about the harmful effects of DDT, he deplored the fact that it was banned rather than better controlled.

In fairness, the industry was not obsessed with *Silent Spring*. Dr. Blair remembered that when he brought the book into his laboratory and showed it to his scientists, "nobody could have cared less about it. Ninety percent of scientists don't pay any attention to these kinds of things." There was also a feeling beginning to take shape among industry leaders that more than anything, they had a public relations problem on their hands. After decades of kid-glove treatment by the nation's press, public opinion about the chemical industry began to chill in the wake of widely reported stories about chemical contamination and lax regulation. Two of the most sensationally reported stories were the 1959 cranberry scare and the discovery in 1962 that a drug prescribed as a tranquilizer for expectant mothers caused horrific birth defects in their children.

Just before Thanksgiving 1959 the government announced that it had destroyed cranberries contaminated with a chemical that produced cancer in experimental rats. The cranberries were not from the current year's crop but had been gathered and frozen two years earlier, when the chemical was approved for weed control in cranberry bogs on condition that it be applied only after harvest. But the animal studies showing a link between the chemical and cancer were not completed until 1959. In the meantime, Congress had passed the controversial Delaney Amendment to the Food, Drug, and Cosmetic Act, which said that no food additive is safe if it induces cancer when ingested by humans or animals. Even though there was no evidence that the 1959 crop was contaminated, the government called a news conference to alert Thanksgiving shoppers to the potential problem. Sales of cranberries dropped precipitously, despite the best efforts of Agriculture Secretary Ezra Taft Benson, who declared that his family would be dining on cranberries come Thanksgiving Day; an Ocean Spray executive, who publicly ate a handful of the fruit; and Vice President Richard Nixon, who ate four helpings at a pre-Thanksgiving dinner in Wisconsin. As it turned out, the 1959 crop was not contaminated, but public fears about dangerous chemicals lingered.

Then, in the summer of 1962, almost simultaneously with the publication of "Silent Spring" in *The New Yorker,* newspapers reported that a drug called thalidomide, prescribed as a tranquilizer or a flu medicine for pregnant mothers, had caused tragic birth defects in European babies. Thalidomide had been kept off the market in America only because a government scientist insisted there be more safety data on the drug. But the nation watched as Sherri Finkbine, a pregnant mother of four who had taken some tranquilizers her husband brought home to Arizona from London, battled the courts for permission to abort her child. Although thalidomide was not a chemical industry problem per se, it strengthened the public notion that science was not controllable and that regulation was not omnipotent.

Some chemical companies launched a rearguard action to counteract the growing public distrust of science and pesticides. It was the beginning of a long industry campaign to convince the country that emotionalism, not hard fact, had caused an unnecessary scare about pesticides, which were essential to human life and to the progress of civilization. The defense of pesticides became a defense of modern life. Dr. William Darby of the Vanderbilt School of Medicine wrote that the application of the ideas in *Silent Spring* to real life would mean "the end of all human progress, reversion to a passive social state devoid of technology, scientific medicine, agriculture, sanitation. It means disease, epidemics, starvation, misery, and suffering."

Dow's Dr. Julius Johnson, speaking in the post-*Silent Spring* environment, talked of "granaries full and overflowing . . . a surplus of fiber for clothes . . . a population where the efforts of one farmer feed 27 of his countrymen. These," he said, "are accomplishments which are the envy of the Communist world and a phenomenon which is the key to survival of all people on the planet." But he warned that "well-fed people are attacking the system and the tools that made them strong." In defense of industry, he cited long-term toxicological testing and safety standards imposed on chemical products. "Fear of the unknown," he declared, "such as in Galileo's day, was a deterrent to progress. This fear can be amplified out of all proportion when vaguely drawn implications are made that pesticide residues are akin to radiation and pose grave problems to generations unborn."

As the debate wore on, the breach between the two sides began to widen. Like British author Cyril Connolly's "river of truth," which was always splitting up into arms around an island of people who "argue for a lifetime as to which is the mainstream," there was no consensus in the chemical debate. And nowhere was the battle more fractious and more burdened with unanswered questions than in the use of herbicides.

chapter 4

In the years following the discovery of the awesomely toxic dioxin, production of the phenoxy herbicides skyrocketed in the United States and other countries. Every year an average 6 million acres of American farm, range, pasture, and forestlands were sprayed, as were highways, power lines, and railroad lines. Herbicide factories in the United States escalated production even more dramatically when, in 1961, the military began spraying in Vietnam.

World War II research in Britain and in America's biological warfare center at Camp Detrick, Maryland, had discovered and developed the herbicides as anticrop agents, but the war ended before they could be used. With the beginning of the American involvement in Vietnam, scientists at the renamed Fort Detrick picked up where they had left off. More than 26,000 chemical compounds were investigated for military use. What was needed in Vietnam were chemicals that would deny the Vietcong guerrillas their hiding places in the dense triple-canopied jungles by stripping the trees and bushes clean of leaves—a process called defoliation. The military also wanted to use herbicides as they were originally intended to be used: as anticrop agents to destroy guerrilla food plantations.

The military conducted its first large-scale aerial defoliation tests in 1959 over four square miles at Fort Drum, New York, using a mixture of the phenoxy herbicides 2,4-D and 2,4,5-T. The success of these tests inspired Secretary of Defense Robert S. McNamara to request feasibility tests for defoliation of jungle vegetation in Vietnam. Teams of scientists from Fort Detrick were sent to Vietnam to conduct field experiments on the susceptibility of Vietnam's vegetation to various chemicals. "Even though wartime conditions interfered with the collection of detailed data," a Pentagon-commissioned study reported, "these tests established that [2,4-D and 2,4,5-T] were active in killing a majority of the species encountered in Vietnam, providing [sic] the herbicide spray was properly applied to the vegetation during a period of active growth." In January 1962 the military formally launched defoliation operations by aerially spraying herbicides on a seventy-mile stretch of road between Saigon and the coast.

Little more than a year later the military and the chemical industry met at the Department of Defense's First Defoliation Conference to review the operation. "The capability of destroying cover and concealment to defend against and fight off guerrilla and other types of tactics is absolutely essential," declared Brigadier General Fred J. Delmore, the commanding general of the army's Edgewood Arsenal chemical, biological, and radiological warfare agency.

Representatives of companies, including Dow and Monsanto, were alerted by General Delmore to the "need right now of chemicals that will do the job at an earlier time, and in a quicker period." Delmore declared that the material used for defoliants must be both "perfectly innocuous to man and animals" and able to "do its job." Albert Hayward, chief of the program coordination office at Fort Detrick, told the conference, "It goes without saying that the materials must be applicable by ground and air spray, that they must be logistically feasible, and that they must be nontoxic to humans and livestock in the areas affected."

Little scientific information was available about the health effects of the defoliation operation in Vietnam, but the chemical industry nevertheless endorsed continued use of herbicides. The endorsement was based on the track record of herbicide use in the United States and assurances from the major herbicide companies that "none of the workmen in their factories have shown any ill health effects." Dow advised General Delmore in 1963, "We have been manufacturing 2,4-D and 2,4,5-T for over ten years. To the best of our knowledge, none of the workmen in these factories have shown any ill effects as a result of working with these chemicals." Dow was telling the truth about its own operations but seemingly ignoring the vast range of health problems experienced by German herbicide workers as well as the chloracne outbreaks at Monsanto's plant in West Virginia and at Diamond Alkali's factory in New Jersey. Neither Monsanto nor Diamond Alkali informed the military about its own problems with the herbicide. Industry representatives dissembled when they told the military at the defoliation conference that there were no reports of problems anywhere in the scientific literature. A judge presiding over a lawsuit filed against the herbicide companies twenty years later wrote: "Plaintiffs argue persuasively that Dow must have known about the explosion at the Monsanto plant in Nitro, West Virginia in 1949 and the resulting cases of chloracne. Dow does not deny such knowledge. Plaintiffs also argue that Dow knew about the outbreak of chloracne among workers at the

Diamond Alkali [now Diamond Shamrock] plant in 1956. Dow does not deny this."

Six chemical compounds were eventually chosen for American military use in Vietnam. They were called Agents Orange, Green, Pink, Purple, White, and Blue after the color-coded strips that girdled the drums used to transport them overseas. Agent Orange was a fifty-fifty combination of the herbicides 2,4,5-T and 2,4-D. It was a stronger version of the same mixture sold by Dow and other chemical companies in the United States as weed and brush killers. They were also the same chemicals that Rachel Carson had called "a bright new toy."

"They work in a spectacular way," she wrote about herbicides in *Silent Spring*. "They give a giddy sense of power over nature to those who wield them, and as for the long-range and less obvious effects— these are easily brushed aside as the baseless imaginings of pessimists." Miss Carson recorded a long list of health problems thought to be associated with the use of 2,4-D and 2,4,5-T and its chemical relatives, including their imitation of X rays in damaging chromosomes. Yet, she conceded, "whether or not these are actually toxic is a matter of controversy."

The government and the chemical industry publicly stressed the safety of the herbicides. Yet the chemical industry undoubtedly knew the enormous problems encountered by workers in herbicide plants in this country and abroad. And the government knew, too, although not as much and not because of any candor from the industry. The scientific report outlining the German physician Karl Schulz's studies on dioxin and its links to chloracne and liver disease was reportedly widely circulated throughout the military. Dr. Friedrich Hoffmann, a chemical warfare specialist and chief of the Army Chemical Corps's Agents Research Branch at the Edgewood Arsenal, was sent to Europe in 1959 to scout potential chemical warfare agents. In his trip report—ten copies of which were sent to Edgewood—Dr. Hoffmann noted that he had received "startling information" about the toxicity of dioxin, including the fact that it had been linked to severe and sometimes fatal liver damage. He also mentioned the 1957 article by Schulz announcing the discovery that dioxin was an impurity in the manufacture of trichlorophenol and 2,4,5-T and was the cause of chloracne. Dr. Hoffmann reportedly told the army that dioxin was too deadly to be used for chemical warfare purposes.

By 1963 there was more cause for concern about the potential threat to human health. An explosion in a herbicide factory in the Netherlands spread the classic symptoms of chloracne among fifty workers, some of whom also developed damage to their internal organs and who complained of psychological problems. The factory was so badly contaminated with dioxin that it was dismantled, embedded in concrete, and dumped in the sea near the Azores. That same year a U.S. Public Health Service physician strengthened the dioxin-chloracne link when his forearm erupted with the disease's characteristic lesions three weeks after he had repeatedly applied a sample of commercial 2,4,5-T to it. In addition, a report by the Institute for Defense Analysis, a Defense Department-sponsored think tank, cautioned that the military was using much higher herbicide concentrations in Vietnam than were common in the United States. The report said military requirements "dictate use of overkill concentrations" of herbicides with "possible toxicological" effects on exposed populations. The report noted the connection between chloracne and herbicides.

Agent Orange was the most heavily used herbicide in Vietnam. The number of pounds applied per acre was three times higher in Vietnam than in America, and contrary to civilian practice, the herbicide was used undiluted. Dow was the largest of nine government contractors supplying herbicides for the war. Starting in 1961, Dow sold Agents Purple, Pink, and Green to the military. Each contained 2,4,5-T. From September 1965 through the end of the decade Dow made deliveries to the government (at a cost of $7 a gallon) of nearly one-third of the total 12.8 million gallons of Agent Orange sprayed by Operation Ranch Hand pilots (whose motto was "Only We Can Prevent Forests"). Agent Orange was loaded onto C-123 cargo and troop transport planes equipped to carry 1,000 gallons of herbicide. The pilots generally sprayed 3 gallons per acre in swaths 240 feet wide for defoliation and crop destruction missions. The herbicide proved so successful against the stubborn jungle foliage that the military recorded a 90 percent drop in guerrilla ambushes in some areas.

Herbicide spraying in Vietnam built slowly through the early 1960's, reaching a peak in 1967 and then declining slightly for several years before it was ordered stopped in 1971. Roughly 17.7 million gallons of herbicides were sprayed across 3.6 million acres during that nine-year period. It has been estimated that 368 pounds of dioxin were in the herbicides released over Vietnam.

To meet the war demand, American herbicide factories worked overtime to make Agent Orange and to supply this country with weed and brush killers. Consequently, dermatologists from Newark Beth Israel Hospital were called into the Diamond Alkali herbicide plant in Newark, New Jersey, in the early 1960's to treat a severe outbreak of chloracne among workers. Years later the herbicide workers recalled that the doctors came every Thursday to lance and drain their boils and give them vitamin and vaccine shots after what they called the rash had spread throughout the plant. Some workers had boils the size of an adult's thumb, and several men had the tips of their blister-ravaged ears removed. The problems had apparently begun after explosions had occurred in the manufacturing reactor in 1955 and in February 1960 (the last one fatal to one worker). Diamond Alkali had reason to worry not only about its workers but about its customers. A corporate official warned the Newark plant manager in a 1962 letter that several large Diamond customers were experiencing chloracne problems, and in one case "we have definitely lost them as a customer."

In a report to the *Archives of Dermatology* in June 1964 the Beth Israel doctors described the typical features of chloracne, which afflicted twenty-nine workers and supervisors. Nearly half the workers also exhibited hyperpigmentation, or a darkening of skin on the head, neck, and hands to colors ranging from mild redness in fair-skinned men to "dark gray intense dusky bronzing." More significantly, they reported that eleven of the twenty-nine men had porphyria cutanea tarda, a disease of the body's blood-forming elements characterized by a darkening of the urine. The doctors speculated it was caused by one of the chemicals in the plant, which exerted a toxic effect on the liver.

In one case, several years after a forty-eight-year-old man had come to work in the factory, his skin darkened, his face, ears, and hands erupted with blisters; his forehead and eyelids sprouted tufts of hair the consistency of his eyebrows; and his urine turned "the color of Coca-Cola."

In another case a welder who had "frequent and prolonged contact" with the herbicidal chemicals was admitted to the hospital after his skin had darkened, his eyebrows had thickened, and his urine had taken on a dark, reddish "strong tea" color. His face, chest, and shoulders were covered with blackheads and boils, and some of his skin had a purplish cast. His feet developed a fungus disease, and his hands were covered with boils.

A third worker had dark and blistered skin, hairy temples, dark urine, and chloracne so severe that company officials removed him from contact with the chemicals.

The dermatologists speculated in 1964 that both diseases, chloracne and porphyria cutanea tarda, were caused by either the finished chemicals in the plant—2,4-D and 2,4,5-T—or one of the intermediate chemical ingredients, including trichlorophenol. They apparently were not aware of the dioxin contaminant. A U.S. Public Health Service physician also visited the Newark plant in 1964 and wrote a report on the workers' problems.

As the war in Vietnam expanded, so did the military's greed for herbicides. In 1962, 4,949 acres were sprayed with herbicides in Vietnam. In 1963 herbicide-sprayed acreage increased fivefold to a total of 24,700 acres. In 1964 it tripled. The herbicide market was booming. Although Dow was not yet supplying the government with Agent

Orange, it was selling 2,4,5-T for use in other herbicides in Vietnam. It also laid claim to the biggest share of the domestic market. To meet the demand, Dow switched from a batch to a continuous operation in making trichlorophenol in its 199 Building in 1963, raising the temperature in the process. "When you make popcorn at home, you put the popcorn in the pot, it pops, you empty the popcorn in a pan, you put some more popcorn in. Each time you do it, you've made a batch," explained Dow's Garry Hamlin. "Balance that against going to the movies, where they have a popcorn popper that's popping all the time in just sort of a continuous feed. Well, that's what we wanted to do. We had continuous production. We also got chloracne."

The operation was pushed to its limit. Trichlorophenol production is usually increased by raising the temperature. The higher the temperature, the higher the amount of dioxin produced as a contaminant. The process Dow used in 1963 required peak temperatures of 225 degrees centigrade, 70 degrees hotter than the limit recommended by the German chemical company Boehringer in a letter to Dow and other companies six years before.

Consequently, some months later, starting in January 1964, two supervisors, three trichlorophenol operators, and one laboratory employee developed facial rashes. Over the spring and summer nearly fifty Dow workers in 199 Building got chloracne. The trademark lesions and blackheads, some of which would persist for several years, spread across the workers' faces, necks, backs, and legs. In August—with the plant in "serious trouble because of chloracne"—the superintendent of 199 Building asked Dow's analytical people to look at the oily waste stream from the trichlorophenol process, which ran in leaky pipes through the plant. He knew from reading the scientific literature that a compound known as tetrachlorodibenzo-p-dioxin was linked to chloracne. He wanted to know if the chemical was in the waste oils.

Dow asked a University of Michigan dermatologist to examine the workers, and the Michigan Department of Public Health was alerted to the chloracne problem. "There was some thought given to contacting the U.S. Public Health Service," said Garry Hamlin, "but the decision was ultimately that this was an occupational problem, and they're not really covered by the U.S. Public Health Service."

In reviewing the 1964 outbreak sometime later, Dow medical director Dr. Ben Holder wrote:

> The clinical picture of the disease is one that primarily affects the skin. In extreme exposures to certain chlorinated compounds, a general organ toxicity can result. This is primarily demonstrated in the liver, hematopoietic [blood-forming] and nervous system. The stimulation of certain skin glands pro-

duces multiple blackheads along with bacterial cellulitis. Many patients demonstrated continuation of the skin lesions several years after complete removal from the exposure.

Dr. Holder noted that "fatalities had been reported in the literature. There is no specific treatment for this disease." Besides the skin effect in the Dow workers, Dr. Holder found no systemic problems other than "mild depression" in about 20 percent of his patients.

Dow scientists had long known that a contaminant produced in the manufacture of trichlorophenol caused chloracne. But even though the Germans had identified the dioxin contaminants in trichlorophenol in the 1950's and told Dow how to get rid of them in their manufacturing process, Dow maintains that its scientists knew only that a contaminant existed, not what it was. They had been monitoring the contaminant with an animal test developed in 1939 as a measurement of skin sensitivity to chemicals. In fact, until improvements were made to analytical equipment during the mid-1960's, rabbits were more sensitive than machines. The animals' ears were rubbed with samples of chemical waste suspected of containing the mystery contaminant. It was a simple process: The ears were shaved, the chemical was dissolved in a solvent and spread over the exposed skin, and the rabbits were put in a box with their heads sticking out through a hole to keep them from scratching their ears. The contaminant was thought to be held at a safe level—one part per million—if chloracne did not appear on the animals' ears.

But when scientists ran a rabbits' ear test on the trichlorophenol wastes in 1964, there was a severe reaction. They concluded that the levels of the mystery contaminant were much higher than normal. Samples taken from around 199 Building itself showed extensive contamination. The building was temporarily shut in mid-1964, while the investigation continued. Welded pipes were installed to prevent waste oil leakage. Automated measuring equipment was installed on the pipes to reduce worker exposure. The reactor room was washed down with detergent and painted with aluminum, equipment valves were sealed, and floor gutters were cleaned out. Dow management asked its medical, biochemical research, and safety departments if 199 Building could be run without injuring workers and vowed to shut it down if the answer was no. The departments answered that the plant could be run safely, but only at considerable cost.

Using mass spectroscopy, which measures the molecular weights of chemical compounds, Dow's analytical chemists discovered that the contaminant had a molecular weight of 320 and that it contained four chlorines. That fitted the general description of the tetrachlorodibenzo-p-dioxins first identified by the Germans, but it did not tell them where the chlorine atoms were attached to the dioxin structure. Twenty-two members of the seventy-five-member dioxin family belong to the "tetra" group; each has the same molecular weight and each has four chlorine atoms. The only way to distinguish among them is by the positioning of the chlorine atoms on the dioxin structure itself. The most toxic dioxin by far has its chlorine atoms attached at the two, three, seven, and eight positions, or 2,3,7,8-tetrachlorodibenzo-p-dioxin.

Initial animal experiments showed that dioxin was enormously toxic. An amount so small as not to be visible on the head of a pin—100 millionths of a gram per kilogram of body weight—could kill half a population of rabbits in the LD50 test. Even less—0.0006 milligram per kilogram of weight—would accomplish the same in a population of sensitive guinea pigs. By contrast, the LD50 for DDT in guinea pigs is 400 milligrams per kilogram.

Eventually advances in analytical technology permitted Dow to identify this most toxic of the dioxins at sensitive parts per million levels. They confirmed what the German scientists had discovered years before. When trichlorophenol was manufactured for use as the raw material in other products, including the herbicide 2,4,5-T, the most toxic dioxin was inevitably formed and transferred to the end product. When measured in December 1964, Dow's trichlorophenol waste oils contained astounding levels of dioxin, ranging from 6,000 to 10,000 parts per million. Far less dioxin was contained in the American herbicide industry's commercial products, but it was possible to find 2,4,5-T being sold with as much as 80 parts per million dioxin before the industry understood its hazards. As Dow's analytical methodology improved in the wake of its in-house chloracne outbreak in 1964, Dow said it was able to hold dioxin in its products down to 0.5 part per million. Later, lawyers would contend that Dow had no proof that its pre-1965 products did not contain more than 1 part per million dioxin.

Industry was the repository of knowledge about dioxin in the early 1960's. Government scientists lagged years behind; they often didn't even have the sophisticated machinery needed to detect low levels of chemicals. Toward the end of 1964 a Public Health Service chemist began making inquiries about 2,4,5-T and chloracne. Dr. David Groth wrote the Thompson Hayward Chemical Company, one of Dow's competitor-customers (it bought trichlorophenol from Dow to fashion into 2,4,5-T), explaining his suspicions about the herbicide and asking for help in developing a method to isolate the contaminant. (Dr. Groth never tested his suspicions because of lack of government funds with which to buy the necessary analytical equipment.) Thompson Hayward's own herbicide plant in Kansas City, Kansas, had been destroyed by fire in 1959 after an explosion. There had been speculation at that time that chloracne-producing chemicals were trapped in the remains, but company officials believed they were destroyed by the heat of the fire. When Thompson Hayward turned to Dow for information about the Public Health Service scientist's request, Dow—in the middle of a chloracne outbreak at its own plant—replied that there might be some methods of production that would lead to the formation of a toxic compound. When pressed further, Dow's research director pointed out to Thompson Hayward the possibility of a "flurry of successful claims by users of the herbicide who allegedly had been injured by it."

The situation had taken a turn for the worse when another herbicide maker sent Dow some trichlorophenol, and it tested out high in dioxin. Dow ran some tests on rival commercial herbicide products and found "surprisingly high" amounts of dioxin. The potential legal problems were staggering.

When Thompson Hayward people went up to Midland in February 1965 to talk to Dow, they found out that the chloracne problem was so severe that Dow trichlorophenol workers were changing clothing and showering in mid-shift. Dow told Thompson Hayward that it was slowing the trichlorophenol production rate to 25 percent of normal in order to reduce chloracne. The company had installed a full-time health supervisor in 199 Building, and extensive cleaning had been undertaken to decontaminate plant equipment. As a result, no new cases of chloracne developed. Dow's people talked openly about dioxin: They said they believed it was systemic and was generally taken into the body by physical contact. Dow people "repeatedly" told their visitors that the company was making an "all-out effort to solve the problem without consideration of the cost, which will be very high."

A few weeks later Dow held an internal meeting to discuss the "exciter" problem and its concerns that products like 2,4,5-T and silvex were contaminated, although company officials insisted that Dow's own products were not part of the problem. "This meeting is to review status of our knowledge of this subject, potential hazards, possible effect on Dow image, legal implications, and need for possible quarantine," the minutes said. "These basic decisions are to be made without consideration of economic impact."

Dow vowed to come clean with the government. It plotted a straight-arrow course to alert health and regulatory authorities to the potential dioxin danger. Dr. Verald K. Rowe, Dow's chief toxicologist, would "move ahead with contact with the U.S. Public Health Service and the Department of Agriculture" as soon as the scientific data were correlated in order to share with government officials Dow's "analytical, testing and toxicological information" about dioxin.

But Dow's good intentions soon degenerated into wrongheaded track covering. The company did not pass information to the Agriculture Department or the Food and Drug Administration, the federal agencies charged with regulating the use of herbicides in this country. Dow did not contact the Public Health Service. Nor did it tell the Department of Defense, which was spraying several million pounds of heavily dioxin-contaminated herbicides across Vietnam annually.

In depositions years later Dr. Rowe claimed he did not remember notifying the government agencies. When asked why he decided not to meet with them, he responded, "I don't know. It was a matter of . . . probably a matter of waiting until things were better correlated . . . then it just died down or something. I don't remember contact." But, he added, "I wouldn't preclude the possibility that I may have had some conversation with somebody there. I can't attest to that. I don't remember." When asked if he thought there was a legal or ethical responsibility to notify the government, he replied, "No. I think we looked at that as an internal problem, mainly. I mean everytime we had a problem we didn't notify the government in those days about it." He said the Michigan Department of Health was notified because it was in charge of worker safety, and the problem was not a "federal situation." He said, "We had concluded as a result of our work that we were not going to market any material that was likely to have a problem in the consumer market. That was our decision." When asked if there was an agreement among the country's herbicide makers not to notify the federal government about the dioxin problem, Dr. Rowe said he did not recollect that there was.

What Dow did do was this: On March 11, 1965, it prepared an internal report detailing the warnings in the 1957 Boehringer letter, which contained the recommendation that the manufacturing temperature of trichlorophenol be kept below 155 degrees centigrade to avoid production of dioxin. Dow had sent three employees to Europe the previous November to talk to trichlorophenol manufacturers whose workers had experienced chloracne problems. They were instructed to buy any technology that would enable Dow to build new, safe trichlorophenol production facilities in the shortest possible time. Not surprisingly Dow made arrangements to purchase that technology from Boehringer.

On March 19 Dr. Rowe sent identical letters to officials at America's major herbicide companies—Monsanto, Hercules, Diamond Alkali, and Hooker. He invited them to come to Midland to discuss "the toxicological problems caused by the presence of certain highly toxic impurities in certain samples of 2,4,5-trichlorophenol and related materials." Monsanto's medical director also got an invitation from Rowe over the telephone. After hanging up, he wrote in a memo to his files that Dow wanted "a crash meeting" with its competitors in order to agree on an industrywide dioxin specification before the Public Health Service got into the act.

On the same day Boehringer secretly sent an outline of its low-temperature trichlorophenol process to Dow.

On March 24 all the herbicide companies but Monsanto met in Midland. Like a Mafia sit-down, at which the ruling don settles disputes among rival families so as to preserve the whole, Dow wanted to convince the rest of the herbicide makers to institute "self-imposed controls" on dioxin production before the government smelled trouble. At the meeting eight of Dow's senior scientists told their colleagues about the recent chloracne outbreak among its workers and shared with them the new analytical technique for detecting dioxin in parts per million levels. They advised reducing the dioxin contaminant below one part per million in the trichlorophenol before it was processed into 2,4,5-T.

Dr. Holder gave an update on Dow's workers and showed slides of the more dramatic cases. Forty-nine of sixty-one exposed workers had chloracne. Five to ten were extremely severe. In the acute stage, as one company representative later described it, "the facial tissues resemble the exaggerated surface texture of an orange." Some chronic cases still lingered two and a half years after the exposure. One bench chemist had been under treatment for two years, and his face was only then starting to show signs of clearing. The small blackheads and eruptions that covered the faces of the 199 Building workers were contrasted with the "Dowicide bumps"—larger boils and bumps—which Dow workers had suffered previously. Dr. Holder noted that a liver biopsy taken on a worker with severe chloracne was negative. Liver and kidney examinations of the workers were normal. Fatigue—"where the employee is completely listless, tired out and nearly incapacitated"—was the only other significant complaint. (In a later report Dr. Holder noted that slight blood disorders in several workers had eventually corrected themselves, except for one man who had chronic blood loss from gastrointestinal bleeding.)

But Dow didn't reveal all the cards it held in its hand. Unbeknown to the rival companies, on the day after the meeting Dow signed a secret ten-year contract that gave the company rights to use Boehringer's low-temperature trichlorophenol process at its Midland plant. Dow was contractually bound to pay for the process only if it succeeded in "eradicating" the chloracne among its workers. In Boehringer's eighteen-page description of the process, which mandated temperatures no higher than 155 degrees centigrade, the company's platitudes about safety stood out in stark relief against the American industry's shoddy methods. "It is certain that the operating methods we use here in

Hamburg... are much more expensive... than is customary for similar methods," Boehringer wrote. "However, the extra costs are justified, in our opinion, in order to have safe operations."

After the Dow meeting the industry representatives went home and wrote memos for their files. Hercules chemist C. L. Dunn wrote:

> Dow says that their [sic] examination of their own and competitors' 2,4,5-T products contain [sic] what they call "surprisingly high" amounts of the toxic impurities. In addition to the [rabbit] skin effect, liver damage is severe, and a no-effect level based on liver response has not yet been established. Even vigorous washing of the skin 15 minutes after application will not prevent damage and may possibly enhance the absorption of the material. There is some evidence it is systemic.

Another scientist in attendance, E. L. Chandler of Diamond Alkali, observed that the meeting "was obviously designed to help us solve this problem before outsiders confuse the issue and cause us no end of grief." He also suggested that Dow believed repeated exposures to one part per million dioxin were hazardous. When Dow scientists exposed rabbits' ears to one part per million dioxin, there was a severe skin reaction after eleven applications. "They conclude, therefore," Chandler wrote, "that 1 ppm [part per million] with repeat exposure can create a real problem." Dow had found as much as ten parts per million dioxin in other companies' commercial 2,4,5-T products.

The rationale for the meeting was transparent. Dow was the biggest producer of herbicides in the United States. It also knew more about their dangers than anyone else. And it had a secret, albeit expensive method of making 2,4,5-T with safe levels of dioxin. Instead of calling down certain government controls on its head, throwing itself wide-open to product liability and workers' personal injury lawsuits, and jeopardizing the industry's multimillion-dollar herbicide market, Dow devised a plan whereby the industry would police itself and keep quiet about its problems until they were worked out.

Dow toxicology chief Verald K. Rowe spelled it out in a private letter to Dow's Canadian unit in June:

> As you well know, we had a serious situation in our operating plants because of contamination of 2,4,5-trichlorophenol with impurities, the most active of which is 2,3,7,8-tetrachlorodibenzodioxin. This material is exceptionally toxic; it has tremendous potential for producing chloracne and systemic injury. If it is present in the trichlorophenol, it will be

carried through into the T . . . and hence into formulations which are to be sold to the public. One of the things we want to avoid is the occurrence of any acne in consumers. I am particularly concerned here with consumers who are using the material on a daily, repeated basis such as custom operators may use it. If this should occur, the whole 2,4,5-T industry will be hard hit, and I would expect restrictive legislation, either barring the material or putting very rigid controls upon it. This is the main reason why we are so concerned that we clean up our own house from within, rather than having someone from without do it for us.

We are not in any way attempting to hide our problem under a heap of sand, but we certainly do not want to have any situations arise which will cause the regulatory agencies to become restrictive. Our primary objective is to avoid this. I trust that you will be very judicious in your use of this information. It could be quite embarrassing if it were misinterpreted or misused.

Dr. Rowe admonished the Canadian Dow official to "practice good citizenship" and warned him that "under no circumstances may this letter be reproduced, shown or sent to anyone outside of Dow."

"I think his remarks were colored by what he had observed in 199 Building, where forty-nine people were suffering because of the overexposure," said Dow's Garry Hamlin. "My guess is that his reaction was deeply emotional, that that kind of stuff was just not going to happen anymore."

Dow may have thought its Godfather-like approach to chemical regulation would work, but it didn't have the unique incentives that the Mafia invokes when faced with recalcitrant lieutenants. Consequently, there was little Dow could do when other herbicide makers continued to produce a high-dioxin product. In July—three months after Dow had admonished the herbicide makers to keep dioxin below one part per million—Dow ran tests on competitors' 2,4,5-T and found that "no one had done anything to remove the acnegen [dioxin] from its products." Hercules was a good soldier; it was making a low-dioxin herbicide. In an apparent effort to enlist his support, a Dow executive called Dr. John Frawley, the chief toxicologist for Hercules, to find out "how serious" he considered the chloracne problem "in relation to the consumer use of 2,4,5-T." Dow knew that Monsanto and Diamond Alkali were flagrantly ignoring the one part per million limit. Monsanto's herbicide usually carried a ten part per million dioxin load; by September 1965 the dioxin had reached an "all-time high" of fifty parts per million.

In a confidential memo written for his files, Dr. Frawley observed that Dow was "extremely frightened that this situation might explode." He wrote:

> They are aware that their competitors are marketing 2,4,5-T which contains "alarming amounts" of acnegen [dioxin], and if the government learns of this the whole industry will suffer. They are particularly fearful of a congressional investigation and excessive restrictive legislation on the manufacture of pesticides which might result.

And that was the end of Dow's experiment in industry self-regulation. Dow officials stopped phoning their competitors and adopted a low profile, keeping silent about the dangers of dioxin for the next four years while simultaneously continuing to supply the defoliation and crop-destruction needs of the Vietnam War and the myriad of domestic uses. Dow did not inform the Agriculture Department because it did not believe the dioxin in its consumer products was hazardous. The silence even extended to the pleas of chloracne-plagued rival companies for help. The Thompson Chemical Corporation was a small company in St. Louis, Missouri, without any testing facilities whose disinclination to make Agent Orange was overruled by the Defense Production Act. When Thompson's technical director, M. S. Buckley, telephoned Dow for help in dealing with its "severe chloracne problem," Dr. Rowe, Dow's toxicology director, did not give him a detailed description of what caused the problem because "it was quickly apparent that Mr. Buckley had little understanding of the toxicological aspects of his problem. Had he asked for methods, etc. I would have agreed to send them to him."

Dow's knowledge about dioxin, however, continued to grow. Dow scientists searched their files and summarized what they knew in an internal company report called "Chloracne-Dow Experience." The scientists raised the possibility that dioxin could cause death in humans. Dismissing chloracne itself as "like teenage acne, but more blackheads," Dow noted that it was usually not disabling "but may be fatal."

The report listed some 100 cases of chloracne among Dow's own workers dating back to the mid-1930's, when there were severe outbreaks among the Dowicide workers. The early Dowicides also caused chloracne in some of Dow's customers during the 1940's. The majority of the worker problems occurred during the 1964 outbreak and during scattered laboratory exposures. In one incident a janitor developed chloracne when he was exposed to some lab wastes.

The Dow report also described the German and French chloracne outbreaks and the "remarkable letter" sent by the Germans to trichlorophenol makers outlining "parameters necessary to avoid chloracne hazard." The report said, "Letter was filed and forgotten!," noting that Dow had finally purchased the process information from the "quite knowledgeable and very cooperative" Germans years after the letter had been sent.

In recalling its own problems in 199 Building, the Dow report took on a strangely comic tone. The section was titled "The Aniline Mountain Story—199 Building." (Aniline is a toxic derivative of one of the chemical ingredients of trichlorophenol.) In 1946, 199 Building had gone into trichlorophenol production, and it had few problems until November 1963, when Dow "pushed process to increase production, raised temperature from 195 degrees C. to 205–225 degrees C." By February 1964 "supervisors, lab man and operators" had chloracne. The process was changed to reduce chloracne, but several months later a 0.25 percent concentration of the waste oil was still killing rabbits. In mid-1964 "Management asked two questions of medical, industrial hygiene and safety. (1) Can plant be operated safely at all? (2) What is needed to ensure safety? Answer to (1) was YES. Answer to (2) Cost many $100,000's and much manpower."

The report on 199 Building was concluded with a joke:

DISPOSITION OF 199 BUILDING
Prize Winning Suggestion—

"Bury it and make a ski resort,
Call it Aniline Mountain."

The building was demolished in 1967 and analyzed for lingering toxicity. All process equipment was cleaned with acid, rinsed three times with soap and water, and buried on Dow property in Midland. Wipe tests showed that contamination of equipment was "very low." Dow set up a ranking of dioxin contamination ranging from "Clean. Congratulations!" to "Clean, but could be cleaner. No hazard" to "Getting dirty. No acute hazard" to "Dirty. Cleanup immediate and re-test. May be hazardous over period of time." A new trichlorophenol plant had been built in 1966. Using the German low-temperature technology, Dow said it had been able to decrease the dioxin load in its herbicides to 0.1 part per million and then 0.01 part per million.

Dow also evidenced a continuing interest in the effect of dioxin on humans. The company contracted with Dr. Albert Kligman, a University of Pennsylvania dermatologist, to test dioxin on the skin of prisoners. Dr. Kligman had been testing mind-control, skin-hardening, and other chemical agents on inmates at Philadelphia's Holmesburg Prison under an army contract initiated in 1964 in order to determine the dosage needed to disable 50 percent of a given population mentally.

After the 1964 outbreak of chloracne at its Midland plant Dow wanted to find out how the effect of dioxin on rabbits' ears compared to that on human skin and if there was a threshold, or minimum amount, that could be expected to produce chloracne. Dow negotiated a $10,000 contract with Dr. Kligman to apply extremely small quantities of dioxin to prison "volunteers" (who were told nothing about the chemical or its possible effects but were offered a small amount of money in return for signing a release form authorizing the "hospital, laboratories or others to perform medical and other tests on me").

The plan called for the application of slight to moderate amounts of dioxin over a period of time. Dow was sufficiently impressed with dioxin's toxicity to caution Dr. Kligman that the dioxin they were sending him was "highly toxic" and that "the seriousness of the consequences that might develop from testing with this type of compound require that we approach the matter in a highly conservative manner." Dow further told Dr. Kligman that an oral dose of one-millionth of a gram of dioxin was "always fatal" in animal experiments and suggested that total exposure of the prisoners should not exceed sixteen times that amount. Because animal tests showed a "typical clinical picture of severe liver and kidney injury," Dow advised kidney and liver tests.

Dr. Kligman's initial tests on sixty prisoners produced no skin effects. The dermatologist informed Dow that "unfortunately, not a single subject developed acne." Dr. Kligman was "grieved that so little had been learned" but hoped that someday he would "shine a light into this cave." He was "encouraged" by the dearth of results to "proceed more vigorously," and Dow authorized more experiments. Dr. Kligman consequently assembled a new panel of ten prisoners, upon whose backs he painted 468 times the maximum dose of dioxin recommended by Dow. Eight of the ten inmates developed persistent cases of chloracne. Dr. Kligman informed Dow:

> In three instances, the lesions progressed to inflammatory pustules and papules. These lesions lasted from 4 to 7 months. Since no effort was made to speed healing by active treatment . . . in no instance was there laboratory or clinical evidence of toxicity. These results implement the conclusions formerly drawn, namely it is much more difficult to produce acne in the human than in the rabbit's ear.

While Dr. Kligman was running his prisoner experiments—which ended inconclusively in 1967 without plans for follow-up studies and without keeping a record of prisoners who were exposed to dioxin— another kind of human experiment occurred in France. This was the more familiar unplanned worker experiment, and it happened on October 4, 1966, when an explosion tore through the trichlorophenol reactor at a French herbicide plant near Grenoble. Twenty-one workers developed *l'acne chlorique.*

In 1967—a time when Agent Orange was at peak use in Vietnam—Defense Secretary Robert McNamara and the Joint Chiefs of Staff were clearly told that the herbicides were endangering the lives of the Vietnamese people. A Rand Corporation report commissioned by the Defense Department's Advanced Research Projects Agency, which assessed the political hazards of high-risk military operations, strongly criticized the use of chemical defoliants as ineffective for crop defoliation and as possibly approaching the "lethal level" of exposure for Vietnamese children and elderly civilians. The study warned that herbicide concentrations twenty times those used in the United States were periodically used in Vietnam, despite the lack of research on health effects. The Army Chemical Corps nevertheless issued a study several months later pronouncing herbicides safe on the ground that the scientific literature was "singularly free" of references to industrial health problems.

It was not the first warning the military high command had received about dioxin. During the spring and early summer of 1965 members of President Johnson's Science Advisory Committee had begun briefing top-level government officials, including Defense Secretary McNamara, about the potential hazards of dioxin. But Donald Hornig, the committee chairman and President Johnson's right-hand man on science and technology matters, did not relay to the President his own feelings that "one ought to be concerned" about dioxin's potential health effects. Yet the State Department confidently issued a report in the spring of 1966 observing that the herbicides used in Vietnam were "nontoxic and not dangerous to man or animal life."

The military was single-minded in its approach to herbicides. It needed them to fight a war; therefore, they were safe until proved guilty. But other government agencies were beginning to have public and private doubts about their safety. The government had launched a sweeping review in the summer of 1963 to test widely used pesticides and industrial compounds, including 2,4-D and 2,4,5-T, for their potential to cause cancer, birth defects, and genetic effects. The review, which was performed by Bionetics Research Laboratories, was a partial reaction to the public storm over *Silent Spring* and the thalidomide scandal. It was known by 1965 that 2,4,5-T had "a significant potential to increase birth defects" in laboratory animals. Bionetics wrote a preliminary report in 1966, but it was distributed only sparingly inside and not at all outside the government during the next three years.

It was not until reports of human injury and ecological damage began filtering out of Vietnam that doubts about the safety of the herbicides were raised outside private industry and scientific circles. Yale botanist Arthur W. Galston warned in an article in *The New Republic* in 1967, "We are too ignorant of the interplay of forces in ecological problems to know how far-reaching and how lasting will be the changes in ecology brought about by the widespread spraying of herbicides in Vietnam. These changes may include immediate harm to people in sprayed areas."

The American Association for the Advancement of Science (AAAS) asked the Defense Department to conduct a fact-finding field investigation of the effects of herbicides in Vietnam. The Pentagon instead commissioned the Midwest Research Institute of Kansas City, Missouri, to review the scientific literature and consult with experts on herbicides. The institute reported in 1967 that it was "highly unlikely" that herbicides were harmful to humans or animals in Vietnam and described "the greatest ecological consequence" of their use as the destruction of vegetation. But the institute delivered no verdict on possible chronic effects from long-term use, citing the lack of conclusive information.

A committee of the National Academy of Sciences, a private organization chartered by Congress to provide scientific advice to the government, declared the report "only a first step" in investigating the consequences of the use of herbicides, and the AAAS reiterated its call for an on-site inspection in Vietnam. Fred Tschirley, a plant ecologist with the United States Department of Agriculture, toured Vietnam in 1968. His observations largely duplicated those of the Midwest Research Institute. "The defoliation program has caused ecologic changes," he wrote. "I do not feel the changes are irreversible, but complete recovery may take a long time." Two independent scientists who went to Vietnam a year later reported, by contrast, that the ecological consequences of defoliation were severe.

As the scientists were probing and squabbling over the effects of herbicides on people and the environment, there was another explosion and outbreak of chloracne among workers in a herbicide factory. At midnight on April 23, 1968, the temperature in a trichlorophenol reactor at the Coalite and Chemical Products Ltd. plant in Derbyshire, England, reached 175 degrees centigrade. The temperature rose continuously for fifty minutes, rupturing the vessel at 225 degrees and causing an explosion "of considerable violence" somewhere past 250 degrees. So forceful was the blast that a chemist was killed by falling masonry and a nearby wall was almost totally demolished. The Fine Chemicals Unit was closed down, and doctors began examining the fourteen men who were in the plant during the explosion or its aftermath.

Eleven of the fourteen men showed abnormalities in medical tests. All but one had abnormal liver function results. Five men had abnormal blood counts. Three men had abnormal amounts of sugar in their urine, and others showed symptoms of kidney disease. But the workers' tests improved to normal within ten days after the accident, and the plant was reopened.

Damaged portions of the plant were sealed off so that they could be decontaminated, while production resumed in the remainder of the building. Over the next few weeks workers began to develop the telltale signs of chloracne. From May 8 through December 8, seventy-nine cases of chloracne were recorded, many of them "severe and extensive." In some men there was such "a liberal smattering" of blackheads that the whole side of the face "bore a dusky grey color." Most of the victims weren't factory operators but were fitters, plumbers, and electricians who entered the plant only occasionally and who worked bare-handed with pipes, joints, and cables. "Obviously," concluded one of the doctors studying the workers, "it was something which had been deposited or embedded in the matrix of the building itself which was causing the reaction and which was being absorbed by contact. Later, it was established that the toxic material was 2,3,7,8-tetrachlorodibenzodioxin."

The workers were treated with antibiotics, zinc sulfide lotions, and ultraviolet light. They were advised to adhere to a "rigorous campaign" of personal cleanliness, which included home steam bath facials and sunbaths. Most of the men recovered in six months' time, although chloracne persisted for several years in a few cases.

In the wake of the chloracne outbreak the plant was closed once again, and no one allowed inside without special protective clothing, including full-face mask, gloves, and boots. On leaving the building, workers were decontaminated in a special room. Rubbings taken from different parts of the building were painted on rabbits' ears, using the technique perfected by Dow in the early 1940's. The rabbit tests showed conclusively that dioxin was clinging to the walls, roof, and other parts of the plant. The entire building was thoroughly cleaned and resurfaced. Heavy and costly equipment that had been contaminated by dioxin was buried 150 feet down an old coal mine. When the decontamination was completed, rabbits left inside the building for a week showed no untoward skin problems, and production resumed. An entirely new and completely automated plant was eventually built for the production of trichlorophenol.

When the British company contacted its German client BASF to inform it of the loss of another chemical normally produced in the Fine Chemicals Unit, the Germans responded with detailed information about their own reactor explosion in 1953 and the subsequent health effects. The Germans would not divulge exact figures, but they told the

British doctors that a number of workers had developed severe liver disorders, some of them fatal; one man "was expected to die soon from this" and others were still under treatment fifteen years after the accident. The Germans also reported that one worker had died of intestinal sarcoma, or cancer of the connective tissue. The Germans were "indeed surprised" when informed that the British workers had developed nothing beyond chloracne. The British rejected the Germans' advice that they run a few liver biopsies on their workers because they did not want to "create undue alarm." During the next several years no new cases of chloracne developed and the workers appeared healthy.

Then, in the late winter of 1971, two pipefitters who were working at an entirely new building that had nothing to do with trichlorophenol developed chloracne. One of the workers, a forty-one-year-old man who didn't change his work overalls before going home, spread chloracne to his four-year-old son, and the second worker, a twenty-four-year-old man, contaminated his wife, who developed the skin disease nine months later. Their only possible link to dioxin was a large metal vessel from the old trichlorophenol plant, which had been repeatedly cleaned by high-pressure steam jets and which had tested out clean. The pipefitters had worked no more than a day or two connecting new equipment to the vessel, yet doctors reported that they developed chloracne more persistent than even the workers who had been inside the plant during the 1968 explosion. "These two men have been very unlucky," observed doctors. "They probably chanced upon a pocket of contamination, tetrachlorodibenzodioxin being almost indestructible and very persistent."

Meanwhile, back in the United States, at the Monsanto herbicide plant in Nitro, West Virginia, where chloracne had erupted after a 1949 explosion, the company's medical director wrote in a brief summary memo in 1968: "I don't want to be cynical, but are there any employees in the Department who don't have chloracne already?"

Monsanto and Dow and the rest of the herbicide industry continued to supply the defoliation and crop-destruction needs of the Vietnam War, although Dow was out of the wartime Agent Orange business by 1970 and declined to bid on a project to manufacture the defoliant in a government plant, citing as one of its reasons the chloracne problem. (Thompson Hayward told the government there was a chloracne problem in the Agent Orange manufacturing process when it negotiated its first government contract in 1967 and even factored the chloracne into the price of the material.) In March 1970 representatives of Dow went to the Pentagon to brief the military about the contamination of 2,4,5-T with the "highly toxic material" dioxin. Although chloracne had been discussed openly with the military, it was apparently the first time Dow informed the military about the link to the dioxin contaminant.

If the appetite for herbicides was declining in Vietnam, it was only increasing in America. One of the biggest domestic users was the United States Forest Service, which sprayed vast stretches of range and forestland with herbicides to clear out weeds and brush that interfered with cattle grazing, timber harvesting, and water drainage. In the spring of 1969 the Forest Service purchased what herbicides Dow could spare and prepared to spray the scrub brush in some unassuming mountains on the edge of Globe, Arizona.

chapter 5

It was the kind of noise that is irregularly heard in isolated places like Ice House Canyon, and that can be the only reason it broke through Billee Shoecraft's swampy sleep early in the morning of the first Sunday of June 1969. Her husband, Willard, was still sleeping when Mrs. Shoecraft slipped outside their bedroom door—a forty-eight-year-old woman in a pink chiffon baby doll nightgown, her hair bleached white and worn in a bouffant flip—and stared with some confusion at a single-engined, piston-powered three-seater helicopter flying directly toward her house.

The house sits in one of several canyons that run like fingers toward the base of the Pinal Mountains in the Tonto National Forest in south-central Arizona. The helicopter was coming from the direction of the mountains, and when it flew over her house, Mrs. Shoecraft was sprinkled with something wet. There was a sweet smell to the air. A neighbor thought it more tangy. He compared it to breathing lightning.

Mrs. Shoecraft hollered to her husband to get up, and then she telephoned the local forest ranger's wife to find out what was going on. The ranger's wife had no radio contact with the helicopter until after 8:00 A.M, and all she could suggest was that Mrs. Shoecraft somehow get the pilot's attention.

Through the sixty-foot wall of glass in her living room that framed the Pinals, Mrs. Shoecraft saw the helicopter swoop low over the aspen grove at the foot of the mountain, brushing the trees with a white fog. She grabbed her boots and a jacket to cover the chiffon nightgown and set out in her blue Toronado, its windshield splattered with the helicopter's droppings. The chopper turned away from the aspen grove and passed over the Shoecraft land again. Mrs. Shoecraft got out of the car and waved her hands in the air at the pilot, then drove after the helicopter as it circled south back toward the mountain. When the road gave out, she headed back to her home, but this time the helicopter trailed her. She jumped out of the car to wave him down, and just as she did, the helicopter passed over her. One of its spray nozzles had jarred loose, and it was spitting out a fine stream of droplets. Mrs. Shoecraft and her car were drenched with liquid.

The pilot noticed the leaking nozzle and flew back to his helispot to fix it. The helispot was three-quarters of a mile from the Shoecraft property, and it was there, the pilot later told the forest ranger, that a "crazy white-haired lady had accosted him—well, had attacked him, more or less." Ranger William Moehn knew immediately whom he was talking about, for Mrs. Shoecraft had what a doctor friend described as a "pressure of speech—she talked too damn much." The ranger had been on the receiving end several times before when she feared herbicides were being sprayed on her property.

The Forest Service had been spraying the Pinals with herbicides on and off since 1965 to kill the chaparral and scrub brush that soaked up mountain rains and prevented water from moving through streams and rivers down to the vast, dry Phoenix valley. Clearing out the stubborn manzanita and mountain laurel also made room for more forage grasses for cattle. The cattlemen liked it, the citrus growers and swimming pool salesmen in the valley liked it, and except for an occasional complaint from people who lived at the edge of the national forestland, the spray project seemed to have public approval. Mrs. Shoecraft was one of the complainers, as was her neighbor Robert McKusick, whose entire family, including two dogs, had been pelted with spray from one of the helicopters in 1968.

Some years the Forest Service did all but post billboards in advance of the spray; others were conducted more quietly. The 1969 spray was one of the quiet kind. The idea was to send the helicopter out early enough in the morning so that downslope winds from the mountain would be light, and even if a gust came up strong enough to carry the spray for a distance, on a Sunday few people would be moving around outside.

The helicopter pilot fixed his broken nozzle at the helispot and took off again, spraying through the morning, part of the afternoon, and into the early evening, finally shutting down at 7:35 P.M. The day had been almost windless when he started but became gusty in the afternoon, grounding the chopper for several hours. The pilot's instructions had been to stay as close to the top of the brush as possible and to leave a 100-foot buffer between the spray area and private houses. If the wind exceeded ten miles an hour, increasing the chances of spray drift onto private land, he was to quit.

While chasing the helicopter that morning, Mrs. Shoecraft had been standing next to her neighbor's clay pit and stock pond when it was sprayed, so she drove down to Robert McKusick's house, an eighth of a mile away in Kellner Canyon, to tell him. Later in the day Bob McKusick saw the helicopter as it passed over the crest of a hill and flew up Kellner Canyon the length of his property, spotting foliage and bed sheets hanging on a line outside. The spray was like a swarm of bees on the wind, hitting one tree and missing the next.

In the same canyon, auto repair shop supervisor Robert McCray and his family were building a house on the land they owned there. Rosalie McCray was making sandwiches for lunch. Her five-month-old baby boy was in his bassinet under a tarpaulin. Her two daughters and oldest son were helping their father. It was close to 11:00 A.M. when the helicopter passed directly overhead, spraying the family with such a fog of liquid that for a moment they could not see.

Farther away Ernie Gardner, sales manager for a plumbing supply business, was working outside his trailer when he felt something on his skin. He worked for a while longer and then felt sick and lay in bed the rest of the day.

Everyone in the canyons would remember the day as being hot and windy and the helicopter as flying too high and too close to private land. After they had been sprayed, the McCrays rinsed themselves off with ice water and stripped off the baby's clothes. They had no idea what was in the spray and made no connection between the helicopter and herbicides. "It soaked us," McCray remembered. "It was all over us like a fog."

As the baby began to gasp for breath, the McCrays got into their pickup truck and drove over to the helispot, less than a mile away. They stood outside the truck, waving their arms at the pilot to attract his attention. The helicopter banked and flew over the truck, soaking the family again.

The spraying continued for three more days, and in the weeks that followed, the troubles began. Nearly everyone who had felt the spray sank into a kind of fatigued, headachy malaise. Some developed blistery, itchy rashes. Others were chronically breathless and, in the worst cases, stabbed by severe chest pains. The McCrays' infant son, Paul, who had immediate respiratory problems, saw doctors sixty-two times during the next year alone. The Satamas, a Finnish family who had not been home that Sunday, ate the fruit from their backyard fig trees several days later. Their thirteen-year-old daughter stopped menstruating. Ernie Gardner's pregnant wife, Mary Lou, drank some water from a public fountain in a recreation area inside the spray area and fell sick.

McKusick's wife, Charmion, an anthropologist who specialized in the identification of prehistoric bird remains, had been recording the disintegration of plant and animal life in her own yard on and off since 1966, one year after the Forest Service had first begun spraying the Pinals with herbicides. After the 1969 spray, the one her husband referred to as "when all hell broke loose," she wrote up a report and called it "A Record of Abnormal Occurrences During the Last Four Years at Kellner Canyon, Gila County, Arizona."

1966—Eleven brown towhees were found in our yard seeking water, unable to fly, had eyes discharging or swollen shut, joints... enlarged. Unable to eat, drink, or stand upright. All died within one to three days.

1967—A change in bird population was quite noticeable. Not only were the towhees absent, but there were no resident cardinals, jays, tanagers, hardly any cactus wrens, no Bewick's wrens which have been plentiful since 1954. The bandtailed pigeons came to eat acorns, but only 12 of them.

1968—The five members of our family were sprayed with our two dogs on our patented land two miles by road from our residence. Immediately on returning home, this was washed off with strong soap and clothes changed. In spite of immediate action, the result was rash, difficulty of breathing, chest pain, muscle spasm. I experienced an annoying painful discharge of milk, unprecedented since I have born[e] no children in 12 years and had a hysterectomy six years ago. We did not think of washing off the dogs—they came down with a pneumonia like affliction within a few hours. There was some damage to the tops of hackberry trees in our yard, apparently due to drift of spray. The leaves fell from the trees and were eaten by the geese, which immediately showed signs of respiratory distress. Only one goose hatched in 1968. No ducks at all hatched in 1968. Many chickens... were also lost.

1969—April: The hackberry trees... shed all their tender leaves in one day and began to disintegrate with bark and branches falling off. Chicken and geese exhibited a depraved appetite, as did the goats, leaving their grain stand while they licked up all the leaves. The fowl became sick immediately, starting a long siege of force feeding. After the afternoon of June 8, when the helicopter passed just above our residential property line... spewing out fluid as it traveled, the remainder of the summer has been spent tending the dead and dying.

Mrs. McKusick described a hen that had become partially paralyzed and sucked its wattles and comb back inside its skin after a convulsion. A peacock was found paralyzed, its feet clenched like fists, she said, and those birds that didn't disappear after the spray were found dead in feeders or on the ground. Mrs. McKusick said the water pond turned bloodred, and an owl and two hawks lost their equilibrium, couldn't fly, and died. No ducks or geese were hatched, the chicken hatch was 8 percent of normal, and those allowed to walk on the ground were quickly dead, according to Mrs. McKusick's report. She wrote:

> As for ourselves, we had a recurrence of all the ills of last year, only more severe, since our yard has been sprayed and we cannot get away from it. Conversations with many complaining residents report over and over the following afflictions which they attribute... to the results of spraying: cramps and diarrhea after eating fruit from trees which later showed leaf damage. Difficulty in breathing, pain in chest, vaginal bleeding, muscle spasm, eye irritation and skin irritation.

The McKusicks, and the rest of the people who lived in the canyons, were by nature isolationists. They purposely lived on the far edge of a town called Globe, which in turn was ninety miles from Phoenix. They moved out to the canyons for different reasons, but the desire they held in common was to live as they pleased. Mrs. Shoecraft was transfixed by the mountains and built an extraordinary house there with four fireplaces, seven doors, and tile hand-painted to match the gray bark of the giant sycamore trees. Bob McKusick mined clay on his property and designed ceramic and mosaic tiles. He and his wife reared three children and all manner of animals.

Neighbors by definition only, Mrs. Shoecraft and Bob McKusick hadn't exchanged five words in the five years before the 1969 spray. It took the intervention of outside forces in their private world to draw them into an uneasy alliance. They both were patriots, in their own way, who believed in the rights of the individual to live a life of his own making and who believed that the government had certain duties to guard those rights. "I'm not a militant; I just believe in my rights," Bob would say, usually in conjunction with a threat to shoot down the next helicopter that tried to spray his land.

Mrs. Shoecraft was a complicated, contentious woman whose hyperboles of speech and dress attracted some people and repelled others. She careened between two selves. One was a quiet, reflective woman who wrote poetry and lived for long periods by herself in a small summer cabin 8,000 feet up the mountain. The other was a bossy lieutenant given to threats and theatrics as she conducted a self-appointed crusade to save the world from itself. Her methods may have been obstreperous, yet everything she did was invested with her own clear sense of right and wrong, and she was more willing than most women of her generation to sacrifice form for purpose.

Bob McKusick was more interested in saving his own piece of the world as opposed to the totality of humankind. He was an artist and a one-handed potter (his right hand had been blown off in a childhood chemistry experiment). At various times he taught public speaking, a Dale Carnegie course, and ceramics. He was a member of Toastmasters International and a practicing psychic specializing in dowsing, which is the detection of hidden water or minerals or missing people with the mind, not with machines. Some thought McKusick was crazy, but he, like Mrs. Shoecraft, was not one to lose sleep over the opinions of strangers.

The two of them came together in a fitful battle that was driven by their own brand of laissez-faire individualism. "Don't Tread on Me" was the unwritten motto; only in this case it wasn't fences or land grabs but toxic chemicals they were worried about. They fought the right of anyone to dump herbicides on their land, their children, their animals, without warning and without giving them the right to choose. At the same time they wanted a strong government enforcer to protect those rights. The basic tenets were outlined in a paper that came to be known as the 9-Point Proposal, which they put together in the weeks that followed the spraying. The first four points were:

1. All spraying or killing by any method of any plant life on Pinal Mountain . . . shall cease immediately.
2. That never again shall the United States Forest Service be allowed to set up a program to destroy without the consent of the people of the entire area involved after all of the hazards or outcome of any such project is known beforehand, and understood elsewhere.
3. That never again shall we be used as government Guinea Pigs.
4. That it be made a law that no spraying or herbicidal use or destruction of plant life can be done until a complete analysis is made public and approved by the public prior to any action.

The final "point" insisted dourly that "we the people do declare that we shall defend our own land from any further desecration." Bob McKusick owned a rifle and several guns, and he had said publicly that he would shoot down any helicopter that dared spray his land again. It was this kind of bravado, coupled with Mrs. Shoecraft's soapboxing, that made the Forest Service people's hair stand on end, but it made absolute sense to the residents of the canyons, who had uselessly flailed their hands in the air as the dripping helicopter passed over their heads.

Twenty-five days after the mountain had been sprayed, eighty residents endorsed the 9-Point Proposal at a Globe City Council meeting. Ranger William Moehn stood up at the July 3 meeting and said, "The stories about ducks' eggs, owls not able to fly, and other wild tales are a bunch of malarkey. Insofar as some leaves falling off trees, in some cases this happens every year at this time." After the council meeting, the *Arizona Record,* the town's normally languid weekly newspaper, woke up with a start. WAR DECLARED ON FOREST SERVICE, its headline shouted. The story said:

> McKusick, Mrs. Shoecraft and others present reported that fruit trees, other types of plant life and even animals had either died or been affected in various and peculiar ways recently, shortly after the Forest Service sprayed close to Globe. Charges were made that the helicopters sprayed from too great a height, sprayed over private land, sprayed on windy days, used a chemical mixture not recommended by the manufacturer, did not know what effects the spray would have on plant life and animals outside the area sprayed, and proceeded with the project without interruption after protests were made.

Ranger Moehn and his staff were startled by the pronouncements of herbicide carnage delivered most loudly—although by no means solely—by Billee Shoecraft and Bob and Charmion McKusick. The Forest Service people genuinely believed the herbicides were not harmful to birds, insects, fish, wildlife, or humans and said so in a 1965 press release about the spray project. Ranger Bill B. Buck even invited people to take a close-up look that year. "If you're as curious as I am, you'll want to drive up and watch the operation! I hope you will," the ranger wrote in his weekly newspaper column, "Tonto Topics." Range and Wildlife staff officer William Fleischman, who was bald, contended testily, "I haven't been to a doctor or hospital for 10 years, and because I don't have any hair doesn't mean I got defoliated."

At Mrs. Shoecraft's insistence, Ranger Moehn and Officer Fleischman went out to the Shoecrafts' place in Ice House Canyon after the spray to take a look around. It appeared to them that herbicides had damaged some native vegetation as well as some peach trees and a small area of yellow blossom sweet clover. They also observed herbicide damage on vegetation in neighboring Kellner Canyon, where Bob McKusick lived, and concluded that both canyons had been hit by spray drift from the helicopter.

The Forest Service visit was followed that summer and fall by an Arizona Fish and Game Department inspection, which was followed by a University of Arizona extension range management specialists' tour, which was followed by a group of officials sent over by the governor, which was followed by assorted and mixed group tours by Forest Service and U.S. Department of Agriculture inspectors. With few exceptions, the inspectors discredited reports of herbicide damage in the canyons. They blamed normal viral and bacterial plant diseases and an overly warm winter for the discolored and dying vegetation, and birth injuries for deformed animals. (One inspector puzzled over the severe headache he developed at McKusick's clay pit but contended nevertheless that no relationship existed between his headache and the herbicide spray.) The Fish and Game Department group found no ill effect to wildlife. In fact, they observed "numerous birds, reptiles and innumerable diversity of insects, together with a couple of ground squirrels, all hale and hearty." Mrs. Shoecraft, who had begun making notes on each "task farce," as she called the visitors, believed contrarily that local wildlife had been seriously affected by the herbicides. The description of animals as hale and hearty seemed ludicrous to her. "What were they doing? Singing and dancing? Smiling or laughing out loud?" she wrote sardonically. "Most of the animals aren't happy. They aren't even sad. They're dead."

The university extension range management specialists rode around the canyons in Mrs. Shoecraft's pickup truck and contended that while much of the leaf discoloration was normal, "an experienced eye might well observe degrees of apparent herbicide symptoms on some plants." But the governor's group flatly declared that "no phenoxy herbicide symptoms were observed."

The group stopped at McKusick's place, where Bob showed them a baby mockingbird with paralyzed legs and feet, a paralyzed bantam chicken with a twisted neck, and a deformed goat kid, which he said was born one month after the spray. The inspectors believed there were natural causes for the deformities, and they did not get on well with McKusick.

"We found it impossible to discuss the damage intelligently with Mr. McKusick," the group wrote in its report to the governor. "He is firmly convinced that all damage to wildlife around the area and plants on his property are [sic] the result of spray drift . . . and refuses to acknowledge the fact that his goats [eating grass and bushes] and other natural agents are causing damage on his property. Mr. McKusick appears to bear a burning hatred of the Forest Service."

Mrs. Shoecraft was stop nine on the tour. The governor's group was in and out in thirty-six and a half minutes. Her lawyer had been talking about a lawsuit, and both sides were wary of the other. The field report reflected the tension. William Warskow of the Salt River Valley Water Users Association, which had co-financed the spray project, directed the governor's evaluation. He wrote:

> Complaint: Saw helicopter pass over four times. "Lawn dying, hackberry trees dying, damage to Sycamore trees, Willard's garden ruined. Look at the film on my fish pond. I'm

just sick!" Complained also of chest and right shoulder pains following spraying.

Observation: We arrived at the Shoecraft residence at approximately 1445 hours. When I suggested to Mrs. Shoecraft that we put the discussion on tape, she countered with a like proposal and we ended up in a Mexican standoff. She first showed us her roses which exhibited leaf browning but no clear symptoms of herbicide. Hollyhocks had died—the cause is questionable.

Mrs. Shoecraft claims the helicopter passed over the Shoecraft property with the nozzles dripping spray solution. She could not remember whether the spray unit was on or off, but could clearly remember seeing the "fog" coming from the chopper over 3/8 of a mile up canyon.

She went to great lengths to show us her mountain mahogany bush about which she had written poems and which she was so dearly fond of and showed us the damage done to the manzanita in the area. The manzanita blossoms did not appear to be properly developed, but exhibited no herbicide damage. The mountain mahogany exhibited slight yellowing of a few leaves.

The report further indicated that Mrs. Shoecraft's oak bushes showed no signs of herbicide damage, the dead branches on her sycamore trees were a natural phenomenon, her failed garden was planted in coarse, hard-baked soil amid rows of unweeded grass, that her peach tree appeared to be suffering from plant disease of an indeterminate sort. The governor was told that no herbicide damage was observed on the field trip with the exception of one property, and almost all the so-called drift damage was due to natural diseases, insect damage, or a combination of both. A tour and report one week later by the United States Department of Agriculture drew similar conclusions, allowing only that spray drift had probably been responsible for "some damage" on private property, although "the amount of damage is far less than some would like us to believe."

Mrs. Shoecraft scrawled across the bottom of the governor's task force report, "Such poor English and I never wrote poetry about mountain mahogany." Then she filed it away with other pieces of evidence she had been collecting since the spray. Her lawyer had learned from the Forest Service that the herbicide used in the June spray project was silvex, which had been purchased from the Dow Chemical Company. The Forest Service had sprayed 970 gallons of silvex across 1,900 acres of forestland during the four-day June 1969 operation. In previous years a half-and-half mixture of the herbicides 2,4-D and 2,4,5-T had been sprayed across 1,238 acres in 1965 and 1,060 acres in 1966. There had been no spraying in 1967, but in 1968—a year when domestic supplies of 2,4-D and 2,4,5-T were rock-bottom because of the demand for Agent Orange in Vietnam—silvex had been substituted on 1,900 acres. The lawyer asked for a sample of silvex from its original container, but the Forest Service said it had only an unopened fifty-five-gallon drum.

Mrs. Shoecraft and Bob McKusick sent away for Dow's toxicology data on silvex, for bulletins from university zoology departments, for user manuals from the U.S. Agriculture Department. They also sent samples of vegetation, soil, and water out for laboratory analysis. GHT Laboratories in Brawley, California, received a box on July 14 containing seventeen vegetables, some soil, and a clump of "creek material" from north of the Shoecrafts' house. The lab found herbicide contamination—not silvex, but 2,4,5-T in Mrs. Shoecraft's bush blackberries, wild cherries, grapes, walnuts, bush beans, corn, and peaches and in the creek. On July 15 a University of Arizona laboratory reported that samples of water taken from Mrs. Shoecraft's kitchen sink and reflecting pool and from McKusick's stock pond were contaminated with low levels of silvex and several similar compounds. The lab advised there was no real public health problem but suggested the results "substantiate some environmental contamination that may be regarded as potentially objectionable."

Mrs. Shoecraft and Bob McKusick were intent on getting the spraying stopped. They circulated a petition, signed by 200 people, which read, in part, "We are convinced that the toxic spray used, especially in the excessive strength applied, represents an unconstitutional attack on our rights" and warned that "necessary measures of self-defense will be employed" if the spraying program was not permanently halted. They sent the petition, plus a four-page record of spray damage complaints, and their 9-Point Proposal, to their congressman, Sam Steiger, and to Arizona's two senators, Barry Goldwater and Paul Fannin, as well as to Senator Henry "Scoop" Jackson, chairman of the Senate Commitee on Energy and Natural Resources. On July 17 the *Arizona Record* ran a story about the petitions, along with a photograph of a black walnut tree with half its leaves gone and a dying mulberry tree with bright yellow leaves. The headline read: AUTUMN IN JULY?

People in Globe didn't know what to think. Mrs. Shoecraft had always been a classic mother and wife: a member of the PTA, a Girl Scout troop leader, a fund raiser for the Globe Community Center, a mother of three boys. Her husband, Willard, was one of Globe's best-liked and most influential residents. He owned radio station KIKO, which he ran from a mobile home trailer next to a trailer park just off Highway 60, the main road into town from Phoenix. Willard had a voice and delivery more suited to the Old Vic than to the broad, twangy Southwest, and everyone loved listening to his deep, masterful phrasings. In the mornings Willard's beautiful voice offered up pieces of furniture and used cars during KIKO's daily on-the-air rummage sale, or moderated interviews, for example, with the nervous stars of Globe High School's spring play, on his Monday through Friday *Open Line* show. Willard thought of *Open Line* as the daily paper Globe didn't have, a kind of "back fence deal."

During the summer of 1969 *Open Line* became suddenly filled with talk of herbicides. Ranger Moehn appeared several times to reassure residents about the safety of herbicides. Bob McKusick used *Open Line* once to reiterate his shoot-down-the-helicopter threat. Mrs. Shoecraft increasingly used her husband's radio station to air herbicide critics' views, although Willard, a radio man since seventeen, valiantly tried to steer an objective course.

By mid-July Mrs. Shoecraft and McKusick had put together a group of ninety-two people who claimed damage to property, animals, or themselves as a result of the herbicide spray. They threatened legal action and laid plans for a big town meeting July 28 to confront the Forest Service with demands for a permanent halt to the spraying. The Forest Service had been fencing with the herbicide issue almost entirely by itself, with some outside assistance from the Salt River Valley Water Users Association.

The Dow Chemical Company had sent Keith Barrons, its director of plant science research, to assess the situation in Globe. Barrons had helped conduct the field trials on 2,4,5-T back in the late 1940's, and he had a good eye for detecting the effects of herbicide spray drift on plants. The Forest Service people took Barrons to the spray area. "It appeared to me that the chap had done a very good job of spraying," Barrons recalled. "Within a couple hundred feet of the sprayed area, I found wild grape, which is a very, very sensitive species, and there was little, if any, effect on this wild grape." He inspected a tree whose owners claimed had been damaged by spray drift. Barrons recognized the symptoms of peach tree curl. "They had a tree that was obviously suffering from peach tree curl," he said. "A peach tree next to it looked perfectly good. How they could think this one tree got all the drift and the one next to it didn't get any, I don't know."

Barrons flew back to Midland to report that he had found no evidence of herbicide damage in Globe. "I left that evening, and I didn't meet Mrs. Shoecraft," he said. "We knew she was the one who was making a lot of noise."

Dow also sent a public relations consultant from California to keep tabs on Globe. Ross Wurm telephoned Dow's Jim Hansen in Midland on July 18, 1969, with the suggestion that Dow send a "technical" man to the upcoming meeting as a show of good faith, to answer questions and calm fears, and as a gesture of solidarity with its friends, the forest and water people. He told Hansen:

> It is obvious that every possible public relations mistake was made in the beginning on this spray job. There are some lessons about this that we want to look at for the future. The Forest Service sprayed on a Sunday morning without any warning to people in the area at all and then the helicopter pilot, although unwilling to admit it at this time, apparently flew back and forth in the process of loading over a residential area. Whether or not his nozzles were dripping remains to be seen. But probably not because there is no sign of it, although the people in the area claim that the wind was blowing and it was spraying on them, etc.

As a result of this, which occurred in early June, Mrs. Shoecraft has really created quite a storm. She got 92 different complainants together. . . . They are not necessarily behind her, but she does have a long list of complaints which has been submitted. These are quotes from the written list. . . .

"Saw helicopter pass over four times. Lawn dying, hackberry trees dying, damage to sycamore trees, complained of chest and right shoulder pains following spraying."

Here's another one.

"Allergy attacks. Went to doctor. Little boy sick. Very bad attacks."

Another one.

"Fruit trees show severe discoloration and wilt. Many large tomato plants with tomatoes have ceased growing and have turned brown."

Here's another one.

"Garden dying. Carrots died. Large Chinese elm by house has been badly discolored on the top. Very unhappy. Peach trees are withering, chickens are dying."

All of these things that normally happen in Arizona in the summer, of course, are being blamed on the spray job. This woman, Mrs. Shoecraft, who is the formidable opposition, has done a tremendous job of pouring out bad information. Here's some quotes from the Arizona Daily Star in Tucson, July 3, and this is about as rotten a story as I have heard or read in my life, and I'll quote from it. "Those leading the protest say that the chemical used is untested in its effect on human and animal life and according to its makers, the Dow Chemical Co., has no known antidote." They have taken information from the label and twisted it very unreasonably and other than a couple of radio appearances by the Ranger, no one here has done much about getting it straightened out. But they have really ground out a formidable amount of lies. They have also announced that they are going to sue. At various times, they have also said that they were going to sue everybody else in sight. Mrs. Shoecraft announced that if the cattlemen had anything to do with it, she was going to sue them also.

The ranger is willing to admit privately after investigation by Barry Freeman, the state range management specialist, that there is very minor damage at the top of the area to a couple of home gardens. That's all they think there is in the way of damage. They claim that all of the other problems that Shoecraft and the McKusicks and the other complainants [mentioned] are just due to bugs and drought and distress; the normal pressure of trying to get plant life to exist in Arizona in the summer.

Let me talk about the meeting on the 28th. They have decided that the normal council chambers which hold about 50 won't be large enough and apparently a crowd of several hundred is expected and it will probably be held in the high school auditorium. The opponents have been talking up the meeting on the radio and they were able to get 80 some people out to a meeting this past week without much notice, so it is assumed that there will be a large crowd. [Robert] Courtney, the head ranger of the Tonto National Forest, says that there are 20 or 30 cattlemen grazers in the forest and he expects quite a few of them to show up, too, as obviously interested parties.

This thing reached one real peak earlier this week when the report came back on a sample of water drawn at the Shoecraft's. They got on the radio and announced that "silvex, a deadly poison" was in the drinking water of Globe so that the public health officer had a constant stream of calls and people racing in the door in a panic. It was good that our medical officer called him because he seemed to have very little real knowledge about . . . silvex. It was revealed later that the report says "less than one part per billion" but they were able to say that there was some there, and it creates a problem.

After having done a real bad early job, the Forest Service has set down and taken a look at this and, one, formulated plans for a P.R. program and a method of going for spraying in the future; two, they are working very carefully on the presentation that they are going to give that night. . . . They expect this to be followed by thousands of hostile questions. I think we should have someone there simply because the Salt River Project and the Forest Service are both important to us and I think that we just have to show that we are with them. I think that we should keep a low profile and not necessarily stand up and testify unless necessary. I also feel that it's important for us to identify the press that are operating in the area and that will be covering it. There's the one nutty radio station but there are some others and even if no one comes out from Phoenix . . . there is still going to be lots of local coverage. Let me think about what I need and we ought to put a [background material] packet out next week.

I can't remember if I told you this earlier, but when this petition was circulated with the 92 names on it, I discovered that Mrs. Shoecraft had sent copies to the Sierra Club, and she sent them to both senators. She sent them to "Scoop" Jackson because they were smart enough to figure out that he is chairman of a subcommittee that has to do with the Forest Service. I saw the list of people, and she has just done a fantastic job of making trouble. As I said, she's talked to the Federal Water Pollution Board and the Federal Air Pollution Board.

So there are two things. One, I think we just have to stand up with our friends and Two, there is a possibility if the Sierra Club or somebody decides to run with this, we ought to know as much about it as we can.

As good an advance man as he was, Wurm did not anticipate the lengths to which Mrs. Shoecraft and McKusick would go to draw attention to their plight. They were feeling frustrated by the lack of interest, unanswered questions, and superior airs of some government officials, one of whom had referred to the people in the canyons as "uninformed." Mrs. Shoecraft and McKusick immediately began calling themselves the Informed Citizens Union (ICU). "Shock and disillusionment set in, because of lies and arrogance," Mrs. Shoecraft later wrote, "and finally I blew Pollyanna a kiss of farewell when she couldn't take it any more, and made room for Scarlett O'Hara. But I guess most of the time the Scarlett O'Hara's [sic] get the job done."

They had stacks of reports that seemed to indicate that not enough testing had been done on silvex prior to its being placed on the market. Mrs. Shoecraft found a Dow Chemical report on silvex issued in 1954, the year it was first sold. The report said that field hazards from herbicide drift to crops had not been fully evaluated and noted that silvex possessed "unique" properties that "warrant further investigation."

They had lists of people, animal, and property damage. Mrs. Shoecraft herself had broken out with a blistery rash after the spraying, and she'd gone to the hospital twice. The first time she had such difficulty breathing and swallowing that she went to Gila General. The second time her labored breathing so frightened her that she went to the emergency room at the copper miners' hospital, Miami Inspiration. She also had chest pains and sharp pains in her arms, legs, and back. Often during that summer she had nausea, and she was losing weight. Doctors took X rays and ran tests on her heart and found nothing out of order.

Her teenage son's guinea pig delivered a one-eyed offspring that summer. Mrs. Shoecraft named him Cyclops. She remembered that many of her son's newborn guinea pigs and white mice had died during the summer of 1968, and when she checked the dates on their graves in his pet cemetery, she realized they had died within forty-eight hours after the mountain had been sprayed with herbicides that year. She remembered, too, that their dog had been sick and lost unusual amounts of his hair during that time.

She tried to think back to her own illnesses. In March 1968, one month after the Shoecrafts had moved into their new house in Ice House Canyon, Mrs. Shoecraft had sudden breathlessness and a feeling that she could not swallow, and then, out of the blue, she could not seem to talk. It was an effort to put words together; she felt confused and panicky and was rushed to Miami Inspiration Hospital in the middle of the night. Doctors thought she was suffering from emotional stress; she and her husband had separated for a time and then reunited —and she underwent hypnosis therapy. Two months later she went out in the forest to collect driftwood, and afterward she couldn't breathe and her throat hurt. Water blisters formed on her hands and spread to other parts of her body; for a time they dripped fluid. When her oldest son received a master's degree, she wore a long-sleeved dress to the ceremony to cover the blisters.

When the mountains were sprayed on the last day of May and the first three days of June 1968, the Shoecrafts were away at a broadcasters' meeting in Tucson. After they returned, Mrs. Shoecraft had trouble breathing again and such difficulty swallowing that she occasionally regurgitated something she was drinking or eating. The right side of her throat felt sore, and a Phoenix doctor thought it was swollen glands. Her chest hurt, and for a time her vision was blurry.

Some people in Globe, including doctors, were suspicious. They suspected that Mrs. Shoecraft and some of the others simply picked diseases from the scientific papers on herbicides and claimed them as their own. Mrs. Shoecraft was indignant: Her skin rashes and respiratory problems had appeared before she even knew that herbicides were being sprayed on the Pinals. Before that she'd hardly been sick a day in her life. "If they wanted to try it on a good specimen to see what it would do," she said, "they picked a good one in me. No outside influences such as drinking or smoking—not because my halo is that shiny; I just never got around to them, I guess." She was tiny (just over five feet tall and weighing 104 pounds) but strong (she boasted that she could hoist a 98-pound bag of cement on her shoulders); she hated doctors and paid as few visits to their offices as possible, and she did not take pills.

Besides health problems, Mrs. Shoecraft and Bob McKusick had proof of soil and water contamination. Mrs. Shoecraft had a $500 telephone bill from calling around the state and country for information about herbicides.

The one thing they didn't have was a commitment from Tonto National Forest chief Robert Courtney to halt future spray projects.

Courtney, whose office was in Phoenix, had sent Forest Service people out to inspect the property damage claims, but he himself had never been to Globe. Mrs. Shoecraft decided that if "Big Chief Sitting Still," as she called Courtney, wouldn't come to Globe, then she'd bring some of Globe to Phoenix.

Mrs. Shoecraft and McKusick convinced Rocky Miller, owner of Rocky's Old Ghost Mine rock store in Globe, to lend them his circa 1899 casket, complete with pleats and tucks and a moth-eaten lid. Mrs. Shoecraft covered the lid with a bolt of black satin she had left over from a failed home decorating project. They put a stuffed bear inside, named it Smokey, pasted on fake eyelashes to make it look dead, and covered it with dead plant material from the stream bed—the same plant material that would later come up contaminated in lab reports. People from town put their dead and rotting plants and vegetables inside a big box draped with red, white, and blue bunting. A malformed century plant and a twisted stool were tied to the top. The local mortician lent them his ancient hearse.

The funeral idea was Mrs. Shoecraft's, and McKusick thought it was "pure genius." The Associated Press reported on July 24:

> Globe area residents are protesting the recent spraying of the chemical called silvex by the Forest Service in the mountain hills south of Globe by organizing a funeral procession. Fruit trees, garden plants and other foliage which were allegedly killed by the spray are being put in a coffin and will travel by a hearse from Globe to Phoenix and will arrive at the Federal

Building tomorrow at 2 p.m. The coffin will be delivered to Tonto National Forest supervisor Robert Courtney's office.

When Mrs. Shoecraft and her "mourners"—teenage girls dressed in black veils and dresses and teenage boys, including her thirteen-year-old son, Robert, dressed up like pallbearers—arrived at the Chamber of Commerce parking lot the next morning, Courtney was there waiting for them. "Courtney was like ice, very steely. He was very tense that day," McKusick recalled. "He told us, 'There's no point in going to Phoenix.' "

The forestry chief tried to talk about the benefits of the chaparral spray program and the harmless nature of the herbicides, but Mrs. Shoecraft grew restless and gave orders for the funeral procession to roll. Her intentions, laid bald by Courtney's surprise appearance, were to generate publicity. She knew the odds were she would get better coverage in downtown Phoenix than in the Globe Chamber of Commerce parking lot.

To her delight, television cameras and newspaper reporters were waiting when the four-car procession, bearing twenty-five mourners, turned down First Avenue in Phoenix toward the Federal Building. The Phoenix *Gazette* ran a photograph captioned: "Mrs. Willard Shoecraft throws dead plants on pile in front of Federal Building as part of protest by Globe residents against Forest Service use of a spray chemical."

"The small but vociferous Mrs. Shoecraft," as the *Arizona Daily Star* called her, lifted from the casket what she described as shriveled, blackened peaches, some with second growths on their sides, some with four seeds; four-sided eggplants; black and twisted peppers, red peppers and green peppers, all grown on a single vine; blackened roses; corn ears two inches long; hollow pumpkins; seven-foot-tall okras with twisted growths on their sides; and an eight-foot-long squash plant that had sprouted tentacles.

Mrs. Shoecraft later said that "it wasn't a publicity stunt for publicity's sake, it was the only way to let them know about what happened and that it must not happen anywhere again. The threat of more spraying was hanging over us. And those who had injured us showed no remorse or regrets."

Another time, she said, "What we were doing was for our country."

The same day as the mock funeral in Phoenix, Dow Chemical sent out a press release on silvex. "As with all such materials currently on the market," it said, "extensive studies have been conducted to determine the effect that [silvex] might have on people, domestic animals or wildlife that might come in contact with it." Silvex was "quite low" in toxicity, causing no ill effect in cattle, sheep, swine, ducks, or chickens that consumed water containing fifty parts per million silvex for four weeks. In an experiment "with human subjects," the herbicide produced "no allergenic effects" even after repeated exposure to the material. It could not "accumulate in the environment and cause long term effects."

Mrs. Shoecraft got hold of the Dow press release and scrawled across the bottom, "Statements in this bulletin (or whatever it is) are ridiculous!" She added the sheet of paper to her growing list of press releases, technical bulletins, and scientific papers on herbicides, which she filed and titled in whimsical ways. File 009-DI, for instance, stood for "Deep Intrigue" and contained Forest Service memos on logistics of the spray operation. The file called USDAS stood for United States Department of Agriculture Stuff, and the title of her Forest Service Correspondence (FSC) file was later revised to Crap. She finally combined the files when she became convinced that the Agriculture Department and Forest Service were operating as one.

Mrs. Shoecraft's methods may have been whimsical, but she was growing more infuriated with each passing day. And those who were subjected to her tirades were turning equally testy. Her fight against the herbicides was escalating into an obsession. She spent hours on the telephone or out terrorizing bureaucrats with her $29.95 Craig tape recorder and requests for an interview. She discovered that by asking for the most important person in a department, "the secretary assumed that 'the general' or 'the president' knew you personally, or you wouldn't be calling without an appointment." She was once denied access to some Forest Service files for lack of a copying machine. "Thirty minutes later" she was back "with a 150-pound copying machine" and "two Boy Scouts out of uniform," and "we just copied up a storm." She carried Cyclops, the one-eyed guinea pig, and Split, a guinea pig born with one eye open and one eye clamped shut, in her pockets to show people. When they died, she stored them in her freezer so that they could be preserved for autopsy.

Mrs. Shoecraft's tactics were unsettling, and even some of those who had also been sprayed by the helicopter and who knew her as a neighbor and empathized with her health problems tried to put distance between their battle and hers. "Some people in Globe hated her," Bob McCray recalled. "People who were sick and really direly needed her help backed off from getting involved because of her. I guess before the spraying she had had some crusades. Her personality was such that I don't think people put any credence in what she did.

"I loved her, though," McCray added. "You could feel how bad [sick] she was." The young auto repair shop manager—he was thirty-three when he and his family were sprayed—had also spent several weeks steam cleaning empty herbicide barrels for the Forest Service in his shop in 1968. His wife, pregnant with Paul, had been with him in the shop as he worked. As his family's health problems mounted—they developed allergies and neurological problems, and the baby began having epilepticlike seizures—McCray pursued his own course against the Forest Service and Dow.

Mrs. Shoecraft, meanwhile, was consumed by details about the 1969 spraying operation itself: the altitude of the helicopter; the speed of the wind; whether or not the herbicide had been mixed with diesel oil as a weight factor to help prevent drift. After a time her investigations broadened, and she compiled data on each of the four times the Pinals had been sprayed and on each of the different but closely related herbicides used—silvex, 2,4,5-T, and 2,4-D. She wanted to know: How potent were they to animals and humans? How long-lasting in the soil and the water? If you got some of the stuff on you, was there an antidote? If vegetation contaminated by herbicides were burned, would poisons be released in the air?

She collected scientific studies. Read cold, without benefit of scientific background or interpretation, they were unquestionably frightening. After skipping over the parts that said "no adverse effects" and "not harmful to fish, birds, etc.," she found words at the bottom of the reports like "pulmonary edema, pneumonia, depression of the central nervous system, liver damage, kidney involvement, testicular degenerative lesions." She read that a teaspoonful of silvex was a lethal dose for a 150-pound man and that a luckless coho salmon lived but twenty-four hours after a fifteen-to-thirty-minute swim in silvex-spiked water. A bluegill, she noted darkly, died immediately. "The one that worried me most," she wrote in her notes, "was the redear sunfish." Chronic exposures resulted in testicular lesions, atrophied spermatic tubules, and production of immature, atypical and abnormal sperm. "I don't know what all of that means," she said, "but it sounds like the redear sunfish is in for a big shock when he tries to reproduce!"

When she obtained a label from a silvex can, it said:

> Don't contaminate any body of water by direct application. Avoid contamination of water intended for irrigation and domestic use. This product is toxic to fish. Keep out of ponds, lakes, streams. To avoid injury, do not use where pond water is being used for irrigation . . . for agricultural sprays, domestic water supply and/or livestock watering. Some fish kill may occur.

A United States Department of Agriculture bulletin told her that droplets or vapors of 2,4-D, 2,4,5-T, or silvex carried on the wind "may injure desirable plants that grow near the area treated" and warned against applying the herbicides when wind speeds were above six miles an hour or the temperature higher than ninety degrees.

"The entire job lot of research garbage tells one thing: these chemicals were designed to kill. They do just that," she wrote. "And no amount of research will make them safe or usable in the hands of fools."

At the July 28 City Council meeting, which was held in the auditorium of the Globe High School and attended by an estimated 150 people, Forest Service officials stressed the harmlessness of silvex. They read portions of the Dow press release without identifying it as such. They said silvex was about as toxic as aspirin or salt. They pointed out that other kinds of herbicides had been sprayed on Tonto in past years; these were chiefly 2,4,-D and 2,4,5-T, which, when combined, formed the herbicide mixture called Agent Orange. Military demand was so high for Agent Orange in Vietnam, where it stripped the enemy of his leafy hiding places, that domestic supplies had nearly dried up, forcing the Forest Service to substitute silvex in 1968 and 1969.

The meeting heated up when Mrs. Shoecraft, clutching a Dow silvex label in one hand and a United States Department of Agriculture bulletin in the other, demanded and was reluctantly given the floor. She grilled the Forest Service officials.

"Are you aware of the cautions listed on the label and in the literature that is published by the USDA as they apply to . . . silvex?" she asked.

She fought about conflicting statements on wind speeds and helicopter altitude the day she was sprayed. Courtney, the frazzled forest chief, tried to lighten things up.

"I am inclined to remember Abe Lincoln's story about the man who was tarred and feathered and was being ridden out of town by a rail," he said. "And while this was happening some wise cracker hollered at him and said, 'How do you like it?' And the fellow says, 'Well, if it wasn't for the honor of the thing, I'd just as soon walk.' "

At the end of the meeting Courtney announced that aerial herbicide spraying would halt in the Globe area pending further study of its safety.

The next day Mrs. Shoecraft went to the ranger station with an aide to Congressman Steiger to get a sample of silvex from two Dow cans stored there. She was convinced that other herbicides besides silvex had been used in the 1969 spray and wanted to have a sample from the cans tested. She had tried to get one before and, failing that, photographed the cans with the Kodak Retina Reflex camera she was now taking almost everywhere she went to document dead plants, deformed animals, and signs in forest recreation areas that she thought lacked proper warnings against drinking contaminated water. Mrs. Shoecraft's camera so jangled Ranger Moehn that he wrote in a memo: "I have attempted to cooperate with these people, but am beginning to lose my cool." But on this morning they agreed to give her what she wanted. Bill Fleischman, the bald Forest Service officer, was removing samples from the cans. As Mrs. Shoecraft later recalled, he told her that he had applied silvex to his head once to grow hair.

"He said it's perfectly harmless, couldn't hurt anyone, even commented that they had been known to drink it in cocktails. He stroked his head and said, 'I have tried it on my head to grow hair.'"

"And I said, 'Well, I . . . it won't hurt anything?'"

"And he said, 'No, it would probably be a fine cosmetic cream.'"

Then she touched her fingers to some liquid pooled on top of one of the cans and rubbed a few drops on her forearms and the back of both hands. She seemed to be trying, in a perverse way, to prove the identity of the chemical in the can. Fleischman (who has subsequently denied the hair-cocktail-cosmetic story) implored her to wash it off, but she refused (although she did wash it off when she got home). The next day, her arm covered with a rash, she went to the Miami Inspiration Hospital emergency room for treatment.

The only way to understand the lengths to which Mrs. Shoecraft traveled was to know that she believed she and her family and her neighbors had been injured by people who were now lying about what they did. "Some people in Globe thought that she was a rabble-rouser," Willard Shoecraft explained. "She was a fighter. If you were her friend and you shot someone, she'd swallow the gun and swear on a Bible that you never did it."

Regardless of her moral imperatives, some people in Globe not only thought Mrs. Shoecraft was a rabble-rouser but also thought she played fast and loose with her facts. It was said that she told a local doctor after the spraying, "Do what you have to in order to prove I have herbicide poisoning."

Her friends dismissed such talk as ax-grinding gossip. Like any town, Globe had vested interests dedicated to preserving what passed as the status quo. There were doctors whose medical books held few, if any, references to health effects of toxic chemicals, and therefore, *ipso facto*, such effects did not exist. There was a small but healthy group of restaurant and hotel owners and Indian jewelry purveyors who depended on tourists driving over from Phoenix or Tucson to take the two-and-a-half-hour guided tour of the Inspiration Consolidated Copper Mine or to visit the Salado Indian ruins or the San Carlos Apache Indian Reservation. There were cattlemen who needed forage for their stock. The cattlemen looked for guidance to the university range management specialists and the Forest Service, which depended on the Agriculture Department, which, as it turned out, was infested with vested interests of its own. And there were 6,000 residents of Globe, most of whom had not seen the helicopters and had not been sprayed. To believe Mrs. Shoecraft and Bob and Charmion McKusick and the other residents of the canyons would be condemning themselves to the same fears and uncertainties.

By the autumn of 1969 those fears were intensifying. News was emerging from Vietnam about health and environmental effects allegedly caused by herbicides, and in this country the government-sponsored study that showed 2,4,5-T caused birth defects in laboratory animals was finally and reluctantly made public. Mrs. Shoecraft and Bob McKusick grew more anxious about their own health and, at least in Mrs. Shoecraft's case, borderline evangelical about stopping the use of herbicides in Arizona and other parts of the country.

The government had contracted with Bionetics Research Laboratories in 1963 to test widely used pesticides and industrial compounds for three effects—carcinogenic or cancer-causing, mutagenic or causing changes in the genetic material of cells, and teratogenic or causing birth defects. Preliminary data in 1965 and 1966 showed that the herbicide 2,4,5-T fed to pregnant mice and rats caused birth defects in their offspring. Over the next several years the Bionetics report was selectively shown to various government agencies—including the National Cancer Institute, the Department of Defense, the Food and Drug Administration, and a National Academy of Sciences/National Research Council committee studying persistent pesticides—but it was not made public. By the end of 1968 a good portion of the government knew that of forty-eight compounds fed or applied to the skin of pregnant rats and mice at different dose levels, two compounds, including 2,4,5-T, produced "sufficiently prominent effects of a seriously hazardous nature to lead us to categorize them as probably dangerous." Mice were fed or injected with doses of 2,4,5-T ranging from 21.5 to 113 milligrams per kilogram of body weight. There was an increased incidence of abnormal fetuses at all doses save the lowest. Three times more abnormal fetuses occurred in rats exposed to 2,4,5-T than in control animals, even at the smallest dose level of 21.5 milligrams per kilogram of body weight and the lowest exposure time of six days. The smallest dose of 2,4,5-T given to rats induced 39 percent abnormal fetuses, in contrast with the usual proportion of 7 to 13 percent. The report further listed 2,4-D—the other component in Agent Orange—as "potentially dangerous" but needing further study.

In July 1969—one month after the Pinals had been sprayed—two "unhappy" government scientists leaked the Bionetics report to Ralph Nader's Center for Study of Responsive Law. Nader's staff in turn gave a copy of the report to Dr. Matthew Meselson, a Harvard University biologist who was concerned that Vietnam was being used as a testing ground for chemical and biological warfare and who had helped gather the signatures of 5,000 scientists on a petition sent to President Johnson to stop the spraying of herbicides in Vietnam. Another scientist who had been made aware of the report was Dr. Samuel Epstein, chairman of a federal panel on the teratogenicity of pesticides. The teratogenicity panel was part of the Commission on Pesticides and Their Relationship to Environmental Health, called the Mrak Commission after its chairman, Emil Mrak, a food scientist and former chancellor of the University of California at Davis. The commission had been appointed by Health, Education, and Welfare Secretary Robert H. Finch to review pesticide use in the United States shortly after the government seized several tons of DDT-contaminated coho salmon in Michigan that spring. Panels composed of government, university, and industry scientists, including Dow vice-president Dr. Julius Johnson, reviewed research on the carcinogenicity, mutagenicity, and teratogenicity of pesticides. Dr. Epstein wanted to see the Bionetics data on birth defects before his panel made its report. He was forced to wait nearly a month for a copy of the report and was finally given it when he threatened to dissolve his teratogenicity panel.

In the fall of 1969 the Bionetics report was still an open secret, and Dr. Meselson went to Washington to talk to government officials about it. Speculation about the reasons for suppressing the report was rampant, but definitive answers were scarce. The government insisted that it was merely putting the data through the required peer review and accuracy checkpoints before releasing it. But some critics suggested that Dow, the country's premier manufacturer of 2,4,5-T, simply applied so much pressure to the government that the report was not released. Others posited a reluctance by the White House to add fuel to the antiwar movement, which might switch its focus from napalm—the incendiary jelly also made by Dow—to herbicides. There was also the matter of international criticism. The United Nations adopted a Swedish resolution that declared the use of herbicides a violation of the Geneva Protocol of 1925, which banned the use in war of "asphyxiating, poisonous or other gases" but which President Nixon interpreted as not covering chemical herbicides, smoke, flame, or napalm.

While all those fears likely contributed to the general atmospherics that surrounded the report, the most plausible reason for its suppression is rather more simple: No one really knew what to do with it. As Dr. Lee DuBridge, the President's science adviser, later told Congress, the heart of the government's control over pesticides lay in the process of registration, and regulations were designed to deal with industry test data, not with outside studies financed by the government. "There has been relatively little thought given to the subject of how to incorporate new, unexpected information which is collected outside the registration process into the regulatory process," DuBridge said. "This was clearly demonstrated after the Bionetics study." Industry had met its burden of proof on safety when 2,4,5-T was registered back in the 1940's and 1950's, even though the early rules required the barest of data and ignored chronic effects. As a result, there was no official regulatory pathway through which to route the report, and it fell between the cracks.

Pressure mounted against 2,4,5-T in the fall of 1969, largely from the scientific community. Suddenly, on October 29, knowing that the secret Bionetics study was no secret anymore and that its results would soon become widely known, the government moved offensively to head off criticism and abruptly announced that it was partially curtailing the use of 2,4,5-T. Presidential science adviser Dr. DuBridge issued the White House directive at a press conference in Washington. Registrations for 2,4,5-T use on food crops would be canceled on January 1, 1970, unless the government could establish a safe legal tolerance for the herbicide on foods by that time. In addition, government agencies would stop using 2,4,5-T in populated areas in the United States, and the Defense Department would restrict use of the herbicide to areas remote from population in Vietnam. "It seems improbable that any person could receive harmful amounts of this chemical from any of the existing uses of 2,4,5-T," Dr. DuBridge said, "and, while the relationships of these effects in laboratory animals to effects in man are not entirely clear at this time, the actions taken will assure safety of the public while further evidence is being sought."

It was a hollow announcement. None of the promises was kept. The Pentagon announced the very next day that it would not alter its use of the herbicide in Vietnam because it believed its present policy conformed to the White House directive despite the fact that the military admitted that its main uses of 2,4,5-T were against enemy "training and regroupment centers." The military was loath to surrender any of its herbicide arsenal and reportedly fought hard against restrictions on the use of 2,4-D, which had been classified as "potentially dangerous" in the Bionetics teratogenic study. Not only was 2,4-D used more widely in Vietnam than was 2,4,5-T, but it was one of the six best-selling pesticides in the United States, with annual sales of more than $25 million. Any curtailment of its use would have provoked huge controversy among industry and growers.

The October directive carried no regulatory clout, but it was the first national airing of what had been up to then a closed debate over the safety of herbicides. News reports that followed the announcement linked the chemicals tested in the Bionetics study to the herbicides sprayed by the United States government in Vietnam. Clement L.

Markert, chairman of the Yale biology department, told *Science* magazine that 2,4,5-T unquestionably exhibited "a high order of toxicity" and posed an "unacceptable risk" to the people of Vietnam, where it might cause hidden, if not overt, birth defects, such as "lessening of brain capacity." In the popular press, an article by the nationally syndicated columnists Tom Braden and Frank Mankiewicz deplored the government's use in Vietnam of a chemical "which has been known for three years to cause birth defects in test animals." The reporters quoted from an army training circular that described Agent Orange as having a "high offensive potential" to destroy food supplies and to deny the enemy food by rendering the soil sterile. "Not since the Romans salted the land after destroying Carthage," the columnists wrote, "has a nation taken pains to visit the war upon future generations."

For Mrs. Shoecraft and Bob McKusick, who read the column when it ran in the Phoenix newspapers in early November, it was "the first admittance we had that the chemicals used in Globe were the same as those being used in Vietnam." They were furious that the government had apparently known since 1966 that 2,4,5-T caused birth defects in animals yet allowed it to be sprayed across their forest and their homes. Their anger and fear intensified in following weeks as newspapers around the country ran stories about health problems allegedly caused by herbicides in Vietnam. The Baltimore *Sun,* in a dispatch from Saigon, reported that vegetation had been damaged and birdlife severely affected. United Press International reported that dogs in Vietnam had been hit by a bizarre epidemic of uncontrollable bleeding. A Saigon newspaper had been shut down by the Vietnamese government the previous summer after it reported that pregnant village women were coming to hospitals in Saigon to have fetuses deformed by defoliants removed. No one could say for certain whether or not the herbicides were at fault, but such subtleties were lost on people in Globe who had been feeling ill, whose gardens had died, and whose animals had delivered deformed offspring.

In November the Mrak Commission issued its report on pesticides. It recommended the immediate restriction of 2,4,5-T on the basis of its ability to cause birth defects in one or more animal species. (The commission also recommended the phasing out of DDT over the next two years.) The prestigious American Association for the Advancement of Science publicly stepped up its pressure against the use of herbicides in Vietnam.

Some chemical industry scientists were puzzled by the Bionetics data, however, because they seemed exaggerated in their potency. They wondered if pressured government officials had not jumped the gun. In short, did they have the makings of another cranberry scare on their hands? Dow obtained some of the 2,4,5-T used in the Bionetics tests, which had been originally supplied by Diamond Alkali, a company that Dow knew had not taken pains to control its dioxin contaminant. Within a matter of days Dow's preliminary analysis of the Bionetics 2,4,5-T showed that it was contaminated with twenty-seven parts per million dioxin. It became clear to them why the 2,4,5-T had produced such a consistent pattern of birth defects at every dose tested. It wasn't the herbicide itself but the dioxin that was causing the problem. The Dow people reasoned that since most 2,4,5-T was now made with less than one part per million dioxin, the herbicide should not be restricted on the basis of animal tests the results of which were skewed by a highly contaminated sample. Dow asked that the animal tests be redone with relatively pure samples of 2,4,5-T—which is to say, those that contained less than one part per million dioxin. But as the end of 1969 drew near and no questions were resolved, no further regulatory action was taken against the herbicide, and it continued to be sprayed both domestically and in Vietnam in great quantities.

Mrs. Shoecraft badgered government officials until she got a copy of the Bionetics study, which was contained in the final Mrak report on pesticides. Mrs. Shoecraft bitterly denounced the time taken to make the Bionetics data available and the money spent on a study ($3.5 million) "just to tell us a lot of things we'd already been told 20 years ago, that these damn pesticides being sprayed all over hell's half acre are going to wipe us out."

In the aftermath of the Bionetics dispute, Mrs. Shoecraft and Bob McKusick redoubled their efforts against the herbicide. They had continued to collect information on herbicides throughout the fall and tried with dismal success to generate concern about their problems. Many years later, remembering that time, Willard Shoecraft had a picture fixed in his mind of his wife cradling the telephone hour after hour, day after day, talking to scientists, congressmen, and newspaper reporters. She rarely had time to socialize anymore; she uncharacteristically took a few days off that fall when a movie crew came to Globe to film scenes for *The Great White Hope* and asked her to be an extra because she was "the only one little enough" to fit into a recycled Debbie Reynolds dress. Whatever gratification she took from her tiny waist was diminished by the clumps of hair that began to fall from her head after the herbicide spray, but she wore clever falls and called it to no one's attention except her doctor's. More in character, she lectured leading man James Earl Jones on pollution and convinced one of the crew to film the plant damage around her home.

Once the film crew had left town, Mrs. Shoecraft went back to work. She dressed up in good shoes instead of cowboy boots and put on her black-rimmed glasses to affect a more serious look, packed her note pad and tape recorder, and covered the Arizona section of the American Society of Range Management's annual meeting in Globe to see what those attending had to say about herbicides. Another evening, she showed her slides of damaged plants and animals to the Globe Rotary Club. The club's *RotoTopics* newsletter reported that "the program was very worthwhile, quite thought provoking, and frightening." The Rotary Club heard the opposing viewpoint at its next meeting when Barry Freeman, president of the Arizona section of the American Society of Range Management and president of the Soil Conservation Society of Arizona, dismissed most of the charges leveled against the herbicides as "absurd." He said the spraying in Globe resulted only in "imaginary damage" to vegetation and people, except for "limited temporary damage due to drift." On the other hand, the ultimate benefits of the herbicide spray would be increased water runoff and better grazing fodder. Freeman wove an allegorical story about pickles: Not only can pickles kill you, but pickles are associated with many diseases. In fact, most people who die in airline accidents eat pickles before the crash. The point of this ridiculous story about pickles, he said, was to show that stories categorically linking herbicides to deaths and diseases were just as ridiculous.

Throughout the fall Mrs. Shoecraft and the McKusicks had been debating the merits of a lawsuit. Mrs. Shoecraft had been heartened by a letter from Professor Frank Egler, a scientist who once promoted herbicides and had come to believe that the short-term benefits of aerial spraying were seriously outweighed by destruction of the environment. "Forestland aerial spraying to kill broad leaved plants," he wrote, "is a gimmick, a fad, a simple technique for simple minded people.

"As for damage to private property from spraying on the Tonto [National Forest]," the professor wrote, "my sentiments are 'give 'em hell' . . . and in the words of Yannacone: 'Sue the Bastards.' " Victor Yannacone was a founder of the Environmental Defense Fund, which used the force of law to fight environmental battles in court and which had waged a public battle against DDT in Wisconsin in 1968.

On Christmas morning Mrs. Shoecraft opened a gift from one of her sons. Inside was a small gray porcelain figurine clutching a lawbook under it arms. Inscribed on the base were the words "Sue the Bastards." She told friends that a lawsuit was necessary because after seven months of dealing with cover-ups and intransigence on the part of both government and chemical industry officials, she finally knew that the only language they understood "would have to be expressed in monetary terms." She lauded Rachel Carson for her gentle, scientific approach to the pesticide problem but pointed out that pesticide production was at record-setting levels and wondered how much effect the author of *Silent Spring* had really exerted.

She felt increasingly alone in her fight and complained bitterly about the apathy of people around her who behaved "like scattered, addle brained sheep with no direction and no leader.

"I have met the enemy," she said, quoting Pogo, "and he is us."

chapter 6

It was the end of an era in a country whose deep, dependable faith in the prowess of progress was being put to a battering, inquisitionary test administered by students—the nation's largest single-interest group—who were both antibusiness and antiwar. A kind of low-grade schizophrenia descended on the country as it switched gears from forward to neutral, then to reverse. *Time* magazine's Man (and Woman) of the Year in 1970 were the Middle Americans, but consumer activist Ralph Nader, antithesis to the silent majority, was runner-up. President Richard Nixon forcefully put his administration on record as defender of the environment, yet he told Ford Motor Company chairman Henry Ford II in a private conversation that his personal views on "the environment or pollution or Naderism or consumerism, are extremely pro-business." Nixon declared to Ford that environmentalists "aren't one really damn bit interested in safety or clean air. What they're interested in is destroying the system. The great life is to have had it like when the Indians were here. You know how the Indians lived? Dirty, filthy, horrible."

Nixon and the nation were largely preoccupied with the Vietnam War and with what would come to be known as the War at Home: the student strikes that crippled 448 universities and colleges; the peace marches on Washington; the shooting of four students by national guardsmen at Kent State University. But there was another kind of war at home—a war against environmental abuses. Environmentalists had spent the bulk of the 1960's in efforts to save forest and national park and wilderness land from development. But the focus began to shift away from conservation with the growing realization that the air and water were polluted with chemicals from car exhausts, industrial smokestacks and discharges, and runoff from agricultural chemicals and waste dumps. Then came 1969 and its attendant disasters: An oil well blew off the coast from Santa Barbara, caking the California beaches and wildlife with millions of gallons of black grime; 21,000 coho salmon were ordered destroyed in the Midwest because they were contaminated with DDT; the wretched Cuyahoga River caught fire in Ohio; emergency smog alerts were issued with alarming frequency in Los Angeles; the moon-walking astronauts brought back pictures of Earth crisscrossed with clearly identifiable bands of pollution; and Floridians battled the construction of a new jetport because of its potential effect on the Everglades.

Time magazine declared 1969 the "Year of Ecology." By the end of the year the federal government had announced timetables for the restriction of the herbicide 2,4,5-T and the insecticide DDT and ordered a halt in the production of the artificially sweetening cyclamates. Each action had been preceded by disturbing reports of laboratory tests on animals, but the most obvious common factor was the growing concern over environmental hazards. It was the worst bear market in eight years, but polls showed that Americans rated environmental protection the third most important domestic problem, and President Nixon reportedly told his aides to put him on the right side of the issue.

An idea about an administrative agency that would be charged with establishing and enforcing environmental standards began to take shape in Washington. Environmental lawyers and professors meeting at a private conference in the fall of 1969 envisioned an agency that would serve not only as enforcer but as a publicist for the environment, "articulating as fully and vigorously the potential costs, risks and adverse consequences of technology as the potential benefits are articulated by industry." It would ferret out, publicize, and press upon Congress and the public the potential adverse consequences and risks of activities that affected the environment.

Yet many people in and outside government doubted that such a radical agency could exist. Even the most optimistic among them were not convinced that a nation of Americans just emerging from the go-go decade of the 1960's would support what was essentially a brake on technology. America's strong social bias in favor of affluence and progress and its almost obsessive national policy in support of technological advance would never permit it, especially if tougher regulations diminished the country's ever-rising standard of living. They knew that similar ideas had surfaced before among policy makers, but either the ideas or the time had not been sufficiently ripe to make changes.

The time was right in 1969. In the spring of that year President Nixon borrowed Roy Ash from Litton Industries to form a group for the purpose of examining ways in which the executive branch of the federal government could be better organized. The President asked the Advisory Council on Executive Organization, called the Ash Council after its director, to take a particularly hard look at the way the government administered environmental protection laws. Pollution control activities were handled by more than eighty federal bureaus. They were, with only a few exceptions, shoved to the lowest rungs in the departmental hierarchy. On the basis of 180 interviews with people in and out of government and other research, the Ash Council developed four alternatives to the federal government's piecemeal management of environmental laws.

The Ash Council's Reorganization Plan No. 3, submitted by President Nixon to Congress in the summer of 1970, cleared the decks for a new and entirely independent agency that would combine the then-scattered areas of water and air pollution control, solid wastes, pesticides, and radiation hazards under its wing. "Our national government today is not structured to make a coordinated attack on the pollutants which debase the air we breathe, the water we drink, and the land that grows our food," Nixon told Congress, adding that he had made an exception to one of his own principles, which was that new, independent agencies should not be created. In this case, he said, arguments for an independent Environmental Protection Agency (EPA) were "compelling."

John Quarles, an Interior Department lawyer who became second-in-command at the Environmental Protection Agency, maintained that President Nixon's zest for the new agency was inspired not so much by his personal concern for the environment as by public pressure. He said the White House had originally favored the creation of a Department of Natural Resources and the Environment, structured around the Interior Department. That plan was quashed after Interior Secretary Walter J. Hickel fell from grace in the administration when a post-Kent State letter he wrote to President Nixon advocating that "the youth in its protest must be heard" was leaked to the press.

Whatever the motivation, the plan for an independent environmental agency received overwhelming support. *The New York Times* called it "a statesmanlike move which could prove to be almost as historic in its own way as the great alphabetical upheavals of the early Roosevelt administration." Historian Theodore H. White writing in *Life* magazine pointed out that in contrast with Nixon's campaign promise "to decentralize Washington and return power to local government," he was now proposing a system that would "enlarge the authority of the federal government even more than did Roosevelt's New Deal."

The most sophisticated observers understood the balancing act the new agency's administrators would be called on to perform. At one extreme were the hard-liners, who advocated strict, punishing controls on industrial production with a nearly pollution-free world as the ultimate goal. At the other extreme were those who prophesied that over-regulation would bring an end to innovation and cause ultimate economic collapse. Straddling the middle was a vast and diverse group of Americans who were both Democrat and Republican, who ranged from the traditional conservationists and fish and game enthusiasts to the younger and more militant breed of new environmentalists, and who had as a common goal the protection of the environment. Even the industry lobby supported the new agency, after some initial protestations about moving pesticide regulation from the Department of Agriculture over to the Environmental Protection Agency. "We think pesticides contribute a great deal more to cleaning the environment than they do to polluting the environment," declared National Agricultural Chemicals Association president Parke Brinkley as he reluctantly endorsed the reorganization plan to Congress.

Despite industry's chest beating about overregulation, chemical companies had been operating in what was essentially a preregulatory America up through the beginning of the 1970's. The bureaucracy itself had grown exponentially since the Steamship Inspection Service, the nation's first federal regulatory agency, opened for business in the 1850's to regulate boiler use on steamboats. By the mid-1960's Washington counted twelve major federal regulatory agencies. The protests of industry leaders sounded no different in the 1960's from those in the 1850's, when the boiler safety laws were enacted over the opposition of steamboat owners. Nor were business policy makers any less guileless in their manipulations of law during the 1960's than they had been in 1909, when Samuel Insull told fellow industrialists that business must "help shape the right kind of regulation" or else face "the wrong kind" forced on them by an angry public. Throughout history, businessmen had openly sought favorable regulation and government assistance, and few expected them to do anything less. The chemical industry was well represented in Washington, beginning when executives were lent to government industrial production and policy boards during World War II and subsequent wars, and continuing through the 1960's, when companies began opening government affairs divisions in Washington to lobby for government contracts as well as for favorable legislation. During World War II and the Korean War, Dow lent executives to the Army Chemical Corps, the Office of Price Stabilization, and the National Production Authority. In peacetime Dow executives were among the corporate leaders who helped shape national industrial policy and who advised the secretary of commerce from their powerful volunteer posts in the Business and Defense Services Administration. Since the early 1950's Dow had posted a man in Washington to act as liaison between military and government research groups and Dow's research laboratories. In the mid-1960's the company formally opened a government affairs department in Washington under the direction of vice-president A. P. Beutel. The initial idea was to lobby for government contracts—Dow had built and operated the Atomic Energy Commission's Rocky Flats, Colorado, plant since 1952; provided engineering support services for the Kennedy Space Center in Florida; and supplied a myriad of wartime needs, including magnesium, picric acid, mustard gas, the herbicide Agent Orange, and the incendiary jelly napalm. Eventually Dow's Washington office took on the broader task of tracking and lobbying for favorable legislation. When combined with their seats on the Business Advisory Council, a group of executives from the largest corporations in the country who tended to have the ear of whatever administration was in power, Dow and the other industrial powers exerted great influence over federal industrial and regulatory policy. As a result, advances in environmental and occupational health and safety legislation were slow in coming until public and labor pressure forced Congress's hand.

Any number of laws had been written during the 1950's and 1960's to guarantee that only the most effective and least destructive pesticides be manufactured, registered, and marketed, but the critical registration and policing powers were invested in the United States Department of Agriculture, a federal agency so myopically attached to its own constituency—farmers and the farming supply industry—that the broader constituency of America was left to fend for itself. The transfer of pesticide regulation to the Environmental Protection Agency was part of a national rejection of the notion that an agency entrusted with promoting and developing a set of resources should be expected at the same time to protect the environment from the consequences of that development. Yet review of the new agency's operations and budget remained in the hands of powerful agricultural committees in Congress.

The history of pesticide regulation in America had been one of subservience to the "more bushels-per-acre" mission of the Agriculture Department. The government had essentially allowed chemical companies to submit scanty data—largely confined to acute poisoning effects —to a loosely monitored, closed-door system of review that did little original testing but relied almost entirely on industry reports, that did not routinely rethink a chemical's registration if new information about harmful effects came to light, and that removed a chemical from the market only under extreme pressure.

The sham quality of pesticide regulation was revealed in 1968 and 1969 through investigations of the General Accounting Office and Congress. The Agriculture Department, which had the authority to grant licenses, or registrations, for chemical products, and to police their use, had simply not enforced the country's chief pesticide law—the Federal Insecticide, Fungicide and Rodenticide Act, which became law in 1947 —until midway through 1967. Even then, the staff was hardly able to cope with the backlog of cases and used its power only *once* to remove a hazardous product from the market pending an investigation of its safety. At congressional hearings, investigators held up to scorn a statement from the 1966 Agriculture Department Yearbook that said "that aggressive enforcement action to take dangerous pesticides off the market has been a significant part of the protection being given the public." The truth of the matter was that in one year alone, the agency had found problems with nearly one-fourth of the product samples on the market, but it made no recommendations for prosecution and was not scrupulous in ensuring that products recalled from the market were, in fact, not being sold anymore. In the case of lindane, a pesticide used in home and commercial vaporizers to control insects in buildings, the Agriculture Department had registered the chemical for use in restaurants and food processing plants over the objections of other federal health agencies for fifteen years. The agency finally canceled lindane's registration two months after the General Accounting Office's critical report to Congress in February 1969. In addition, the agency had registered 252 chemical products over a year's time despite the objections of the Public Health Service, which questioned their safety. Investigators also found that the Agriculture Department ignored conflicts of interest among consultants and failed to make sure that label directions were adequate. The public scandal that ensued made certain that pesticide regulation would be transferred from the Agriculture Department to the new Environmental Protection Agency when it opened for business in 1970, with the Food and Drug Administration retaining power to remove from the market food contaminated with excess pesticide residues.

On the first day of the new decade the President signed the National Environmental Policy Act (NEPA) into law, with the admonition that the 1970's "absolutely must be the years when America pays its debts to the past by reclaiming the purity of its air, its water, and our living environment." NEPA compelled the government and industry to consider the effect of their every action on the air, water, and soil through environmental impact statements. The Council on Environmental Quality (CEQ) was formed, setting up a sort of three wise men of the environment who had the President's ear in much the same way that the Council of Economic Advisers did on financial matters and who wrote an annual report measuring the changes in America's environment. "The keeper of our environmental conscience, and a goad to our ingenuity," President Nixon called the CEQ, which had advisory but no regulatory powers.

In mid-February Nixon followed through with a major environmental address to Congress, which his staff had been preparing for six months. He outlined a thirty-seven-point program for environmental improvement, including twenty-three specific proposals for legislation.

"Like those in the last century who tilled a plot of land to exhaustion and then moved on to another," Nixon said, "we in this century have too casually and too long abused our natural environment."

The President's proposal to expand government action in water and air pollution control and solid waste disposal and to increase national parklands did little more than strengthen legislation that was already on the books, but it signaled a shifting of power away from the states and back to the federal government. It also exacerbated the growing competition between the President and Congress to take the lead on the environmental issue as they scrambled to keep up with the great wave of public support for environmental protection. "For a president whose primary interests are saving a buck and not rocking the boat," observed the sardonic *New Republic,* "the ecology craze is heaven-sent. Mr. Nixon can make a lot of noise on the stairs and never have to come into the room."

Artful or not, his message served the symbolic purpose of putting the environment on the national agenda and paving the way for broad bipartisan support of environmental reform. And with his advocacy of nationwide federal pollution standards, Nixon effectively laid the first brick in the construction of a vast regulatory house that, with government intervention as its foundation, stood in direct antagonism to the Republican party's traditional business constituency.

From January through March 1970, Congress introduced 273 pieces of environmental legislation. The environmental fever ran so high that Gaylord Nelson, the senator from Wisconsin, proposed a constitutional amendment guaranteeing every American "the right to a decent environment." The press marveled at the cooperation between the Nixon White House and the Democratic Congress, but as television commentator David Brinkley observed, "Nobody's for pollution. Everybody's against it."

Each day's news seemed to bring new word of environmental problems. *The New York Times* noted ominously in a story headlined WE LAY WASTE THE WORLD the "widespread belief that man's rapid alteration of his own biological environment threatens his long-term survival unless it is monitored far more closely than now. The greatest concern is for effects too subtle to be immediately apparent." By the spring of 1970 the environmental movement had picked up enormous steam, partly because it was a popular student issue. On April 22 teach-ins were held at 1,000 colleges and 6,000 schools across the country to celebrate "Earth Day." Leaders of Environmental Action, the group that promoted the celebration, hoped that it would "serve as a catalyst in the values of society."

Much of the business community, while privately suspicious of the movement's motives, either took a conciliatory stand or ducked behind a low profile. The Dow Chemical Company put itself aggressively on record as a proenvironment business. Dow formed an "ecology council" in 1970 to "guide and expand the company's work in environmental improvement"; it had as one of its goals the persuasion of others "to join the crusade for a clean environment." At the annual shareholders' meeting that spring Dow president Herbert Dow Doan talked briefly about the company's 9 percent increase in sales and earnings and then devoted the rest of his speech to the environment. Doan characterized environmental improvement as a "fundamental problem" and one that "won't go away when the shouting dies." Uncharacteristically he declared that regulations, which "are becoming tougher, and necessarily so," can be "good for everyone." He reiterated his faith in Dow's scientific knowledge and techniques but outlined plans for the upgrading of toxicological testing and waste disposal facilities. Doan said, "We recognize that public attitudes are changing fast as awareness of pollution grows. Not so long ago, smoke from a factory meant bread on the table and a table to eat the bread on. Industry has been successful in supplying human material needs—and I view it as a sign of progress that today smoke is a dirty word." The Dow president talked directly and obliquely about recent problems with the herbicide 2,4,5-T. He acknowledged that the company "ran into difficulties" with an impurity in the 2,4,5-T manufacturing process in 1964. It shut down the plant, identified the problem, redesigned the plant, and started producing again in 1966 after spending an extra $5 million to keep "the impurity" at a level considered safe, which was less than one part per million. Doan explained how 2,4,5-T containing twenty-seven parts per million of this "impurity," which he never mentioned by name as dioxin, had caused birth defects in animal tests. He asserted that herbicides containing less than one part per million dioxin have a "large margin of safety."

Herbert "Ted" Dow Doan was the grandson of founder Herbert Henry Dow. His father, Leland Doan, who had married one of the founder's daughters, was a forceful president who led Dow through the spectacular growth years of the 1950's. The younger Doan was just forty when he took over the presidency of Dow in 1962. A reflective man with a special interest in the ethics of business who worked at Dow by day and spent ten years of nights reading Will and Ariel Durant's eleven-volume *Story of Civilization,* Ted Doan seemed the right man to steer the company through the troubled decade ahead. Rail-thin and compactly built, Doan had attended prep school at Michigan's prestigious Cranbrook Academy, where he ran the student council, and studied chemical engineering at Cornell. His heritage may have pointed him ineluctably toward the executive offices known as Mahogany Row at Dow's corporate headquarters in Midland, but his parents let him make up his own mind. "Instead of being raised to go to Dow," Doan recalled, "my mother was always telling me, 'I think you'd make a good preacher or teacher.' " But by the time he finished college "I had no idea of going to any other place," Doan said. "I came straight to Dow."

As president of Dow, Doan replaced the hierarchical organization of management with a more specialized system, which structured the company into a dozen or so separate businesses according to product line—agricultural chemicals, commodity chemicals, and so on. He had been sold on the team approach ever since a team he helped form within the research division built a pilot plant for a new product line in a record-setting three months. Doan adapted the entire Dow organization to team leadership—a system that critics said diluted corporate responsibility by increasing the distance between the top executives and their mid-level managers and scientists. But the company was extraordinarily adept at making money—profits rose 80 percent from 1962 to 1967—and there was little outside scrutiny until Dow suddenly found itself the target of international vitriol for manufacturing and selling napalm to the government.

The air force contracted with Dow in 1966 to supply part of its napalm needs for the Vietnam War. A mixture of benzene, gasoline, and polystyrene plastic, napalm was packed into canisters and dropped by plane on enemy targets in Vietnam. A dozen Dow workers filled the $5 million napalm contract at the company's Torrance, California, plant. It was there on May 28, 1966, that more than 100 pickets first demonstrated against Dow's manufacture of napalm. On the same day a similar protest took place in front of Dow's offices in Rockefeller Center in New York City. Protesters carried signs that said NAPALM BURNS BABIES—DOW MAKES MONEY.

The napalm protests continued for the next three years, paralleling the growing intensity and violence of the antiwar movement itself. If it were true that Dow was swept up in the wide net of opposition to the country's most unpopular war, it was also true that napalm was a cruel weapon. Antinapalm groups trapped and harassed Dow recruiters on college campuses and picketed Dow offices. The war came home to Midland in the summer of 1966, when protesters passed out leaflets to Dow's factory workers and held an antiwar rally during which an Episcopal chaplain from the University of Michigan announced that napalm "turns people into blackened, screaming shards and if you work for Dow, you must know that or you wouldn't be in the business." Roman Catholic priests and nuns broke into Dow's Washington office in the spring of 1969, throwing files onto the street and pouring what they said was human blood on company records. (The files revealed that a Dow lobbyist had made contributions of $100 and up to all but three members of the Michigan congressional delegation in 1966 and 1968 in apparent violation of federal campaign finance laws.) Protestant and Catholic pastors read a "thesis of protest" against napalm from their pulpits. In the autumn of 1969 eight people, including the twenty-two-year-old son of a prominent Midland businessman, vandalized Dow's computer research center. The group destroyed computer tapes and punch cards that it believed contained information about napalm, nerve gas, and defoliants. "Beware!" the protesters said in a mimeographed statement. "Your brutality has been recognized and we won't be stopped. Stop producing weapons, stop pollution, stop exploiting the poor." As it turned out, they missed the military projects records and instead destroyed tapes containing information on blood banks and workers' health records.

In defending its manufacture of napalm, Dow argued that it was simply exercising good citizenship in filling a $5 million contract for the military that far from being profitable, represented only half of 1 percent of the company's sales. "Why do we produce napalm?" Ted Doan wrote in *The Wall Street Journal* in 1967. "In simplest form, we produce it because we feel that our company should produce those items which our fighting men need in time of war when we have the ability to do so." Doan despaired that Dow's contributions to health and peace (measles and rabies vaccines, artificial human organs, and the twenty-seven-foot silicone rubber hose that supplied astronaut Edward White with oxygen during his space stroll in 1965) were so darkly overshadowed by its contributions to war. The normally placid Dow stockholder meetings turned acidic with the accusations of outside protesters. The Dow board itself was said to be deeply divided over napalm, and when the contract ran out in May 1969, the company lost its bid for a new contract to another company.

If Saran Wrap hadn't done the trick of making Dow a household name, napalm surely did. Like America itself, the napalm era forced Dow into a position of both self-examination and public confrontation. In another sense, it served as something of a training ground for future showdowns over toxic chemicals. New moral standards were applied to business and the making of profit. Corporations were held responsible for the social consequences of their products. "The creation of a responsive and responsible corporation," wrote economist Robert L. Heilbroner in 1972, "becomes an indispensable step in the creation of a responsive and responsible state: perhaps the central social problem of our age."

Dow's Ted Doan tried to bridge the growing chasm between business America and antibusiness America. He held "dialogues" with students about profit, growth, and service to society. During the napalm years he invited twenty-five college newspaper editors to Midland. "They were convinced we were doing this for the worst of reasons—profit, mainly—and they thought they were going to meet people they could flail at here. They knew we were going to wear horns and would be easy to trap," Doan said with his characteristic twangy, no-nonsense delivery. "So here we were all innocent, babes-in-the-woods fellows, and it really shocked them. We got twenty-four good editorials out of those twenty-five people. One guy lied through his teeth." Another time Doan's speech to a business group was cut short by napalm protesters at Notre Dame, and the university suggested the situation might be defused if Doan talked privately with the protest leaders in his hotel room. He was smuggled through a tunnel and back to his hotel. "And here's these three leaders, these hippies," Doan recalled. "And so we sat down and talked for half an hour, and I said, 'Now I've got to go to dinner.' Here's this dinner for a hundred and fifty people at the business conference. And I said, 'Why don't you guys come with me?' So in I walk with these three hippies, and I make them clear us a space, and I sat down at dinner with these guys, and they're drinking our good wine. As you know, these people were all good people. The leader of the thing wound up trying to get a job with Dow." In his more formal discussions with students, Doan emphasized Dow's statement of corporate objectives, which he wrote with C. B. "Ben" Branch, who succeeded Doan as president in 1971. "They were cribbed from a book of John Gardner's called *Excellence,*" Doan recalled. "A lot of stuff about people doing things well and honesty. It was real motherhood, and it was real good motherhood. The last statement was, 'We'll do business in a way that leaves the world better because we were in business.' That was Branch, direct quote. That was the way he felt. He couldn't stand having a product go out if he thought it was going to do anybody harm. He would go a long way toward putting a company out of business and restructuring it in some way if he thought he was really doing damage. Let me not be so modest. I think all of us were the kind of guys who would say, 'If we have to go out of business, fine, we'll go out and we'll find a way to rebuild the damn thing some other way. We won't put arsenic in babies and kill 'em.' You know, that kind of thing."

Doan believed, above all, that maximum contribution to society was consistent with maximum long-term profit growth. If he saw an ethical reason to clean up Dow's plants and to invest in the pollution control business, he also saw it as a way of making money. During the confrontational 1960's, when industry was first forced to reckon with the side effects of its pollution, Doan was on the board of one of the industry lobbying groups. "I can remember debating these things with people from Du Pont and Monsanto," he said. "We were trying to see where we were vis-à-vis the government, struggling to see what they were going to do. The consensus was the environmentalists are probably dead wrong—[but] we probably ought to do something and sort of move slowly, plus it's going to cost us a bundle. So I would sit there and say, 'It might cost you a bundle, but we're going to make money doing it.' " Dow had been in the forefront of hazardous waste technology at its own plants and had taken pains, after a rocky start, to clean up the Tittabawassee River. "The whole industry," Doan said, "would have been ahead if they'd followed Dow's lead."

But Dow's lead didn't always point in the right direction. For years —until mercury residues appeared in fish and the government clamped down—the company dumped mercury wastes into Canada's St. Clair River. "We decided to shut down the only mercury cell we operated and write off forty million bucks," Doan said. "I'll lay this at Ben Branch's feet. I believe he was the guy who said we were going to write the thing off. The reason was it was a bad thing. It doesn't make any difference whether it's not killing anybody. It's easy to call other people bad when you can afford to do something, and one of Dow's advantages is that it's big and it's technically good and it can afford to be virtuous. I think I invented the idea that we should make a policy that said we would live with and exceed—and then we had to put in the fudge word 'reasonable'—federal regulations. The reason we could do that so enthusiastically was we convinced ourselves we could make money doing it."

In his speech to the 1970 stockholders' meeting—his last as Dow president—Doan addressed the growing public fear of toxic chemicals. He suggested that scientific tests using extreme doses of chemicals on sensitive animals could be designed to "suggest almost anything." When combined with increasingly sensitive analytical methods capable of detecting minute amounts of chemicals in the environment, the tests were "capable of creating unnecessary public fear." He suggested that the question of whether or not there are threshold levels of chemical safety below which no damage occurs was a matter of "great national importance."

"Without this knowledge and a greater certainty than we have today," Doan concluded, "it is understandable that policy makers in government may be inadequately informed and that we will see a continuation and perhaps escalation of public fear. And perhaps also an escalation of speculative stories. These are stories whose vote potential the politicians will not miss. They are exciting news on television and make beautiful copy for newspapers and magazines."

President Doan was likely conjuring up a private vision of Globe, Arizona, as he spoke, for by the spring of 1970 Billee Shoecraft and her antiherbicide colleagues had finally and spectacularly commanded the world's attention. The ground swell had begun in January with a telegram from Globe Mayor E. Ross Bittner to President Nixon. STOP SPRAYING THIS AREA, the mayor pleaded. THIS IS A POPULATED AREA.

The restrictions on the use of herbicides in so-called populated areas, which had been announced the previous fall by Dr. Lee DuBridge, the President's science adviser, had never been enforced. The government did not determine a safe residue level for 2,4,5-T on food crops by its January 1, 1970, deadline and so allowed spraying to continue, both domestically and in Vietnam, while studies continued. Both Dow Chemical and government scientists concluded their duplications of the Bionetics tests by that spring. Dow's studies showed no untoward health effects in lab animals exposed to 2,4,5-T containing less than one part per million dioxin. The government's tests showed just the opposite. While the preliminary data from all the rerun tests were being analyzed, the government stalled for time. The Food and Drug Administration rarely found more than trace amounts of 2,4,5-T residues on food crops and so was not inclined to cancel registrations if the herbicide tested out safely. But people who remembered the promises made by Dr. DuBridge the previous fall were left confused and suspicious. "The ban is so full of loopholes," Mrs. Shoecraft told a Tucson newspaper, "it makes me want to laugh."

The herbicide fighters in Globe had quickened the pace of their battle that winter after new health effects had come to light. Mrs. Shoecraft learned in February that a sample of her fatty tissue contained 35 parts per million of the herbicide silvex and 2.5 parts per million of the herbicide 2,4-D. Indeed, her tissue looked like a repository for every persistent pesticide marketed during the past twenty years. She had trace amounts of DDT and its derivatives, plus lindane, endrin, and chlordane. The laboratory noted that the 126 parts per million of endrin found in her fat "is by far the highest concentration of any endrin we have ever encountered in any sample."

Also that winter, Bob McKusick's anthropologist wife, Charmion, compiled a list of what she considered herbicide-related reproductive problems in animals and humans in Globe. Several pregnant women had miscarried, she said, and there were a host of reproductive problems in animals.

> Neighbors complain of small calves, born prematurely, some blind, some with deformed backs. I have raised milk goats for 17 years and until the application of herbicides, had only one kid born dead in 12 years. In the seasons of 1965–66 through September 1, 1969, over 60 percent of our kids were lost through miscarriage, or born dead.
>
> On 15 December, 1967 a doe was born with displaced eyes —the left one low, the right one high—to the best of our judgment, a defect in the joining of the two halves of the embryo similar to "harelip." We butchered this monstrosity

as soon as possible, as it looked pretty bad and we did not wish to offend our customers. On the same day, another doe was born to a different mother with a short right foreleg, humped back, and hollow chest; this doe is so crippled that she walks on three legs or crawls on the hind legs and front knees.

On 15 December, 1968 she dropped two male kids, one dead, one with a short right foreleg. All deformed goats have been red (deformity transmitted by genes?). This same crippled doe's health (after switching to hauled city water) improved greatly, and on 3 January, 1970 she dropped three kids at least 10 days prematurely. One buck had been dead about 5 weeks; two does were alive with long bones but not filled out yet.

In February of 1969, a female kid was born to our hair goat. The kid is 1/4 milk goat and 3/4 angora. Her horns have never been normal, but soft and painful to the touch, breaking and flaking continuously.

Mrs. McKusick also detailed the deaths and deformities in the Shoecrafts' guinea pig litters. As for the children in the canyons, she said they were "very badly dragged down by this—apparently because of their low body weight." Robert McCray's son, Paul, who had been just five months old when the helicopter passed over his family's property, continued to have severe health problems, according to his father, who said the child was racked by epileptic types of convulsions. "Before the spray, I was never sick," McCray said. "Afterwards I was allergic to everything." He said that allergies and problems of concentration plagued his family, and that he himself had continuing episodes of chest pain, muscle spasms, finger and toe numbness, eye irritation, and skin rash. His goats, cows, and horses also fell ill. "We lost everything," McCray said.

Mrs. McKusick said her own children were tired all the time—especially after they had played in the brush along the mountains. She said her son, Randy, sometimes slept for the better part of the day after his excursions into the brush, and all three of her children were bothered by headaches. The two girls had skin rashes, she said. Kathleen's were so severe across her face that they oozed white liquid. She also had painful menstrual periods and abdominal cramps. Stephanie's eyes were irritated, and she had intermittent muscle spasms, according to her mother. There were bloody spots in Stephanie's hairline. Kathleen's hair fell out in patches. When Kathleen went away to school, her health improved dramatically, her parents said.

The McKusicks had a long list of their own health problems. Directly after the family had been sprayed by the helicopter in 1968, Bob

McKusick had severe pains in his chest, and he found that the pains returned whenever the weather was hot or after a brief rain. He said his back was racked by muscle spasms, which, combined with the chest pains, kept him from working for six weeks. He said he'd also had a rash on his arms and some blistering of the skin, and for a time he was losing his hair, but it began to grow back. The McKusicks' dogs had been sprayed, too, and they had forgotten to wash them down afterward. They said one animal developed pneumonia and eventually died, hemorrhaging from the mouth.

In previous years, before she knew about the herbicide sprays, Charmion McKusick had been plagued with a particularly virulent vaginal infection. It felt, she said, as if she were "being eaten by worms." Later, after the 1969 spray, she said she'd had headaches, nausea, intestinal cramps, and a painful discharge of breast milk. Then she said her hands grew mysteriously weak—a consternation since she used them so much to type her scientific reports, sketch illustrations, and help her husband with his tile-decorating work. She kept working by wrapping layers of material around pens, which made them bigger and easier to hold.

Attention to their problems had been long in coming, but come it did, finally, in February 1970. In typical snowball fashion, what started as a few local newspaper and television reports became headline news on the three national television networks and in national magazines. The British Broadcasting Corporation flew in a reporter, and when ABC's Learjet landed in Globe, Mrs. Shoecraft was there to greet it, snapping photographs with her camera. At the center of the commotion was Democratic Congressman Richard D. McCarthy from New York, who had written a book about chemical warfare and who was an outspoken critic of the Nixon administration's chemical and biological warfare policies, especially as they related to herbicides used in Vietnam. McCarthy came to Globe the first week of February to hold informal hearings on herbicide use there as he built a case for halting the defoliation of Vietnam.

He was the only official of the dozens Mrs. Shoecraft had pursued that fall and winter who actually came to see her and investigate for himself the facts of the case. She accorded him rare approval. Others were less convinced of his intentions. Edwin Fitzhugh, an editor of the conservative Phoenix *Gazette,* wrote of McCarthy's visit:

> McCarthy is a standard New York politician of the leftist persuasion. One of his specialties is riding off in all directions against germ warfare, the industrial military complex, and the United States presence in Vietnam. Globe's silent world donnybrook started when a small group got the idea that the

United States Forest Service had poisoned land, plants and people by spraying herbicides. With no exception that I know of, men who have dedicated their studies and their careers to various of ecological sciences and who have looked at the Globe scene say there is nothing there to justify the clamor of politics, publicity and distrust that is being worked up.

McCarthy's visit was hailed, on the other hand, by a California group called the National Health Federation, the motto of which was "Civilization May Be What Will Finally Eliminate the Human Race." The group's frequently overwrought communiqués about an epidemic of dead military dogs in Vietnam, for instance, sometimes obscured the real issues. But Mrs. Shoecraft was willing to talk to just about anybody who might help in her fight against herbicides and so opened the door for groups like the National Health Federation and even something called the Society for the Prevention of Environmental Collapse to get into the act.

McCarthy opened his two-day hearings in Globe with a terse summary of his concern about herbicide use in Vietnam, his puzzlement and distress over the American government's failure to enforce the herbicide ban announced the previous fall, and his fears about the continuing use of a chemical that produced birth deformities in test animals. "True science teaches us to doubt," the congressman said, quoting the French scientist Claude Bernard, "and ignorance to refrain."

Yale biology professor Arthur W. Galston was the first witness. He contrasted the mild acute toxicity of 2,4,5-T with its potentially more dangerous chronic effects, including cancer, interference with genetic material, and birth defects. He quoted from the Mrak panel's report, which recommended the immediate restriction of pesticides that were found to be teratogenic in two animal species in valid tests. "I believe that is the safe policy when you think you may be doing harm," Professor Galston said. "You stop until you find out whether you are in fact doing harm." He suggested that a little 2,4,5-T went a long way in provoking birth defects in animals and estimated that in Vietnam, where higher concentrations of herbicides were sprayed than in Arizona, a peasant woman need drink or use in food preparation less than three quarts of rainwater a day to receive a teratogenic dose.

"As a biologist, I am terribly concerned about this because I believe in herbicides," the professor declared. "I want to see that they continue to be used. I'm afraid there may be overreaction on the part of the public." He suggested that only hard data would allay doubts and recommended that the government's pesticide review policy be revamped so that after a product was registered, it would come under mandatory review at an initial two-year interval and then subsequent five-year intervals.

The second witness represented the Forest Service. John Pierovich, an assistant regional forestry official, said no decision on the fate of future herbicide spraying in the Pinals would be forthcoming until studies on its safety were concluded. In response to questions from Congressman McCarthy, the witness conceded that the ten-mile-an-hour wind limit invoked by the Forest Service during the Pinals spraying was double the agency's generally recommended limit.

Dr. F. I. Skinner, a Globe veterinarian, testified that he had not seen any clinical effects in local animals after the herbicide spraying, but he said he would not be in favor of continued use of herbicides without further research.

Mrs. Shoecraft was the fourth witness, and as was her fashion, she showered the hearing room with a spray of earnestly offered but wildly disconnected facts about herbicides. Quoting from press releases, newspaper articles, toxicology reports, and government memos, she outlined her chief complaint: that a fine mist of quickly vaporizing herbicides had been sprayed recklessly from high altitudes on a hot, windy day across populated areas by a government agency that took a chemical company's word that the spray was harmless when, in fact, scientific studies had shown that major questions about its safety were unresolved.

She also offered up her personal analyses of government ineptitude and cover-up. "I refer further to the Department of the Army's Circular 33661," she said. "I have a letter here from Representative [Sam] Steiger's office, to apply back in 120 days, but I didn't choose to apply in 120 days. I called the Adjutant General's office, I said we needed it now, I'm one of the victims. I was informed by the Department Office that they sent it out to the printer's. My suggestion was that you either get it from the printer's, or you get a copy, I need it now. I received it in 3 days."

In conclusion, she dropped a well-timed and precisely aimed bombshell. "I suppose I am the first human to go on record to be able to say that they have now found 2,4-D in my pound of flesh, and that was as of this morning from two different laboratories," Mrs. Shoecraft announced.

Bob McKusick ended the day's testimony with a summary of his family's and his animals' medical problems, and later that afternoon he and Mrs. Shoecraft led the congressman and his staff on a tour of the canyons. Testimony continued on the second day with Dr. Paul Martin, a University of Arizona geologist and ecologist, who declared that herbicide drift had occurred and that in four trips to the area he had observed "some things that I have not seen in Arizona vegetation before. Such as the presence on century plants of flowering way out of season, and immature new plants going on the old stocks of old ones without normal seed being set." Evidence was further introduced by a college student from Massachusetts who had conducted a door-to-door survey of fifty-six families in the canyons. The admittedly unscientific survey showed that twenty-three of fifty-six people had been ill during the previous two years, twenty-one of them with respiratory problems, four with chest pain, five with serious diarrhea, three with uterine hemorrhaging, one with miscarriage, two with pains or numbness in arms, and there were other ailments.

Dr. Galston summarized the field trip he had taken the day before with Mrs. Shoecraft and the McKusicks. At the helispot, where the helicopters fueled up before the spraying operation, he could smell herbicide residues. In the canyons he observed plant damage from herbicide drift. "There is no doubt about it," he said, "it has occurred. I have seen it, and as a plant physiologist, I could testify that this is typical damage due to herbicide drift." He also saw "damaged animals" and humans who "alleged that they were adversely affected."

The professor concluded: "All I can say here is the damage is there, and spray operations did occur, but I know of absolutely no scientific evidence which would link the spray operation to the damage, and I think the people who showed me the damaged animals showed it to me in the spirit that this could be a consequence of spray operations, but they weren't sure, and certainly I'm not sure, but unlike some people I would not immediately offhand say this is ridiculous."

Dr. Galston reported a case he had come across in the medical literature that described two young girls who played for several hours in a yard that had been heavily sprayed with 2,4,5-T to kill poison oak. The following day both girls had reddened skin and swelling of the mucous membranes in the mouth and vagina. On the third day there were signs of kidney damage. Other reports, he said, described toxic effects on the human nervous system after exposure to herbicides.

In his concluding statements Congressman McCarthy reiterated his concern about using a chemical that might cause birth defects in humans and his outrage that the burden of proof seemed to be on government, not on the chemical industry, to prove or disprove its safety. "... the attitude is, 'We'll keep using it until you can prove it unsafe.' Well, I quarrel with the basic assumption, I think that it should be just the reverse. I don't think that any toxic substance whether herbicide, pesticide, drug, whatever, should be used, sold in the United States until it can be shown that it is not harmful to human beings, that it doesn't produce cancer, or birth defects, or genetic effects. One would think that we have learned from the Thalidomide experience, but apparently we haven't."

McCarthy was also incredulous that "the Dow Chemical Corporation could have succeeded in helping reverse an order from the White House." It was his theory that the White House science adviser's October directive on herbicide use had not been enforced because of pressure brought on the government by Dow, the nation's largest producer of 2,4,5-T. It was his hope, the congressman said, that the experience of Globe would teach a lesson and "we will stop using 2,4,5-T around the world until we can run a series of tests that show that it is not harmful to this generation, and to the next generation."

During the hearings United Press International wrote a story about Mrs. Shoecraft, who was "doused with the chemical as she stepped outside her forest edge home last June. Since then, her conviction that 'my life is expendable but this mountain isn't' has carried her through several illnesses which she blames on exposure to the chemical." Television stations across the country broadcast footage of Mrs. Shoecraft and the McKusicks and their allegedly herbicide-damaged animals. Two of McKusick's animals—a crippled goat and a deformed duck—became media celebrities.

The residents of Globe were divided about the McCarthy hearings. A Globe doctor told the Phoenix newspapers that the herbicide issue was "the biggest farce he had ever encountered" and suggested that it was "a political question designed to make headlines for Mr. McCarthy." The physician dismissed allegations of herbicide-caused health effects. "I doubt if it will ever become a medical question," he said.

The doctor's negativism only reinforced Mrs. Shoecraft's general distrust of the profession and her conviction that doctors closed their minds to any but the most established and conservative of medical facts.

After failing to find a doctor who took her seriously in Globe, Mrs. Shoecraft obtained the name of a physician in California who believed that toxic chemicals played a decisive role in modern disease. Dr. Granville F. Knight specialized in allergy and nutrition in his private practice in Santa Monica. Dr. Knight had been interested in the effects of pesticides on humans for more than twenty years. He had testified about pesticides before local and federal committees, particularly regarding his belief that toxic reactions to pesticides were much more common than the mainstream medical profession suspected. He had observed in his own practice many cases of peculiar types of illness—often characterized by headaches, chest pains, nausea, vomiting, and diarrhea, preceded or followed by upper-respiratory infections—which he associated with the use of pesticides. Dr. Knight had come to believe that some pesticides attack the body's peripheral sensory nerves as well as its genetic material, or chromosomes. Dr. Knight's radical views on pesticides, coupled with his megavitamin therapies and his general unwillingness to be strapped by convention (he had his official stationery printed up with the requisite name, address, and telephone number but added an Edmund Burke quote across the bottom: "The people never give up their liberties but under some delusion"), tended to make him a pariah in the established medical community. The people of Globe, however, who made pilgrimages to his little office on Wilshire Boulevard in Santa Monica believed that Dr. Knight saved their lives. "I think he was crusading," Bob McCray recalled, "but I thought he was a real good doctor."

In the wake of the national publicity stirred up by the McCarthy hearings, the United States Department of Agriculture decided to send a new task force into Globe to investigate conflicting claims about herbicide damage. This "government interdepartmental panel" of scientists was directed by Dr. Fred Tschirley, the assistant director of the USDA's agricultural research service who had surveyed herbicide damage in Vietnam two years before. The task force spent four days in Globe, conducting on-site inspections, interviewing local residents and physicians, attending briefings by the local Forest Service and Department of Fish and Game officials, and collecting samples of soil, human and animal blood, and animal and vegetative tissue for testing. Mrs. Shoecraft called these scientists, according to her mood, "the Big Eight" or "the jet set" or "government boys sent out to investigate the government blunder made by their own branch of the government." She compared their visit to a judge who, in seeking to learn the truth, sends criminals to investigate their own crimes. She thought it absurd that they were launching an investigation eight months after the spray. She also considered the winter season an odd time to study plants since most were shorn of their leaves.

Dr. Tschirley had phoned her in advance to say that "dedicated scientists" were coming to Globe "with an open mind" and to suggest that she set up a meeting with residents. She promised to cooperate if he would bring her answers to some questions she'd been unable to get, including the results of Agriculture Department testing on soil, water, and plant samples taken from her property. Dr. Tschirley did not meet Mrs. Shoecraft's demands, and so on "the big day," when the task force met with concerned residents at the American Legion hall downtown, Mrs. Shoecraft stayed home "and counted ant hills." Later that evening, when two task force members drove out to her home in Ice House Canyon, Mrs. Shoecraft had a temper tantrum when they declined to look at certain animal and plant specimens she thought important.

After leaving town, the task force reported preliminarily that the herbicides had not caused ill effects in humans or animals in Globe. The government scientists blamed plant damage on a variety of natural causes, including woodpeckers, sapsuckers, and root rot, a fungus found exclusively in the American Southwest that attacks woody plants. While no major herbicidal effects had been found, the task force members acknowledged how frightened some Globe residents were and declared that from a purely psychological standpoint, they would hesitate to recommend resumption of the spraying program. Mrs. Shoecraft was enraged. "Those USDA boys work fast when a few million dollars belonging to the chemical companies are at stake," she wrote later. "Just three days after they galloped into and out of Globe, and with no results from any analysis on plants, blood, water or urine, the newspapers carried the story around the world: Herbicide Spray Cleared of Animal and Plant Damage."

The final single-spaced, twenty-nine-page task force report, issued some months later, stuck forcefully to the initial decision acquitting herbicides of blame, even though the scientists conceded that herbicide drift "undoubtedly" had caused some problems. The report also clarified the types of herbicides used in the spray. As originally reported, silvex had been the primary herbicide used during 1969, but subsequent investigations showed that roughly 3 percent of the total spray was 2,4,5-T, with additional trace amounts of 2,4-D. Analysis of the herbicides used in 1969 showed they contained 0.5 part per million dioxin, well within the range of safety recommended by Dow. The report also confirmed that the herbicide had been mixed with water instead of oil but discounted fears that the rapidly evaporating water made for a more concentrated and potent herbicide spray or for wider drift. Investigators conceded that winds could have exceeded the recommended ten-mile-per-hour limit during the spray operation.

On the basis of their limited observations and those of earlier investigations—and with the caveat that a complete investigation was impossible because of the season of the year—the task force members reported that spray drift had caused herbicidal damage to some plants on private properties, but that most damage was attributable to natural diseases, insects, drought, and air pollution from the area's copper smelters. They reported that almost all plants in the Globe area seemed to be under "moisture stress" because of below-normal rainfall during the first eight months of 1969 and that "abundant damage" from many twig insects was readily apparent. They found Chinese elms riddled with thousands of woodpecker or sapsucker holes and others, without holes, that had apparently died from root rot.

Richard Lewis, who lived down the road from the McKusicks in Kellner Canyon, showed the task force his small vegetable garden, which he said had died after the spraying. Lewis remembered seeing the helicopter circling over the hill behind his place on a night when near gale-force winds were blowing down the canyon. The investigators doubted his memory on the ground that a gale wind exceeds thirty-nine miles an hour. They also thought his peach trees were in fine condition despite some dead branches and concluded that another peach tree had died from transplant shock. When the task force visited McKusick's place, they determined that a walnut tree had probably died of root rot. They concluded that canyon residents were trying to raise fruit and ornamental trees in soil that was too dry and alkaline to support them and that this, not herbicides, caused their trees to turn yellow and decline.

While noting dioxin's suspected teratogenic effects and the lack of information about how herbicides are metabolized in the body, the investigators emphasized the "notably few authenticated incidents" of acute harm to humans or animals in herbicide exposures throughout the world. Pushing the possibility of chronic effects to the background, they asserted instead that herbicides do not present a "direct toxic hazard" when correctly used. Since herbicide residues persist in the environment for a short time in comparison to more stable chemicals like DDT, there was little reason for concern about potential cumulative health hazards. The investigators acknowledged, however, their ignorance about the persistence of the dioxin contaminant.

As for herbicide effects on local wildlife, the task force cited the observations of fish and game officials and a local cattle rancher, who said coyotes and jackrabbits were more numerous than ever. The local wildlife ranger could substantiate no reports of fish or wildlife injury, despite claims by canyon residents that fewer birds, skunks, and ground squirrels had been spotted since the spray. He said it was not unreasonable that bird species such as the spotted towhee and Arizona jay would disappear after the spray, not because they were injured by herbicides but because their chaparral habitat had been disrupted. The task force further noted that no animal distress or deaths were observed after the spraying of nine national wildlife refuges in the southwestern United States with 2,4,5-T and silvex between 1964 and 1969.

The task force report found few links between human health complaints and herbicides and suggested that the residents were victims of something close to mass hysteria. On the basis of interviews with residents who had health complaints and with nine Globe doctors, the task force concluded that except for skin rash and eye irritations, "it is highly unlikely" that the ailments were caused by herbicides.

Two teenage girls living in Kellner Canyon reported irregular and painful menstrual periods, the flow so heavy they used twice the normal number of sanitary napkins and so painful that one girl had passed out several times. The task force suggested their problems were possibly due to stress from a death in the family and preparations for a trip. The report noted but did not comment on their mother's worsened asthma, chest pains, and arm numbness or on their brother's muscle spasms and weight loss.

In another family in Kellner Canyon, a thirty-year-old man had lost his sense of taste and smell and developed worse allergies. His wife had two miscarriages—one in the spring of 1969, one during the following winter—and was "quite frightened about attempting to have additional children." She also had increased menstrual bleeding and cramping.

A fifty-year-old laborer who had worked on houses in Kellner Canyon reported that he'd been rushed to the hospital in 1968 with severe, crushing chest pains and shortness of breath. The left side of his face turned numb, but laboratory tests showed nothing abnormal, and he was discharged. Since then he'd had numerous episodes of chest pain and dizziness, especially if he visited the spray sites, and he was routinely getting injections of a local anesthetic in his back, plus taking pain and nerve medications.

The one-year-old son of a family living in Kellner Canyon—this was Paul McCray—had an eye infection for three weeks after the 1969 spray as well as frequent upper respiratory infections, ear infections, and episodes of vomiting and diarrhea. His father told investigators that he had performed mouth-to-mouth resuscitation on the child when he went into respiratory and cardiac arrest at home. "Son has also apparently been somewhat anemic, is receiving gamma globulin shots, and is currently being treated with phenobarbital for nervousness," the report noted.

In describing Mrs. Shoecraft, whom they identified only as "female, about 50 years of age, living in Ice House Canyon, interviewed in presence of attorneys," the investigators noted that her fat biopsy had reportedly showed herbicide contamination. "Subject apparently has had numerous physical complaints and has visited many local physicians," the report said. "Dr. Randolph reported a visit at which she said to him: 'Do whatever is necessary to prove that I have herbicide poisoning.'"

After reviewing four pages of reported sicknesses—which included recurrent respiratory infections in children, irregular menstruation, heavy vaginal bleeding during menstruation, chest pain, miscarriages, diarrhea, and eye and skin irritations—the task force declared the "emotional peak" of Globe's inhabitants more significant than their exposure to herbicides. "The psychosomatic effect of an aroused public very likely has played a role," the report said. "It is also important to note that except for three subjects, all of the complaints dated only from the June, 1969 spraying, despite the Forest Service having sprayed the same area three other years."

The task force also asked nine physicians in Globe to answer three questions:

1. Have you been aware of any disease which could be related to herbicide spraying?
2. Have you noted any increase in the incidence of miscarriages since the spraying?
3. Have you noted any increases in any other human illness, such as birth defects, unusual skin rashes, or muscular weakness during this time?

One doctor declined comment, saying his experience was too limited. The rest answered willingly. To the first question, one doctor answered yes, that he attributed several episodes of chest pain and "vague female complaints" to the herbicide spraying. The rest said they knew of no disease caused by the spray. To questions two and three, all the doctors replied in the negative, although one did note an increase in the number of complaints "as the emotions of the citizens became aroused."

As for domestic animals, the task force reported that the local veterinarian believed that no animals were directly affected by the spraying operation, even though he had treated an increased number of dogs with respiratory disorders from the entire Globe area during 1969. The task force also noted that cattle grazed in the sprayed area had no abnormal loss of calves, noticeable increase in disease, or birth defects.

The parade of complaints from residents told another story. Richard Lewis's wife had two dogs who had died after the spray application, one with an enlarged abdomen and respiratory discharge. Ruth Steinke reported that her chickens had lost weight and died, and several had paralyzed legs and wings and distorted necks. The task force said this was similar to a common poultry disease. One of Mrs. Steinke's cows that had wandered on a hill where there was drift from the spray produced milk with a red color and stringlike attachments—a condition attributed by the task force to an unsterilized milk container contaminated by bacteria.

Another of Mrs. Steinke's cows that developed a labored breathing condition showed a malformed lung when it was slaughtered for meat. The task force blamed pulmonary emphysema.

The McKusicks came in for particularly skeptical treatment by the task force, which painted a picture of a family that badly neglected its animals. "The general husbandry practices," the report noted sourly, "left much to be desired." The task force reported that vegetation, including poisonous snakeweed, in the McKusicks' yard was heavily grazed by their goats. The animals were "in poor flesh."

The task force scientists were apparently most incensed by a goat named Big Red, which had become something of a television celebrity. Big Red was a mess. She had an arched back and stiff legs and walked on her two front knees, dragging her hind legs behind. The McKusicks said the goat had been born after the 1968 spray and deformed by herbicides. The task force said the goat's condition was a typical case of malnutrition. It also scrutinized the goat's teeth, which "contained a full set of permanent incisors." This would indicate "without doubt that this animal was at least five years of age." The task force also criticized McKusick's goat-breeding program, pointing out that one billy among a herd of females would make inbreeding unavoidable and would result in high mortality of offspring.

Another animal celebrity named Charly—a deformed duck with a slipped tendon—also was the subject of steamy debate among the task force members. They concluded that the duck had been hatched four miles from the spray area and given to the McKusicks and that its slipped tendon was a common condition in fowl that were lacking choline, which is essential in fat and carbohydrate metabolism. The task force concluded:

> The deformed goat and the duck with slipped tendon had been shown on television with the inference that the conditions were caused by exposure to the herbicides. These cases were clearly misrepresented. The duck was crippled when given to the family, and at that time already suffered from a slipped tendon. The goat was born before any herbicide was applied. We do not believe the goat was born deformed. Its condition is typical of severe nutritional deficiency.

Soil samples taken from the helispots used for the spraying operation were analyzed down to trace parts per million amounts for dioxin without success; it did not mean that dioxin was absent, but that analytical equipment could not detect less than parts per million amounts. No 2,4,5-T was detected in the soil either, but silvex was found in amounts as high as 5,500 parts per million in one sample. Attempts to detect dioxin in blood, fruit, and animal muscles and livers were not successful, the investigators said, because reliable methods for analysis had not yet been developed. Analyses of animal tissues for 2,4,5-T and silvex were inconclusive, leaving unanswered the question of whether or not herbicides were present.

The McKusicks and Mrs. Shoecraft were predictably upset about the report. Her distaste for the Globe medical profession knew no bounds. She remembered a time when she had swollen eyes, discolored skin, and swollen arms and legs and was losing patches of hair and eyelashes, and a local doctor had suggested that she was "nervous" because of the deaths of her animals. She had blurted out, "My God, doctor! But why did the animals' hair fall out and their skin look like mine just before they died?" She thought the Globe doctors were ignorant about the effects of pesticides, and they thought she was overreacting. "There's a good possibility some of the human cases are related to spraying," Globe Dr. William Bishop conceded in an interview that winter, "but symptomatic connections aren't connections and I'm no toxicologist. People here are emotional and each morning wake up with new nails pounded into their palms. What's needed is solid scientific investigation."

Mrs. Shoecraft telephoned each member of the task force to find out what special herbicide expertise he brought to the investigation, and none satisfied her tough standards. She also telephoned a University of California herbicide specialist when she read in the *Los Angeles Times* that he thought the stories about Globe "were extremely misleading" and that they "pollute the emotions of unknowing people, making logical analysis impossible." She was resigned to people's calling her uninformed and unknowing, but what upset her most was the specialist's assertion that a task force member had openly called the Globe incident "one of the greatest hoaxes perpetrated on the American public." Off she flew to the telephone again; the task force member in question told her that what he'd really said was that people in Globe were honest but misled.

Mrs. Shoecraft's attention was diverted from the task force by a new outrage—a story in *Time* magazine that winter that rather casually dismissed people in Globe as a collection of nutty misfits. The group Environmental Action, which was preparing for its Earth Day teach-ins that spring, had sent representatives to Globe to assess the situation and talk to the residents. Neither side felt comfortable with the other. Mrs. Shoecraft tried to check the credentials of some members of the group and confusedly accused some of them of being representatives of chemical companies. Worried that burning might release poisonous fumes, Mrs. Shoecraft and McKusick threatened to shoot one of the group scientists when he attempted to set some herbicide-sprayed brush on fire. "She called him a fake and a fraud," recalled Bob McCray, who was not opposed to the experiment. The environmental group quickly left town.

According to *Time,* the "silvex-touched residents" had a "paranoid outburst." The story said, "Mrs. Shoecraft is convinced that her phone is being tapped, her mail opened, her every movement watched by lurking spies."

The story described the aftereffects of the herbicide spraying—eggplants turned as "orange as pumpkins," pumpkins turned as "black as charcoal," and desert yucca that developed S curves. But "outrage became an ecological crusade when some of the people who were exposed to the spray began to have odd complaints." Is silvex, or any other defoliant, the real culprit? the magazine asked, and answered its own question by quoting a Globe physician who said, "I keep trying to see the relationship between the spraying and the illnesses but I have simply not found anything." Another doctor told *Time,* "Old troubles have been given new names." *Time* concluded, "So far, only one thing seems clear. Environmental concern can do odd things to some people's minds."

Mrs. Shoecraft was momentarily dumbstruck by the article, which was accompanied by a photograph of her in her old range boots standing next to a twisted plant that looked like Jacob's staff. When she recovered, she tried to turn derision to her advantage. "Don't you see?" she told her friends, who were urging her to sue the magazine. "Time Magazine has done a lot of research we didn't know about at that time, and they've found how all this spray stuff can damage your brain and do all kinds of horrible things to people's heads." She also told them with tongue planted firmly in cheek that she was going to start a business called Paranoia for Hire. She made up a poem to hang on the wall:

> *If your taxes are high and you can't save a bit*
> *Just call on me and I'll throw a fit*
> *If your butcher now charges two more dollars for meat*
> *My paranoiac behavior will make him retreat*

Mrs. Shoecraft's impulsive behavior made most scientists and doctors uneasy, even when they were in basic agreement or sympathy with her goals. She railed against the conservatism of science and the reluctance of its practitioners to draw conclusions that she thought obvious. Her equation was simple: $1 + 1 = 2$. Person A is exposed to Chemical B, which has been shown to cause certain harmful effects in laboratory animals. Person A develops some of those symptoms. Chemical B is to blame.

But scientists knew there was no such thing as a simple cause-effect equation because the variables of life made it geometrically complex. How many other harmful substances had Person A been exposed to over the course of his years? Cigarettes? Air pollution? A bad diet? Radiation? When any of these toxins were mixed together with Chemical B, what kind of synergistic effect resulted? What kinds of effects did Chemical B produce in animal studies? Were the animals more susceptible to those effects than other species? Were they dosed with unrealistic levels of Chemical B? Was a no-effect level found? Was there a good control population used as comparison? Should these results be extrapolated from animals to humans?

But Mrs. Shoecraft wouldn't buy it. Nor could she countenance the traditional reluctance of scientists to get involved in public brawls and political squabbles over the regulation of chemicals, food additives, and drugs. She wrote:

> Why are so many scientists as apathetic as the general public in their reaction to many of the alarming facts regarding what is really happening to man? The majority of them leave the burden of informing those who should be doing something about it to a handful of their more courageous members. Why must the few always fight the battles for the many?

Yet increasingly, as the dioxin debate wore on, concerned scientists were stepping out to speak publicly against the continued use of herbicides. Dr. Matthew Meselson, the Harvard microbiologist who had been appointed by the American Association for the Advancement of Science to head a 2,4,5-T evaluation project and had been partly responsible for the release of the Bionetics study, kept up a steady and unfaltering pressure against herbicides after he had returned from a fact-finding tour of Vietnam and Cambodia in late 1969. Dr. Meselson and colleagues at Harvard had developed a breakthrough method of analyzing dioxin down to parts per trillion levels. Using this method, they detected dioxin in fish and crustacean samples collected from herbicide-sprayed areas in South Vietnam and concluded that the chemical could possibly have accumulated to biologically significant levels in the country's food chains. In a report on his Vietnam studies, Dr. Meselson noted that doses of dioxin measurably smaller than the acute lethal dose produced birth defects and death in animals exposed over a period of time. Rats were far less susceptible to dioxin than other animals, and as a result, birth defects occurred in rats at doses substantially lower than those required to kill. In addition, feeding studies in monkeys showed that dioxin poisoning was cumulative. Repeated intake of tiny doses of dioxin that were far smaller than the lethal dose caused serious poisoning and death in monkeys. As a result, Dr. Meselson believed there were sufficient questions about dioxin's chronic toxicity, to both animals and humans, to justify a ban on production of 2,4,5-T. "I say when in doubt, stop it," he told a medical magazine in 1970.

Another scientist reported preliminarily in 1970 that dioxin was teratogenic in chickens. Dr. Jacqueline Verrett, a Food and Drug Administration toxicologist, had been comparing the effects of highly contaminated 2,4,5-T against a Dow sample containing less than one part per million dioxin. When injected into the chick embryo, both samples produced birth defects, including eye and beak deformities and slipped tendons. In one study a clutch of twenty-five fertilized eggs were injected with fifty parts per trillion dioxin. Fifteen of twenty-five chicks hatched, and of those, eleven were born with slipped leg tendons, which caused them to walk on their knees. Cleft palates and beak deformities were found in other survivors. In the unhatched chicks there was pronounced evidence of swollen tissue, or edema, as well as structural deformities. In another brood, Dr. Verrett produced a similar pattern of birth defects using just two and a half parts per trillion dioxin, or 1/400,000th of the one part per million limit proposed as safe by Dow.

But the evidence that tipped the scales and forced the government's hand came that winter, when the federal government's National Institute of Environmental Health Sciences and the Food and Drug Administration weighed in with the results of their duplications of the Bionetics tests. The Bionetics tests had used 2,4,5-T samples contaminated with twenty-seven parts per million dioxin. The new tests used samples that contained less than one part per million dioxin. But the herbicide continued to produce birth defects in some animals—including cleft palates, crooked tails, missing limbs, and skull defects—even when it contained less than one part per million dioxin. The major differences between the earlier Bionetics tests and the new studies were the reduced intensity and range of birth defects and the absence of certain defects, particularly cystic kidneys, in mice and rats. Additional studies made it clear that dioxin by itself, in half part per million quantities, was capable of inducing birth defects.

Critics declared that the studies were untrustworthy because, as in the Bionetics tests, massive doses and susceptible animal species were used. The opposition also claimed that any chemical given in sufficient amounts at the correct time in the pregnancy cycle to the right species would increase birth defects. They noted that mice of the same breeding strain used in the studies also developed birth defects if the mother was fed a diet of nothing but raisins for one day, or was flown to a new location while pregnant, or was subjected to loud noises or a five-degree change in laboratory temperature. They pointed out that aspirin and vitamin A were potential teratogens under certain conditions.

The chemical industry was skeptical of the tests and discounted their applicability to humans. Dow, which had run its own tests on purified 2,4,5-T without finding the birth defects detected in government tests, was convinced that something of a fear-and-smear campaign was being waged against herbicides. "If we thought 2,4,5-T was harming anybody, we'd take it off the market tomorrow," Dr. Julius Johnson, Dow vice-president and director of research, told a reporter that winter.

Dow's central argument was that humans are not exposed to harmful concentrations of herbicides under normal conditions of use and that herbicides degrade relatively rapidly in the environment and so do not bioaccumulate in the food chain. The contaminant dioxin was immensely toxic, but it could be held at safe levels under proper conditions.

"It is difficult to imagine a situation short of a person treating his food with 2,4,5-T in which man could ingest as much 2,4,5-T proportionally as was given to the test animals," Dr. Johnson said.

Supporters argued that high doses were necessary because the statistical chances of detecting effects in the small numbers of animals used in a typical study were very low. A pesticide might cause birth defects in 1 out of every 10,000 humans after a certain exposure, but the probability of detecting that statistically significant effect in 50 rats or mice exposed to the same amount of pesticide for the same time would be slim. Just because a chemical shows no increase in birth defects in animal tests, they argued, doesn't mean it can be considered safe. Further, humans are sometimes more sensitive to teratogenic chemicals than are animals. Humans are 60 times more sensitive to thalidomide than mice, 100 times more sensitive than rats, 200 times more sensitive than dogs, and 700 times more sensitive than hamsters. As scientists pointed out after the thalidomide scandal, inestimable human deformity could have resulted in this country if the government had tried to set a safe human exposure level for the drug based on animal studies. They argued that the dioxin-contaminated herbicide might well be responsible for less obtrusive birth defects than thalidomide's missing limbs and that these deformities, such as cleft palates, were simply absorbed into the national pattern of birth defects and so not recognized. They also disputed the contention that any chemical was a teratogen under the right conditions. A nonteratogenic compound may kill the animal if the dose is high enough, they said, but it won't cause birth defects.

Although the test results were disputed, and although the industry began to argue paradoxically that animal tests were not good predictors of human health effects, the new round of results seemed to show with some clarity that 2,4,5-T was a teratogen even when it was contaminated with less than one part per million dioxin, which is the level Dow claimed it had maintained since 1965, when it learned how to manufacture a cleaner herbicide.

In April 1970 Mrs. Shoecraft went to Washington. The herbicide debate had built to a crescendo, partially as a result of a critical *New Yorker* series about the hazards of 2,4,5-T written by Thomas Whiteside. Congress finally took notice and scheduled several days of hearings. Mrs. Shoecraft showed up in Washington with several bulging boxes of files, including photographs she'd taken of herbicide-damaged plants in Ice House and Kellner canyons. She carted her boxes over to the office of Michigan Senator Philip Hart, who was preparing to chair the Senate hearings on the effects of 2,4,5-T on humans and the environment. The senator had his staff help Mrs. Shoecraft carry her files to the hearing room.

On April 7 Senator Hart's Subcommittee on Energy, Natural Resources and the Environment opened hearings on 2,4,5-T. In an eloquent opening summary that went to the heart of the conflict, Senator Hart conceded that the debate about herbicides "may in the end appear to be much ado about very little indeed. On the other hand, they may ultimately be regarded as portending the most horrible tragedy ever known to mankind. What does emerge clearly from this uncertainty is that we must take steps to eliminate it."

In the manner of expert witnesses testifying against each other in a trial, herbicide supporters and opponents squared off at the hearing over the central question of whether or not 2,4,5-T should continue to be sold while its safety was a matter of dispute. Congressman Richard McCarthy reported on his investigations in Globe and on the continued use of 2,4,5-T in Vietnam, despite the fact that no field studies regarding its long-range effects had been conducted by the military. The congressman attacked the Forest Service for its negligent enforcement of regulations, the Food and Drug Administration for not establishing safe residue levels for herbicides on food crops, the Agriculture Department for not adequately policing herbicide use, and the White House for backing down from its announced ban on herbicide use the previous fall. "I am now told boldly by [science adviser] Dr. Lee A. DuBridge in a letter of March 2, 1970 that 'we anticipate, indeed we will insist upon final action of 2,4,5-T before its period of principal usage in late spring,'" McCarthy said. "I will not hold my breath."

The congressman called for an immediate five-year ban on the use of 2,4,5-T pending a determination of its safety and an organized effort by the federal government to collect statistics on birth defects in the country, including those that might be caused by herbicides.

Two public-interest attorneys from Ralph Nader's Center for Study of Responsive Law testified that 2,4,5-T posed "a grave and unnecessary danger to public health." They quoted scientists' opinions that 2,4,5-T "may represent the ecological equivalent of thalidomide." They questioned why the government was still stalling on a herbicide ban when duplications of the Bionetics tests by the Food and Drug Administration and the National Institute of Environmental Health Sciences continued to show that even 2,4,5-T containing less than one part per million dioxin caused harmful effects to animals. "Whether or not human beings are more or less susceptible than test animals to these chemicals, we do not yet know," they said. "But clearly on the evidence now available, the burden of proof should be on the industry to demonstrate that they are not harmful. In the meantime, all uses of these herbicides around the home and in populated areas should be immediately suspended." They raised the specter of a nation of parents grieving over children born with birth defects "with no knowledge of the cause" because 2,4,5-T "leaves no unique fingerprints on the fetus" but produces instead the most ordinary of defects, such as cleft palates and kidney abnormalities.

"The parents will gain no compensation for their loss," the attorneys testified. "Moreover, there will be no lawsuits to force the chemical companies to test more thoroughly their products for teratogenic effects before they are released on the market or to maintain strict quality control standards which will keep the level of contamination of dangerous dioxins as low as possible. In the absence of legal remedies for private citizens, protection must come from the Federal Government."

The attorneys reported that an informal survey they'd made the previous week in ten Washington area stores showed that more than 60 percent of all herbicides sold to the general public contained either 2,4-D or 2,4,5-T, or a combination of both. Some were so badly packaged that they could not be handled without spilling dust, and others did not bear even the minimum federally required warning "Keep out of Reach of Children." Scott's Kansel weed killer, in a salt shaker-like container, had no warnings on the front, and only "Avoid contact with skin, eyes, clothing, etc." written inconspicuously on the back. "In short," the lawyers said, "the general population is being exposed to 2,4,5-T herbicides in use in residential areas."

The attorneys used the hearing as something of a soapbox to broadcast their concerns about potential health effects to all Americans from the thousands of chemical products they were exposed to in daily life and the growing concern of scientists that cancer and genetic damage were related to chemical exposures. "The regulatory history of 2,4,5-T and 2,4-D serves as an example of the governmental failure to protect the public from potentially dangerous chemicals," they warned, citing as evidence the ineffectiveness of the Agriculture Department in regulating pesticides and its reliance on tests conducted either by the chemical companies marketing the products or by independent testing facilities hired by the companies that "tend to seek the answer most advantageous to their clients."

The Agriculture Department, in the person of Dr. Ned Bayley, its director of science and education, and his assistant director, Dr. T. C. Byerly, took an opposing view. While agreeing that there had been some problems in the registration process, Dr. Bayley admonished that as a civilization without pesticides, "we would be in a very serious situation from the standpoint of our ability to produce food and fiber for this country."

Rebutting earlier testimony on the potential of 2,4,5-T to accumulate in the environment and on the lack of a safe residue level for the chemical on food crops, Dr. Bayley declared that the herbicide degraded in the environment in a matter of months (although he conceded the paucity of information about dioxin's persistence in the environment) and that residues on foods were so rare that among 5,300 crop samples analyzed during the previous four years, only 25 contained trace amounts of 2,4,5-T and only 2 were higher than that, although both were still well below one part per million. Despite the findings of Harvard biologist Matthew Meselson and reports from medical people in the country itself, Dr. Bayley asserted that Agriculture Department scientists had not seen evidence of irreversible ecological damage from herbicide use in Vietnam. The same was true for Globe, Arizona, where a team of Agriculture Department scientists found that the only effect of herbicide use was "damage to some susceptible plants near the treated area."

As for the Bionetics study and its aftermath, Dr. Bayley explained that the birth defects were in fact due to the high levels of the contaminant dioxin, not to 2,4,5-T. Because the chemical industry was now able to limit the levels of dioxin in 2,4,5-T to below 1 part per million, there was no cause for concern. A government survey released at the hearings, however, showed that Monsanto's 2,4,5-T contained 2.9 parts per million dioxin.

The repeat tests on 2,4,5-T, which still showed birth defects, were suspect, he said, because the herbicide was fed to test animals at overly high-dosage levels. The Agriculture Department official further pointed to repeat tests run by Dow, which produced no birth defects. "Therefore," he said, "we do not believe this in any way changes the hypothesis that the low level of dioxin is safe."

Senator Hart, the subcommittee chairman, asked Dr. Bayley if the basic conclusion of his testimony was that the registered uses of 2,4,5-T did not constitute a sufficient hazard to order production halted.

"That is correct," Dr. Bayley replied.

"And yet," retorted Senator Hart, "this morning we have heard testimony that preliminary tests suggest that 2,4,5-T when contaminated by dioxin comparable to that found in currently produced 2,4,5-T is teratogenic in three species; that the Mrak commission or a panel advisory to it said that the teratogenic effects in one or more such species should be grounds for immediate restriction of pesticide use; that residues of 2,4,5-T are now found on approximately one out of every 200 food samples analyzed by the FDA; that we can't be sure of the amounts of dioxin in 2,4,5-T now being sold, nor do we have as yet clear ideas on the amount of other dioxins in pesticides . . . ; that no evidence suggests that these dioxins are not persistent or cumulative in human tissue, and that some evidence [exists] which would indicate perhaps they are.

"If you accept that as premise, in view of all of this, would you say that you are sure that registration of 2,4,5-T for use directly on food crops does not constitute a hazard to man?"

"I would say," replied Dr. Bayley, "that the information we have does not give us indication that it is a hazard to man in accordance with the registered uses."

Mrs. Shoecraft sat in the hearing audience taking notes on testimony. She was angered by the intransigence of the Agriculture Department (she called it "Washington's Sacred Cow") and by its officials (Dr. Bayley and Dr. Byerly were tagged the "Bobbsey Twins"). When the hearing adjourned that afternoon, Mrs. Shoecraft went back to her hotel room and wrote a poem:

> Humans deranged and gasping for breath
> Accept unprotesting this chemical death
> These poisons are easy to buy any day
> All stamped and approved by the USDA
> They'll kill off the world with these chemical tools
> Nothing makes sense to some doddering fools!
> So let's vote them all Laugh-In's 'Finger of Fate'
> And spray them all under before it's too late.

She went to bed early, counting not sheep but the "unbelievable number of government employees whom I met whose minds appeared to be so full of 'inspired ignorance' that there remained no room for knowledge—or understanding." The next day, during a break in the hearing schedule, Mrs. Shoecraft, dressed in her Indian moccasins and Navaho squash-blossom jewelry and carrying a tape recorder in a satchel, presented herself unannounced at the Department of Agriculture, where she demanded to see either Agriculture Secretary Clifford Hardin or one of the "Bobbsey Twins"—Dr. Bayley or Dr. Byerly. "Since no one could tell by the size of my carry-all whether I had brought not only my lunch but my clothes as well and planned to stay awhile," Mrs. Shoecraft was quickly taken to Dr. Byerly's office. After rattling around in her bag ("everyone was afraid I was hunting a sandwich"), she was even more quickly escorted in to see Dr. Byerly, the agency's assistant director of science and education.

Mrs. Shoecraft asked for, and was given, some of the studies that the Agriculture Department officials had used to support their hearing testimony on the safety of 2,4,5-T. Dr. Byerly admitted that spray drift had done some damage in Globe—"errors do happen"—but he reminded her, too, that private citizens have proprietary rights and those rights are limited. "There are no unlimited property rights for anybody in the United States," he said. Reconstructing their conversation later, Mrs. Shoecraft remembered that it was like a shooting match with aphorisms instead of bullets.

Dr. Byerly peppered his conversation with Latin phrases, admonishing Mrs. Shoecraft to "write them down and look them up and ponder them."

"After the fact, therefore because of it," Dr. Byerly said, explaining that the quotation captured the reason why he as a scientist had to remain skeptical.

"Seek the truth and the truth shall make you free," countered Mrs. Shoecraft, quoting from the Good Book on her mother's library table.

"Seek wisdom," she added after a while.

"And above wisdom, seek understanding," Dr. Byerly replied.

Then he said, "Just as I try to seek the truth, I try to keep this in mind from time to time because I'm capable of error . . . but be assured I'll do my best."

"This land is mine," Mrs. Shoecraft said, quoting from an Apache saying as she picked up her carryall bag to leave. "No one but I can lose it for me."

The hearings resumed with testimony from Dr. Jacqueline Verrett, the Food and Drug Administration toxicologist whose experiments had shown dioxin to be an extraordinary teratogen when injected into chicken embryos. Dr. Verrett brought two trayfuls of deformed chickens to the hearing room; some were crippled by edema or by slipped tendons and could not stand up. She testified that while the heavily dioxin-contaminated 2,4,5-T used in the Bionetics study was a more potent teratogen, the Dow sample—with just half a part per million dioxin—also caused birth defects in the hatched chicks, including eye defects, beak defects (the most predominant of which was cleft palate), short and twisted feet (the result of slipped tendons), and edema in various parts of the body. Some scientists discounted the applicability of Dr. Verrett's findings to human risk because of the extreme sensitivity of chick embryos to toxic substances. The World Health Organization did not recommend chick embryos for the screening of drugs for teratogenicity because they were too sensitive to a wide range of agents and were anatomically different from pregnant mammals and their embryos. But Dr. Verrett was convinced of their worth. Since the human placenta is a highly permeable membrane, which easily permits the passage of chemicals to the fetus, she reasoned that a teratogen in a chicken had a very high probability of being a teratogen in a human.

Looking back on her work some years later, Dr. Verrett recalled that of 1,200 or so chemicals her laboratory had examined over the previous two decades, dioxin was "the most toxic thing we have tested. That is without exception."

Her testimony on teratogenicity was supported by Dr. Samuel Epstein, a toxicologist with the Children's Cancer Research Foundation in Boston who had directed the Mrak Commission's teratogenicity panel. He reviewed the wide range of birth defects found in animals exposed to 2,4,5-T and dioxin and suggested that a clear and potentially costly risk to humans existed. "The financial cost to society of one severely retarded child, computed on the basis of specialized training and custodial care alone, approximates to $250,000," Dr. Epstein said. "Continued use of these herbicides in the environment constitutes a large-scale human experiment in teratogenicity."

On the morning of Wednesday, April 15, the lead-off witness, United States Surgeon General Dr. Jesse Steinfeld, advocated a moderate approach to hazardous chemical regulation—an approach that balanced both the benefits and the risks—and then, without warning, he read an announcement:

> Agriculture Secretary Clifford M. Hardin, Interior Secretary Walter J. Hickel, and Health, Education and Welfare Secretary Robert H. Finch today announced the immediate suspension by [the Department of] Agriculture of the registrations of liquid formulations of the weed killer, 2,4,5-T, for use around the home and for registered uses on lakes, ponds and ditch banks.

The surprise announcement stunned the hearing-room audience. Not only was 2,4,5-T being restricted from use around the home and in waterways, but the Agriculture Department also intended to cancel registered uses of the herbicide on food crops. Herbicide opponents had been lobbying for this order for six months, but the Agriculture Department had only a few days ago told this same hearing that such drastic restrictions were not warranted. As it turned out, new information had become available only two days before. Statistical analyses of the preliminary data from the National Institute of Environmental Health Sciences studies showed that 2,4,5-T currently on the market—with less than one part per million dioxin—as well as the dioxin contaminant itself each produced birth defects in laboratory mice.

The surgeon general continued to read from the announcement:

> In exercising its responsibility to safeguard public health and safety, the regulatory agencies of the Federal government will

move immediately to minimize human exposure to 2,4,5-T and its impurities. The measures being taken are designed to provide maximum protection to women in the childbearing years by eliminating liquid formulation of 2,4,5-T use in household, aquatic and recreational areas. Its use on food crops will be cancelled, and its use on range and pastureland will be controlled.

The suspension of herbicide use around the home and waterways was the most drastic enforcement tool in the government's armamentarium, because its effects were immediate and final. The proposed cancellation of 2,4,5-T registrations on food crops was less drastic because by law, chemical companies could appeal the decision, forcing hearings to be held on its risks and benefits, and could continue to use the pesticide pending a final determination on its safety. As for the use of herbicides on nonagricultural range, pasture, and forestlands—the use that most concerned Mrs. Shoecraft—the order specifically allowed the continued spraying of 2,4,5-T for brush and weed control. Silvex was also exempted from the order.

On the same day Deputy Secretary of Defense David Packard ordered the immediate suspension of the use of 2,4,5-T within the defense establishment, pending a safety review.

Later that afternoon, when Dr. Julius Johnson, vice-president and director of research for the Dow Chemical Company, appeared at the hearing, he steadfastly defended the safety of 2,4,5-T.

"Now, do you agree with the position that has been taken, as announced this morning by the three Secretaries?" Senator Hart asked Dr. Johnson.

"In a matter of practical hazard, as an imminent hazard to health, I do not agree," the Dow executive replied. "Under the climate of pressure today, it was a wise decision."

"That sounds like you are planning to run for reelection, but you do not want to announce it," retorted the senator.

Dr. Johnson was a skilled public speaker who moved easily between the worlds of industry and government. He had been a member of several government panels investigating pesticide use, including the Mrak Commission. His testimony before the Hart hearing clearly showed the depth of his feelings about the safety of Dow's product and his disdain for the dioxin doomsayers. The word "dioxin," he said distastefully, as if speaking about a relative gone wrong whose escapades are spread across the tabloids, has become "almost a cause célèbre." He lectured, as had others before him, that even vitamin A is a teratogen at high doses. He counseled the need to replace "emotion, rumor and misconception" with a "clear explanation of the facts."

As evidence, he offered the results of Dow-run studies that showed that 2,4,5-T containing less than one part per million dioxin did not cause birth defects in rats or rabbits unless they were "stressed to the limit of their tolerance" or were given extremely high doses. Additionally, Dow physicians studying the health of 130 men who worked in Dow 2,4,5-T plants for periods up to twenty years found no evidence that exposure to the herbicide resulted in adverse effects.

In discussing the history of dioxin, Dr. Johnson disclosed that Dow had been aware since 1950 that the Germans had isolated a contaminant in the manufacture of the intermediate product 2,4,5-trichlorophenol, one that they believed caused chloracne in exposed workers. Dow monitored the level of the contaminant in its 2,4,5-trichlorophenol by rubbing the chemical on rabbits' ears and gauging the intensity of the reaction. By keeping the contaminant at reduced levels in the intermediate product, Dow was convinced that its end product, or 2,4,5-T, was also safe. Dr. Johnson maintained that Dow was not aware until October 1964 that dioxin was in its 2,4,5-T.

It was then that Dow herbicide workers developed such severe chloracne that the plant was shut down. In contrast with graphic descriptions of chloracne in the medical literature as "comedones, retention cysts, nodules, pustules, boils and patches of pigmentation" that sometimes gave the skin "a dirty grayish-brown appearance" and that lingered on some men's faces as "closely arranged pitted scars which have a disfiguring effect," Dr. Johnson described chloracne as "similar in appearance to severe acne often suffered by teenagers."

Rabbits' ear tests run on the 2,4,5-trichlorophenol waste oils showed that the chloracne-provoking contaminant was building up to a "danger point"—a result, Dr. Johnson testified, of manufacturing changes made to improve production capacity. The temperature of the manufacturing process, in other words, had been pushed so high that greater quantities of dioxin were formed. The plant was shut down, and Dow scientists, using sensitive analytical equipment just then being developed, were able for the first time to detect dioxin at parts per million levels in its 2,4,5-T.

Dow dismantled the herbicide plant and began designing a new one to manufacture 2,4,5-T containing reduced levels of the dioxin contaminant. In the meantime, Dr. Johnson testified, Dow notified the Michigan Department of Health, the Institute of Industrial Health, the University of Michigan, and various other health groups in private medicine and industry about its concerns over dioxin. "In addition," he said, "we called a meeting which was held in March 1965 to notify other manufacturers of 2,4,5-T of the difficulties encountered. We described to them the nature of the health hazard and shared our test procedures and analytical standards." From the day the new plant came on stream in 1966, Dow met a self-imposed manufacturing specification of less than half a part per million dioxin.

Senator Hart asked the Dow executive why the company had notified so many health agencies and other chemical companies, but not the Agriculture Department or the Food and Drug Administration.

"At that time, Senator Hart, we considered our obligation discharged by removing the dioxin from our product, by notifying health authorities in the state, and we thought we had the problem solved," Dr. Johnson said. "In retrospect, it would have been much preferred had we notified the U.S. Department of Agriculture, the agency that has statutory authority for the registration."

Dow reacted angrily to the restrictions on its herbicide. Meetings were held around the country to convince state officials that the herbicide was safe, and Dow scientists and executives began making a concerted effort to defend the chemical publicly. Dow saw itself threatened by a "battery of 'instant ecologists' and those who swing with current fads" as well as by a "population of underinformed, honestly concerned people who have not heard the whole story."

The entire industry went on alert to this new threat. University of California agricultural scientist Boysie E. Day told the annual meeting of the Society of American Foresters in 1970:

> The problem is one of increasing public distrust of modern technology, as evidenced by a rising wave of critical re-evaluation of all our institutions and practices, in which everyone seems to be dissatisfied with everything and all objectives and motives become suspect. There is a view going around that science and technology are somehow the cause of all our problems. This view is particularly widespread in relation to the use of agricultural chemicals, especially pesticides.
>
> After a long period of national self-congratulation for seemingly endless progress and mounting affluence, we are undergoing an inevitable period of reappraisal and disillusionment. After the binge of optimism comes the hangover of realization.

Yet the scientist saw hope for 2,4,5-T. "I grant that it is of no political significance whether the alleged hazards of 2,4,5-T are real or imaginary," he said. "It is sufficient and justifiable to ban it when enough people want to ban it. The fact that this chemical has been extensively used in an unpopular war, combined with other doubts and suspicions, may yet turn the tide in its disfavor. However, in view of all we know about the chemical, and our long experience in using it safely, this seems unlikely."

* * *

After the Senate hearing, Mrs. Shoecraft called her attorney in Phoenix and told him he might as well go fishing because Senator Hart had saved them a lot of time and money in getting answers to questions they needed from Dow—"questions that we would never have known the answers to," she said, "or how to ask."

When she returned home to Globe, Mrs. Shoecraft decided to file a damage claim with the Agriculture Department for alleged property loss from the herbicide spraying. Her house and the land it was built on meant a great deal to her. She and her husband had bought the rugged twenty-acre plot at the foot of the Pinals for $1,500 an acre and spent two years building their house and working the land into shape. She had an appraiser calculate the loss: fifteen of twenty acres covered by vegetation in various stages of death, and two acres of destroyed gardens.

She filed a claim for $4.5 million. The claim was not written but was orally delivered by Mrs. Shoecraft on tape, against a background of organ music performed by her husband, Willard. She thought the organ music appropriate "in case anyone dropped dead from shock." She also included some of her poems.

The tape began with Mrs. Shoecraft's unmistakable voice: "You can reach over and turn off this tape, but you won't, for you'll probably want to hear what I'm going to say." She addressed her remarks to Agriculture Department Secretary Clifford Hardin, White House science adviser Dr. Lee DuBridge, and President Nixon "if he cared to listen.

"I am not speaking to you about Vietnam . . . but about the United States of America . . . and your continuing to allow the use of 2,4-D or 2,4,5-T or silvex on rangelands, forests, and food crops. All three have been proven hazardous and deforming. I cannot help but wonder how helpless a mother in Vietnam must feel when she can't even complain to you about these chemicals as I am able to do.

"You know as well as I do, Mr. Hardin, that the order to restrict the use of 2,4,5-T in America to rights-of-way and rangeland is totally meaningless, while still permitting the use of 2,4-D and silvex. And since these chemicals can drift for hundreds of miles, where are your 'unpopulated areas' for application, Mr. Hardin or Mr. DuBridge or Mr. Finch or Mr. Hickel or President Nixon?

"The disappointment I feel in finding such feet of clay in high places has added to a general disillusionment with your entire department. I gave my word that I would do all within my power in whatever time is left to me, to see that these things which are wrong are stopped. I gave my word to the trees that are left, and to the Arizona desert which I love, and I will keep that word.

"Because I am a realist, I have reduced the hurt to the only terms I believe can be understood by those who would continue these plans of desecration anywhere in the world. And since, as Dr. Bayley, USDA, has so eloquently stated, it must be for 'economic reasons,' I now present the extent of my damage economically in terms of a monetary figure. This figure is far short of the actual loss I could express, and while it will never cause me to forget what has happened, it may help you to remember."

She closed the claim with a poem and a flourish of organ music.

This be my prayer, I ask tonight, Oh God!
Bring back the flowers . . . and start my world anew
Help move the rocks! I'll use what strength I have
Don't let me fail in what I have to do!

I will avenge each needless useless death
And share within my heart their silent pain.

chapter 7

Billee Baldauf met Willard Shoecraft when she traveled west from Indiana in 1947 to visit her sister in a raw Arizona mining town called Globe. Mrs. Baldauf was a pretty twenty-six-year-old divorcée with snow-white hair and two small sons. "She was extremely vivacious, a very attractive woman. She burned up fuel fast," said a doctor who met her on that first trip to Globe. She had been born to Etta and William White on a farm in Pleasant Ridge, Ohio, in 1921. They named her Wilma. She preferred Billee. She married when she was eighteen and had one year of college and one course in radio training before teaching herself interior design and decorating. When she was nineteen, she bought a set of Audel's *Carpenters* books and learned how to be a builder. She wrote poetry, rode horses, and knew how to use a rifle. Frank Lloyd Wright, whom she met once, was one of her few heroes.

Mrs. Baldauf's older sister, Esther, moved to Globe after World War II, when it was still a rough-hewn place tucked in a basin deep between the Pinal and Apache mountains and named after a globe-shaped piece of almost pure silver found nearby. Globe had already been through two booms—silver and copper—and the Apache who terrorized the first prospectors had long since been herded onto the San Carlos Reservation four miles to the east, where they raised cattle, danced the Devil Dance, and sold baskets to tourists on land where Geronimo once roamed. The region fed off its natural bounty of copper and a rare hollow-fibered asbestos tougher than piano wire. It was spectacular country, half desert and half pine-carpeted mountain, filled with mesquite, cat's-claw, cottonwood, and prickly pear. At dusk the sky hung dense and salmon-colored over the canyons. The sound was of crickets. The smell was the cool, perfumed assault of a florist's icebox.

Billee Baldauf from Indiana fell under Arizona's spell. It was a place that gave her a sense of being alone, but not lonely. She also fell under the spell of Willard Shoecraft, the charming and loquacious manager of the local radio station. He and Mrs. Baldauf's sister were longtime friends. They met when a boy fired a BB gun at a bird sitting on a telephone wire. The pellet missed the bird but hit the telephone line, fusing it into the radio network line, which was transmitting the *Catholic Hour* to Willard Shoecraft's radio listeners. The gossipy telephone chat that Mrs. Baldauf's sister was conducting with a friend suddenly replaced the liturgical music on Globe's radios. Her husband, who listened in amazement as his wife's conversation danced across the radio, stormed over to the radio station to complain. The couple had been friends with Willard Shoecraft ever since.

Willard Shoecraft had no legs. Both had been amputated at the hip after he suffered third-degree burns in a house fire when he was seven. He refused to wear artificial limbs or spend his life in a wheelchair. As a child he rigged a skateboard for transportation and eventually forced his muscular arms into service as legs, getting about on the palms of his hands. He learned how to fly a plane and how to hunt from a jeep. He swam and played the piano, and he got into the radio business when he was seventeen. He was handsome, witty, and articulate. "Willard could charm the horns off the devil," a friend said. When Billee Baldauf came to town, he courted her so assiduously that she collected her things in Indiana and moved back to Globe to be his wife. They were married on New Year's Eve in 1948. Harry Truman had just been elected President. Mrs. Shoecraft brought to her new marriage "an apartment house in Indianapolis and a Frazer automobile." Willard adopted her two sons; their own boy, Robert, was born on New Year's Eve in 1955. Sometime later Willard built his own radio station, KIKO, in Globe, and he bought twenty acres of untamed land at the foot of the Pinals in Ice House Canyon so his wife could build her house. But Willard's devotion to his radio work, coupled with the forcefulness of both his and his wife's temperaments, chewed at the marriage. After a string of domestic trouble, they separated and were briefly divorced. They remarried, ran into trouble again, and divorced again. Mrs. Shoecraft took the boys to California, then to Phoenix. After a few months she and Willard reunited. "I think that every minute she was gone, I did everything I could to get her back, which I finally did," Willard said. They were remarried in Las Vegas and were never apart again. "They fought like cats and dogs, but they truly loved each other," explained Mrs. Shoecraft's daughter-in-law, Mari, who laughed and said, "I don't know how two Aquarians can live in the same house anyway." Mari Shoecraft was called as a character witness for her mother-in-law at one of the divorce hearings. "Yeah, she's a character," Mari testified. "Bill and Willard were both strong-willed people. They argued a lot. But he was madly in love with her, and she with him," said Mary Lou Gardner, who was one of Mrs. Shoecraft's closest friends. "Willard was the one who was the star here in Globe. Everyone in Globe adores Willard. He's so charming. Bill didn't want to take the time to be charming. She had to hurry. She had a premonition of time cut short."

Mrs. Shoecraft set to work building her house. Working with a crew of "cowboys, mine workers, alcoholics, preachers, a Mormon bishop, an auto mechanic and some of the boys from the county jail," in the midst of giant sycamores and cottonwoods, next to a stream that coursed down from the mountains, she built stone walls from 165 tons of rock; laminated together 36,000 board feet of lumber with giant beams from the first theater built in Arizona to form the ceiling; and carved out four fireplaces, seven entrances, and walls of glass. It took her more than two years to finish. In the winter of 1968 the Shoecrafts moved into their new house in Ice House Canyon. Fifteen months later Mrs. Shoecraft stepped out of her bedroom in a pink chiffon nightgown to investigate a noise coming from the mountains.

If Mrs. Shoecraft surprised people with the force of her passion after the herbicide spraying, it was because she had led such a relatively conventional life until then. (Although as one friend pointed out: "She was one of the most litigious human beings I've ever known.") She reared her sons and sent the oldest two off to college. Her youngest, Bobby, absorbed his mother's love for animals. He had twenty-five pigeons, some collie dogs, black cats, and a horse. Every September his mother and he would drive from Globe to Phoenix at three in the morning so they could get a good place in line for a big rummage sale. She scoured the local stores and built up a fine collection of good turquoise and silver jewelry. She spent long hours on her poetry. Willard had found scraps of her poems on envelopes and pieces of paper from the time he first met her, and she had a slim volume of her work titled *Moondust and Other Poems* published in 1960. She collected manzanita in the canyon beyond her house and twisted it into decorations for her interior design business. She helped design Mary Lou Gardner's house—a sumptuous 3,500-square-foot aerie on a ridge overlooking the Pinals with a huge living room open to the panoramic view.

No one outside a small circle of family and friends really knew Mrs. Shoecraft. She was not a publicly religious person but had what her husband called a self-serving, functional personal religion. She picked her own friends and, as a result, did not have many. In private she was warm and irreverent. When her children had children, she could not countenance being called by any of the sugary grandmother names. Her grandchildren called her Mom Bill. "She was totally different from other women," Mari Shoecraft said. "She was all open arms and love from the first time I met her. I came over to their home to spend Thanksgiving, and she welcomed me at the door like I was a long-lost friend. She was a bubbly, lovable person. But she said things with such authority that if you didn't do it, you knew the wrath of God would come down on your head." But in public she was not gregarious and kept to herself. "She was very much a recluse. No one in Globe even knew her," Mary Lou Gardner said. "She hated parties. She wanted to go into a corner and read or write."

She had never been an environmentalist and was closer to Phyllis Schlafly than to Betty Friedan in her philosophy of women's rights. She claimed she was going to write a book called "Women Are a Mess" about that "overfed, overpaid, overindulged part of the human race called woman." She believed in femininity "all the way" and declared that she had "all the equal rights I want, and so does any other female that has enough brains to really be a female, and not a facsimile of one." But the spraying of her home and family and friends drew fighter's blood from Mrs. Shoecraft. "One hundred pounds of love and adoration for her family, her pets, her garden, her home, her mountains—finding all this jeopardized—was [sic] metamorphosed into such an energized force that I even believe an irate mother grizzly bear would be put to shame," observed Frank Egler, a scientist who came to be one of Mrs. Shoecraft's allies. "It is the sort of syndrome that lesser mortals bow and bend before and try to preserve their own egos by calling it emotionalism and paranoia."

When someone called her an iconoclast, she looked up the word in the dictionary and decided it was exactly what she was. "A breaker of images," the dictionary said. "One who attacks cherished beliefs." She wrote: "If those beliefs are wrong and if on close inspection they will not stand the spotlight of truth, then I believe they should be exposed and destroyed." She may have believed that a woman should be a woman, but "if there are times she can't stand the way things are, I believe in her having enough guts to do something about it."

And so she did. In the spring of 1970 Mrs. Shoecraft and twenty other Globe residents filed suit against Dow and three other chemical companies that made the herbicides that were sprayed over the Pinals, the helicopter company that sprayed them, and the state agency that helped finance them. In a separate claim, they sued the federal government under the direction of which the Forest Service defoliation project had been undertaken. That summer Billee Shoecraft began writing a book to expose "the true facts of the spraying of Globe, Arizona." Through the rest of that year and into the next, she spent long, isolated stretches in her cabin 8,000 feet up the mountain, writing a sardonically emotional manuscript that was published under the title *Sue the Bastards* by a small Phoenix press in 1971. In matters of fact, the book was prone to exaggeration and was often flat wrong. In matters of the heart, it was lethally on target. From the beginning disclaimer ("Regarding the characters in this book: Any similarity to persons now living [or dead] is purely intentional. . . . Their true names have been used to protect the innocent, and condemn the guilty") to the back-cover photograph of the author resplendent in a Tammy Wynette-style blonde fall superimposed over a poem "For My Son's Guinea Pigs" ("I'll close your eyes now that are swollen . . . I'll close your eyes now that you're dead . . ."), the book was pure Billee Shoecraft. She wrote:

> "Of the people, for the people, by the people" should not be just a parrot phrase. Whatever it was that caused this sleeping sickness to slowly engulf many of those in our country and numb them into silent acceptance does not need to be perpetuated forever.
>
> What happened in Globe, Arizona is part of what has happened to you. You cannot shut yourself away and say, "It does not involve me," for you are involved. Those who seek to delude themselves, even in government offices, that the incidents which happened here in no way affect them, are only lying to themselves. The lies which they told about a mountain town in Arizona will ultimately cause their own destruction. And in their attempts to suppress the truth about what these and other chemicals are doing to an expectant mother in Vietnam, a worker in a factory in Germany, a reindeer in the forests of Sweden, or a Mexican American in Globe, Arizona—they injure not only themselves, but you also.
>
> When I am asked, "What are you?" I reply, "many things: A poet, designer, architect, writer, mother and friend." But most of all I am an American, and a citizen of the world. And as such, I am free to say that no one, not even those who call themselves "Government Agencies" have a right to destroy what is mine.

Mrs. Shoecraft shopped the book around to some New York publishers, who were put off by her refusal to allow any editing of the manuscript. She eventually paid about $15,000 to have it published in Phoenix. The press run was 5,000 books; most were bought by her new friends in the antiherbicide movement. After a time the book was almost impossible to find. The Dow Chemical Company had to buy a copy from Willard Shoecraft.

The next several years of her life were spent in virtual crusade, with Willard as an uneasy ally. The Dow Chemical Company spent the next several years in a crusade of its own, publicly defending the safety of its herbicides and privately maneuvering in the courts to keep them on the market. "We are deep in the game of political expediency, environmental fadism, and trial by press, and it is beginning to seriously hamper scientific progress at a time when progress is vital," Dow's Dr. Etcyl Blair wrote in the company's *Down to Earth* trade magazine. "If we want to lead the world in technology that up to now has fed the world, we had better re-assess our priorities, unhitch science from politics, separate fact from emotion, and get back to work."

Through legal appeals and injunctions, Dow was able to continue selling 2,4,5-T and silvex for a period of nearly nine years following the ballyhooed "ban" announced in the spring of 1970. Dow's efforts were made easier by the federal government. Although the media widely announced the 1970 action as a ban on 2,4,5-T, the order restricted only about 20 percent of the actual use of the herbicide in America. In its order the government completely ignored silvex, which carries as much dioxin contamination as does its cousin 2,4,5-T. And even though restrictions were slapped on 2,4,5-T, the government gave the chemical industry room to maneuver by suspending some uses of the herbicide and canceling others. Suspension is more harsh than cancellation—it immediately halts interstate shipment of the pesticide. Cancellation allows the manufacturer to appeal the order and to ask for an independent review or a public hearing. The critical difference is that the manufacturer can keep selling the pesticide throughout the entire appeals process. In 1970 the government *suspended* liquid 2,4,5-T for use around the home and both liquid and nonliquid mixtures of the herbicide used to kill vegetation around lakes, ponds, and irrigation ditches. But it *canceled* the use of nonliquid mixtures of 2,4,5-T around the home and on food crops—opening the door for court appeals. Dow exercised both its rights—to appeal and to ask for an independent review. One of Dow's major herbicide markets was southern rice growers, who used 2,4,5-T to kill weeds in their rice fields. Dow's lawyers filed an appeal of the cancellation order in federal district court in Arkansas—the heart of rice country. A public-interest group sued in Washington, D.C., meanwhile, to force the government to replace its cancellation order with the tougher suspension.

The 1970 order also left untouched the aerial spraying of forests, rangeland, and pastureland, as well as power line, highway, and railroad rights-of-way. Mrs. Shoecraft's files bulged out of control with information on herbicide spraying that people sent her from every part of the world. She learned that forests were being sprayed in Oregon, Arkansas, Michigan, and Wisconsin. Invited by several civic groups to attend public hearings in San Diego on the planned herbicide spraying of a national forest there, Mrs. Shoecraft made a command performance. "One of the water quality boys spoke and made a real smart speech . . . how this stuff only lasts a day or two," she recalled. "I shot him out of the saddle with the report I was holding in my teeth." The Forest Service announced after the hearing that it would substitute another herbicide for 2,4,5-T in some of its sprays "because of the extreme emotionalism that has been produced." When the use of Agent Orange wasn't immediately phased out as promised in Vietnam, Mrs. Shoecraft sent a letter to General Creighton Abrams. "I wanted to believe you would keep your word after the dangers of those chemicals were exposed in Senate hearings in Washington," she wrote, and tucked a poem inside about Agent Orange ("Orange, bright orange, is the shroud that I wear. . . . My leaves once were green that I wore for my hair").

The last Agent Orange helicopter spraying operation under American control was flown in Vietnam on Halloween 1971. The military ended the spraying with reluctance. "I would say purely as a sop to the political, this [Agent Orange] was one of the programs we felt should be removed to decrease the opposition to our involvement [in Vietnam]," Air Force Major General Garth Dettinger said years later. In the aftermath Vietnamese doctors reported increased birth defects, stillbirths, and liver cancer. "In the abominable history of wars," wrote Vietnamese professor Ton That Tung, "have we ever seen such an inhuman fate reserved for the survivors except in the case of atomic war?" The Vietnamese reports were ignored by most American scientists because they were too difficult to validate.

Mrs. Shoecraft spoke to science classes at Eastern Arizona College. She attended herbicide teach-ins in Portland, Oregon, and in Canada, doing radio and newspaper interviews and testifying at hearings. In Portland she dropped in unannounced at the branch office of the Agriculture Department and was upset by the dearth of information on herbicides available to the public. So she held a teach-in to supply the USDA with updated material. When someone casually asked her a "So what do you do in your spare time?" type of question, she replied, "I educate myself to be the most knowledgeable individual there is on the subject of herbicides." She told an Arizona reporter that after an appendicitis attack at age nine, "They didn't think I'd live but I did. All of my time is borrowed." She seemed to thrive on this new larger-than-life persona, and some people began to resent her. She shot and killed a bull that barged onto her property acting "like he was drunk"—she thought he'd been rendered senseless by the herbicide spray—and the bull's "prochemical" rancher owner sent her a bill for $450, which Willard paid to keep the peace. If she alienated people, it hardly mattered, for she was making new acquaintances left and right. There was the woman in Malibu who also thought she'd been injured by herbicides, who sent 100 copies of Mrs. Shoecraft's book to friends and libraries. There was the woman in South Dakota who believed that pilots had purposely sprayed a field of edible millet with herbicides and who commiserated with Mrs. Shoecraft that "It is a lonely struggle I feel against the chemical companies." Robert Reynolds, the technician from the lab in California that tested the samples of her fat, even stopped by to visit her while on vacation in Arizona.

When she wasn't writing or proselytizing, Mrs. Shoecraft was answering lawyers' questions. From the fall of 1970 through the summer of 1974 she was depositioned seven times. During her first depositions, she told the lawyers about the book she was writing. "In fact you will chuckle over parts of it and parts of it you will cry over, I guess," she told them. The lack of a proper warning on the herbicide product label was one of the claims in the Shoecraft lawsuit. When Dow's local counsel asked her what *she* would have put on the label, Mrs. Shoecraft fired back: "Do not use this product unless you are aware of the hazards connected herewith, which may include central nervous depression, loss of memory, paralysis, loss of hair, psychological symptoms, needing psychiatric care. No antidote is known. Milk of the cattle may be contaminated. Period of residual properties in soil are unknown. Method of removing from contaminated water unknown. Do not use unless you are prepared to face the consequences. Toxicity studies are very limited. No human experimental studies have ever been done to completion. Research by Dow very incomplete . . . the hazards relative to what dioxin can do should be listed. This can be very toxic to dogs —I would appreciate that—and guinea pigs. They might also put on it—Can cause deformities, in very small amounts, to the test animals."

"Does that exhaust the list?" Dow's counsel asked.

"No," she replied. "I would like to see on there—This is not to be used by anyone unless he is a qualified graduate toxicologist and chemist, and that its use be forbidden for the ordinary type applicator that I have met at the moment. I would like to see put across the front of it—by prescription only."

Another time she showed up in the Phoenix high-rise office of Dow's local trial counsel wearing a minidress and "pushing a grocery cart obtained with my Green Stamps. The cart contained three tape recorders, a grocery box of unedited tapes, five empty slide carousels, and 1,069 slides in a brown Payless grocery store bag." The tape recorders were set up so she could make a recording of her own deposition. The unedited tapes were interviews she'd conducted with various government officials and scientists about herbicides. The slides were pictures she'd taken of herbicide damage and just about anything else that struck her fancy along the way. When the lawyers shut off the lights to view the slides, they found them upside down and sideways. Interspersed with the slides of one-eyed guinea pigs, deformed goats, and an eight-foot-tall okra plant without leaves were photographs of Santa

Claus and Shoecraft family friends waving from the beach. Mrs. Shoecraft's three-year-old granddaughter, who liked to help her grandmother collect specimens and take pictures, had "sorted" the slides and included some of her own favorites from the family collection while Mrs. Shoecraft was asleep one day. "I lost count," Mrs. Shoecraft later recalled, "but I believe we screened all those pictures in one and a half days and all of the boys seemed very polite and most anxious to get me, my recordings, my paper bags, my squeaky grocery cart, and my lawyer's control of the situation quietly out of their office, as they brought to a close 'The Deposition of Billee Shoecraft.' "

Mrs. Shoecraft may have been winning some minor skirmishes in her war against herbicides, but Dow was winning the major battles. An independent advisory panel reviewing the government's cancellation order gave 2,4,5-T a clean bill of health in the spring of 1971. The committee essentially recommended that restrictions on 2,4,5-T be lifted as long as it didn't contain more than 0.1 part per million dioxin. As reported by one of its members, Auburn University botany professor Donald Davis, the panel believed the only possibly sound basis for banning the herbicide would be if the dioxin contained in it entered the food chain of humans and animals. They found no such evidence and so concluded it was safe. But the advisory panel report was roundly criticized by a group of scientists and consumer representatives, who said the committee had accepted the idea that low-level doses of dioxin will not affect animals, even though scientists had never found a "no effect" dosage; had dismissed the possibility that dioxin can accumulate in body fat when there was animal evidence to the contrary; and had concluded unreasonably that the herbicide quickly breaks down in the environment.

The fracas was one of the first skirmishes on what one participant called the "battleground of opposing philosophies about the relationship between technological risk and human safety." Public-interest lawyer Harrison Wellford looked at the science advisory panel, for instance, and saw a group that reflected the classical and troublesome opinion that "the public should be exposed first and the experiments done afterwards." But Professor Davis, who served on the panel, looked around and saw a public whipped into what he considered an irrational "climate of fear" about pesticides. Davis drew a parallel with the abuses of the McCarthy era, when a politician was able to feed the frenzy of a Communist-fearing country because "the media gave complete coverage and an air of credence to his wildest accusations." As Davis put it, "We seem now to be suffering a similar paranoia concerning agricultural chemicals."

In the seesaw battle that played out over the next decade, one side would take a step forward, then backward. In the summer of 1971 EPA

Administrator William Ruckelshaus rejected the panel's recommendations and said the agency would continue to press for cancellation of 2,4,5-T on food crops. He called dioxin "one of the most teratogenic chemicals known," and he warned that even nearly pure 2,4,5-T might be a hazard to humans and the environment. The next month the federal district court in Arkansas granted Dow an injunction halting the EPA's plans. But Ruckelshaus refused to discontinue the cancellation order, saying it was "abundantly clear" that Dow had not met the burden of proof that 2,4,5-T was harmless. The agency appealed the Arkansas court decision to a federal appeals court, which upheld the EPA's cancellation plans. In the summer of 1973 the EPA scheduled a hearing the following spring to determine if all uses of the herbicide should be canceled.

In the interim, new facts about dioxin were coming to light. Doctors who had followed the medical fate of seventy-eight Czech workers exposed to dioxin in a 1965 industrial explosion reported eight years later that two workers had died from lung cancer (although they said there was no evidence the cancers were caused by dioxin). All but two of the workers had developed chloracne, twelve had liver lesions with symptoms of porphyria, and there were numerous reports of nervous, mental, and sexual disorders. Scientists also reported at a government-sponsored symposium on dioxin in 1973 that it produced multiple organ damage in animals and acted as an "extraordinarily potent teratogen" in a number of animal species. Although there was a wide range of animal susceptibility to dioxin (guinea pigs are 1,000 times more sensitive to it than dogs), studies showed that regardless of dosage, animals died weeks after a single administration of dioxin.

There still was little evidence of dioxin's toxicity outside the laboratory and the factory. In the spring of 1971, though, a salvage oil company sprayed a riding arena on a horse-breeding farm in eastern Missouri with waste oil sludge to keep the dust down. Three days later sparrows and other birds that normally populated the barn rafters were found dead on the arena floor. Over the course of the next several weeks hundreds of birds, several cats and dogs, and numerous rodents died around the farm. More than half the eighty-five horses exercised in the arena died, some of them less than a month after the spraying, some of them nearly three years later. Before they died, the horses' weights dropped dramatically, and they had loss of hair, skin lesions, intestinal colic, dark urine, inflamed eyelids and feet, and stiff joints. Several children who frequently played in the arena soil had chloracne. A six-year-old girl developed headaches, nosebleeds, diarrhea, and urinary tract problems. Tests showed the walls of her bladder were inflamed and bleeding, and portions of the kidneys were also inflamed.

Three of eight rabbits exposed to the soil died within a week. When the rest were autopsied, they had skin lesions and damaged livers. Scientists guessed from the rabbits' response that dioxin was in the waste oil sprayed on the arena. Tests showed thirty-three parts per million dioxin contamination—a level substantially higher than the one part per million industry limit on dioxin. Further investigation revealed that the sludge came from a company that manufactured hexachlorophene. A disinfectant commonly used in soaps and detergents, hexachlorophene is made from trichlorophenol. The production of trichlorophenol—which is the first step in making 2, 4, 5-T and silvex—inevitably produces dioxin by-products. The federal Center for Disease Control subsequently linked the illnesses of the six-year-old girl—who recovered uneventfully—and several domestic animals on the farm to dioxin.

In England, meanwhile, scientists in two government laboratories who were experimenting with dioxin developed the same kind of chloracne that had plagued industrial workers in trichlorophenol factories. In one instance a scientist wearing overalls and disposable gloves who was well aware of dioxin's toxicity and took pains to avoid skin contact nevertheless developed chloracne two months after his exposure. In another case a young scientist who took equally rigorous precautions noticed, five weeks after his experiments, that his skin was extremely oily. He said it felt as if his skin had been spread with melted butter. Soon after, chloracne appeared on the sides of his nose and spread to the lower parts of his cheeks, his ears, the front of his neck, his chin, and finally behind his ears and the back of his neck. The chloracne eventually subsided, but two years after the experiment, in the spring of 1972, he developed colicky abdominal pains and gas. He lost weight, had "oppressive" headaches, was exceedingly tired and irritable, his concentration diminished, and he fell into "episodes of uncharacteristic anger." Abnormally long and dark hair began to grow on his shoulders, the upper part of his back, and around his nipples. Larger, dark hairs grew in around his eyebrows. His mental and muscular coordination deteriorated so badly that he felt as though he were not fully in control of his limbs. His writing was labored; his vision, blurred. With the passage of a year and a half his symptoms disappeared, save for a raised serum cholesterol count. One of the scientist's colleagues who had not participated in the original experiments but had been exposed to a diluted form of dioxin developed nearly identical medical problems. He did not have chloracne, but he reported an inexplicable loss of energy and drive, marked loss of concentration, vague indigestion with gas, loss of appetite, and a peculiar sensation of flickering of vision. His skin was unusually oily, and abnormally long and coarse hair grew on his shoulders, on his upper back, and around his eyebrows. Writing in a British medical journal in early 1975, a doctor who examined the scientists concluded that "such profound long term effects suggest that the dioxin molecule is highly active when absorbed into the body in minute quantities."

In June 1974 Dow won its battle with the Environmental Protection Agency. The EPA called off hearings on 2,4,5-T and withdrew the cancellation order against the use of the herbicide on rice crops. The agency said it needed more time to collect evidence. Environmental groups said the understaffed, overworked agency was simply outgunned by the chemical industry and withdrew rather than take a weak case into the hearing room and lose. Russell Train, the new EPA administrator, was reported to be under tremendous pressure from the Nixon White House to go easy on regulatory matters, and congressmen from corn and cotton districts were already fuming about suspension actions against two hugely popular pesticides, aldrin and dieldrin. But industry supporters believed the agency had come to its senses and finally realized that the herbicide was in fact one of the most studied and safest of modern chemicals. By canceling hearings, the agency had in effect shifted the burden of proving that 2,4,5-T was hazardous from the manufacturer to itself. For the next four and a half years Dow made and sold 2,4,5-T and silvex with minor interferences from the courts and state legislatures, while the federal government stood out of the way.

By 1974—the year the government backed down on 2,4,5-T—Mrs. Shoecraft was seriously ill. Her chest and side pains had increased, and her eyes swelled up while she was writing her book, so she moved up to the mountaintop cabin, which sat above the sprayed area. Her health improved there. It diminished every time she came down to Ice House Canyon. She began trying to spend half her time on the mountain and half in the canyon. The week before one of her depositions in the fall of 1970, some of her eyelashes fell out and she was nauseated and vomiting. Twice during that fall she flew to Santa Monica to see Dr. Granville Knight, the elderly nutritionist-allergist who believed that her illness was herbicide-induced. "The only kind of illnesses that have put me inside a doctor's office or hospital until after 1966 were cleancut, easy-to-see things like having babies or a nail through my boot or a scorpion bite," she wrote, "but never any of those 'what's wrong with me—I don't feel good' kind of visits. None of those 'I feel so depressed I could die' kind." Dr. Knight prescribed nutritional supplements and told her to avoid sugar and food contaminated with pesticides.

Her hair thinned and came out in patches, which she covered with her blonde falls. "She would always say she had a pain in her side. That was the main thing, and she complained about losing her hair," recalled her youngest son, Robert. "She took very good pride in herself, and she had hairpieces. But I can remember a couple of times getting into her closet and seeing the hairpieces. Now that I think about it, it seems like the hairpieces got bigger and bigger as time went on." Unlike his mother, Robert hadn't been directly sprayed, and although he had a few minor problems, his health was good. Willard had been doused once by the helicopter, but his health problems were not nearly so severe as his wife's. He downplayed the ulcer that his wife said developed on his stump as a result of the spray. "That was Billee," he said. Willard made no claim for damages in the lawsuit but was joined as a plaintiff because of community property laws.

Mrs. Shoecraft traveled to California to see Dr. Knight because most of the Globe doctors were skeptical of the herbicide claims. In notes on her visit in January 1971, Dr. Knight observed that she was "greatly fatigued, and she was also gathering evidence on pesticides. She's worse. She was worse after returning to Globe. She has spells of severe weariness, moderate depression and irritability. Working hard on a book." A year later he wrote: "Hair still falling out. Extreme fatigue. Son, now 17, has wiry, kinky hair (not curly before spray) and after '69 spray, he grew seven inches in a few months time. She's had no rash this past year. Weight has dropped. No zest. Wants to just sit. Says pigeons are dying and they lay eggs on the ground. Some ducks have two extra wings." In the summer of 1972 she had "numbness of left shoulder girdle and left side and anterior left thigh increasing in frequency and extent. Recently had menstrual bleeding lasting three weeks (on returning to Globe) (already menopaused)."

She spent much of 1973 giving speeches and interviews and hearing testimony. In the summer of 1974 she gave a forty-minute speech to 200 people at a convention of the International Association for Cancer Victims in Los Angeles. But she canceled a speech she was to have given against a proposed herbicide spray project in another state because she felt so tired and ill. She called Dr. Knight from Globe. "Exposed to smoke of forest fires—marked throat and esophageal congestion, swallowing pain," he wrote in his notes. "Menstruating again. Irregular areas of partial numbness (all toes, outer aspect of legs, parts of neck). Patchy pigmentation on arms below short sleeve line. Says her cowboy (handyman) lost because of brain tumor. Animals still dying, deformed, hair falling out markedly, wearing a wig. In spite of marked fatigue and disability, she's giving talks in many parts of the country." A month later he wrote that she had "been up to campground used as helicopter spraying station and she and son became sick with nausea, vomiting, diarrhea, fatigue." Tests he administered showed traces of 2,4-D, 2,4,5-T, and silvex in her blood.

The lawsuit was transferred out of state court and into the federal district court in Phoenix in the fall of 1974. A companion claim against the federal government stayed in Tucson. The sole remaining defendant was Dow. The rest of the chemical companies and the helicopter company had made small out-of-court settlements along the way. The state water agency had been dropped from the suit on a technicality. As things began to heat up with the lawsuit, Dow went to court and got an order requiring Mrs. Shoecraft to be examined by a doctor of its choice. He was Dr. Charles Hine, the San Francisco toxicologist and industry consultant who had done some of the original research on the nematocide DBCP for the Shell Chemical Company. Dow had used Dr. Hine on and off for the past twenty years as a consultant. He was, Dow said, "a medical doctor with a specialty in clinical toxicology and is particularly qualified to examine Mrs. Shoecraft." Before the examination Dow sent Dr. Hine a copy of the court order; digests of depositions taken from Mrs. Shoecraft, the McKusicks, and Dr. Knight; Mrs. Shoecraft's medical records; the test results on her fat, blood, soil, and water; a summary of the Forest Service chaparral spraying program; the twenty-nine-page task force report on the Globe spraying; and a copy of *Sue the Bastards*.

Mrs. Shoecraft resisted mightily and finally flew to San Francisco in September only after her lawyer said her recalcitrance might cause the claim to be dismissed. Her daughter-in-law, Mari, went with her. "She had a phobia about hospitals that wouldn't quit," Mari said. "She didn't want anyone ever to be in control of her. Also, one of the very bad side effects of these chemicals which Bill was exposed to is just extreme nervousness, and you get very suspicious, and you think the world is coming down on you and you can't do anything about it."

Mrs. Shoecraft put everything on tape, and this would be no different. She secreted a miniature tape recorder in her purse and kept it covered with a scarf in the doctor's office. (She made so many trips to the bathroom—to change tapes—that the doctor remembered and remarked about it in a later deposition.) Dr. Hine told her that he was a toxicologist experienced in pesticide intoxications, and he understood she'd been exposed to a weed killer. The examination went downhill from there. "He started asking questions that had nothing to do with herbicides—how much money she made as a writer and so forth," Mari said. In deposition, Dr. Hine said Mrs. Shoecraft was two and a half hours late for her appointment, that it took four and a half hours to extract a history from her, that she refused to be hospitalized, refused a fat biopsy, refused X rays, and refused a gynecological exam. Fur flew; attorneys were called. Mrs. Shoecraft located a gynecologist in San Francisco to take a Pap smear. "She was crying to me in the hotel room," Mari said. "She didn't want Hine to have anything to do with her private parts. He had dirty fingernails, and his suit looked like he'd slept in it."

Dr. Hine seemed as put out with Mrs. Shoecraft as she was with him. "Taking the history was the laborious, difficult part," he said in deposition, "and she just simply did not respond or wanted to know why the question was being asked, what the relevance of it was, and what my experience was in that matter, and it was a very difficult situation." He asked her what she did for a living, how much money she made the previous year, and what organizations she belonged to. He asked her if she was an active environmentalist because he believed that would play a part in her attitude toward her sickness. She said she was not.

During the examination Dr. Hine noted that she had an approximate hair loss of 50 percent, with thinning and areas of baldness and severe patching. At one point during the exam Mrs. Shoecraft told the doctor that she believed chemical agents should be thoroughly tested before they were released on the market and that all conditions of their application should be understood before use. "She was very sly when she was saying it and she knew that she was saying it for a purpose to me," Dr. Hine recalled. "She was attempting to establish her position with me on that basis."

Dr. Hine extracted ten cubic centimeters of blood from Mrs. Shoecraft. It was deep-frozen and shipped airfreight, twelve days later, to Dow's Ag-Organics Residue Research Laboratory in Midland for analysis. When questions were raised about the propriety of sending blood from a plaintiff to a defendant for objective testing, Dr. Hine declared that Dow had the only laboratory capable of doing chemical analysis of trace concentrations in the blood. He said the blood could not be sent to two California labs that had previously analyzed trace concentrations of chemicals in Mrs. Shoecraft's blood and fat because they "were already involved." (Dow took more than a year to complete its analysis of Mrs. Shoecraft's blood sample. No traces of the herbicides or dioxin were detected.)

"My diagnosis," Dr. Hine wrote, "is that, one, she has a delusion which is a false belief... and secondly, she has a great deal of psychosomatic complaints which in my opinion arise from her ... and are associated with and caused by her delusion. She had no other pathological thought processes than her delusion. She was oriented in time and space and she didn't think she was Queen of the May and she didn't think she was in a sphere circling the earth. But she had a persistent delusion." He concluded that Mrs. Shoecraft's health problems were associated with "degenerative diseases of her age group" (she was then fifty-three), which included arteriosclerotic heart disease, calcification of the aorta, and degenerative disease of the spine. Those diseases accounted for part of her complaints. Some were due "to the anxiety which she has as a part of her delusion."

Mrs. Shoecraft was so distraught after Dr. Hine's examination that she flew down to Santa Monica to see Dr. Knight. Willard flew over from Globe to be with her. "She was in a frightful state," he said. "I believe there was no earthly purpose for sending her up there other than harassment."

Already thin, Mrs. Shoecraft lost another seven pounds in the nine months between her fall 1974 and summer 1975 visits to Dr. Knight. "Lost weight," he wrote. "Is living mostly above the area where her house was contaminated and where things grow nicely without any evidence of herbicide poisoning. Whenever she gets down into her house she gets symptoms. She's having menstrual bleeding and is wearing an emergency plastic diaper (for incontinence). Hair is still very sparse, dry, stiff and abnormal and of course she must wear a wig in order to be respectable. Still complains of hoarseness, roughness in throat. Difficulty in swallowing, sore throats, fatigue, depression. There is no reasonable doubt but that pesticides have caused her troubles."

In the late spring of that year Dow brought new defense counsel on board for the Shoecraft case. He was Rudolf H. Schroeter, a partner in the Los Angeles firm of La Follette, Johnson, Schroeter & De Hass, which had represented Dow in product litigation in southern California and other areas for the past dozen years. Schroeter was a diminutive man with a small mustache and impeccable taste in clothes who played concert piano and was an expert on herbicides and their effects. He charmed the plaintiffs; he charmed the plaintiffs' lawyers; he even charmed the nearly uncharmable Bob McKusick, who affectionately called him weaselface and Herr Schroeter. "Just a gentleman, a pure gentleman," said Mary Lou Gardner. "He had to come to Globe once, and it was his wedding anniversary, so he brought his wife with him. They were staying over at the Copper Hills Motel, and we sent them champagne and a fruit basket.

"I told him once that I hoped we'd meet again under better circumstances," she continued. "He was such a good lawyer. He was ten times better than any attorney we had. We forgot that he was coiled and ready to strike."

Mrs. Shoecraft had driven one attorney from the case, and now she had a falling-out with her new attorney and fired him. She was increasingly impatient with court delays. She seemed to have, as her friends pointed out, a premonition of time cut short. Willard thought the San Francisco fiasco was the "trigger"—she had been slipping ever since the trauma of examination. In the spring of 1976 she made an appointment to see Dr. Paul Singer, a Phoenix urologist whom she had known for thirty years. He was very fond of her, foibles and all. "You never have to ask Billee any questions," he said. "She would tell you. She was a very vivacious, very active and very intelligent person . . . who was not herself that day." He found a nodule in the twelve o'clock position above the nipple on her left breast and a nodule in her left armpit. She had a hard uterus and was bleeding from the cervix.

Dr. Singer sent Mrs. Shoecraft to see a Phoenix surgeon that same day. He found small cysts in both her breasts, a fairly common condition in women. But X rays showed a suspicious area in her upper left breast. He recommended she undergo a biopsy with immediate surgery if there was a malignancy. Her Pap smears had come back negative, and he thought the breast problem the most pressing concern.

She put it off. She was acting as her own attorney now and wanted to petition the court to speed up her trial date. In May Dr. Knight wrote a "To Whom It May Concern" letter. He said she was suffering from rapid physical degeneration. Spine X rays indicated that several vertebrae had collapsed. She also "apparently has developed a large fibroid of the uterus in the past six months or so," which indicated "an urgency which has not heretofore been present as far as the time of the trial is concerned. If justice is to be done," he concluded, "time is of the essence."

Mrs. Shoecraft and Mari went to Tijuana, Mexico. "She had a tumor in her left breast, so I said, 'Well, I'll go with you to Tijuana,'" Mari recalled. "We went to a clinic, and she got a shot of enzymes and a shot of Laetrile. The enzymes act as a catalyst for the Laetrile. She got them one a day for ten days. I took them with her. My husband doesn't even know that. Tijuana was a zoo, but people had received very good results from these injections. We brought back two weeks' worth of Laetrile with us. But she still had something in her uterus, and I was more concerned about that."

In July the pain in her side became so severe that Mrs. Shoecraft finally surrendered to exploratory surgery. Mary Lou Gardner fixed her favorite meal the night before—shrimp and vegetables. "Bill never ate," Mrs. Gardner said. "She was just one of those people. So we traded off: She'd do artistic things, and I'd cook for her." In surgery the next day Dr. William Bishop, the family doctor who had delivered her son Robert, was so disturbed by what he found upon opening her abdomen that he left the operating table and went out to the waiting room to talk to her family. Everywhere he felt there were tumors—in her abdomen, in her liver, in her pelvis. Dr. Bishop told Willard, Mrs. Shoecraft's sister, and Mary Lou Gardner that it was impossible to operate. "I can't tell where the cancer starts and where it ends," Mrs. Gardner remembers the doctor saying. "She's just full of it." Biopsy showed that cancer had widely invaded her ovaries. A sample of fatty tissue sent to GHT Laboratories in California revealed traces of 2,4-D, 2,4,5-T, and silvex. Lab director Robert Reynolds had been testing Mrs. Shoecraft's soil and plants and fat for seven years now and he had grown so fond of her he didn't even send a bill for this latest analysis. Instead, along with the chemical results, he sent her a bouquet of flowers and a note that hoped she was "on the road to being completely well" and told her about the "wonderful, new gas chromatograph" the lab was buying.

"I'm not going to look into your eyes," Mrs. Shoecraft said when Mrs. Gardner came into her room after surgery to tell her she had cancer. "I'm going to beat this," she told her friend.

Four days later the Shoecrafts left Globe in search of Laetrile. "We were frankly looking for anything that would help," Willard said. "We just kind of went into a panic situation from July 5 on and packed up everything and moved to San Diego. We went to Mexico first, and she almost died down there. It was bad. She had just gotten out of surgery with this enormous incision in her abdomen, and she was in very sad shape, and that first night in Mexico was bad. We got her out of there and moved to San Diego and put her in a motel the next day. We were there about two months."

On July 10 an explosion in the reactor at a 2,4,5-trichlorophenol factory north of Milan, Italy, vented a white cloud up into the sky and across the nearby community of Seveso. The cloud carried in it several kilograms of dioxin. People who were outside when the cloud descended had burnlike sores on their bodies, and they developed headaches, dizziness, and diarrhea. Within two days scores of small animals died. On autopsy most showed extensive liver damage. Pregnant women petitioned doctors for therapeutic abortions, and others obtained them illegally. Seven hundred people were evacuated from their homes. It was a mass human experiment in dioxin contamination, and scientists from all over the world watched Seveso closely in the aftermath of the explosion.

Willard rented rooms at the Bahia Motel in San Diego. From there he made regular runs to Mexico and smuggled Laetrile back across the border. He gave his wife three shots a day. Mari was there with her for a while, and her children came to see her. She had good days, and she had bad days. Mari's husband, John, spent a week with her. He and Mari lived half the year in Hawaii, and he put Mrs. Shoecraft's granddaughter on a plane and sent her to San Diego. The Gardners came over from Globe several times.

"She had a good time in San Diego," Mrs. Gardner said. "It gave her a well feeling. I dragged her through Old Town—we played *turista*—and we took her on a boat called *The Bahia Bell*. We went through Ramona's Marriage House, which is an 1800's Spanish house. She lay down on the grass in the park and said, 'I can smell the chlorophyll. It's helping to cure my body.'"

In their room at the Bahia Motel the Shoecrafts had other kinds of visitors. "She was quite well known in this business of chemical sprays and cancer . . . and a lot of her friends gravitated to her over there and suggested other cures or medicines," Willard recalled. "It seems there are hundreds of people who suddenly come to you and they've got the cure . . . it's everything from Zen to a religious thing to a diet. Take your choice."

The Shoecrafts visited "Edie somebody," who "cured herself" with something called, "believe it or not, wheat grass juice. It grows high and green and comes in plots and costs five dollars a plot, and you cut it off with scissors and buy a machine that costs five hundred dollars and put grass in there and out comes this real thick green juice . . . and they swear by it. Then there's watermelon juice." They visited a couple of quacks: "One fellow kept coming down from Los Angeles. He didn't have a phone. He kept coming down with an old, dirty milk bottle with what looked like tea in it and kept insisting that it would make her well. He would knock on the door at three in the morning with the bottle of tea, whatever it was. We finally ran him off. We saw a lot of people, enough to drive you nuts."

By September, her health no better, the Shoecrafts returned to Globe. She bounced back and forth between Mari's house in Phoenix and Mary Lou Gardner's house in Globe. It was that fall at Mrs. Gardner's house that she met Dow attorney Rudolf Schroeter for the first time. "Bill read the Bible every day, and we were just reading about 'Love Thy Enemy' when Mr. Schroeter was coming to meet us," Mrs. Gardner said. "She said, 'We can kill him with kindness.' So we cooked up a storm, and she got dressed in all her Indian jewelry; she looked brilliant, even though she was so sick, she could hardly walk. She could put up such a good front. We had potato salad, shrimp, and deviled eggs. Bill said, 'Usually they bring the food over after the funeral. How wonderful that I am going to get to eat at my own feast.'

"After the meal, Bill was getting weaker, and she decided to take a hot, hot bath. She pulled a towel over herself and had Mr. Schroeter sent in," Mrs. Gardner said. "When he came out, he was in tears. He could see the pain. He said, 'If anything has caused this pain, I hope you win.' That was the first time and the last time they met."

Of Mrs. Gardner's account, Schroeter said: "Certain facts are correct; half is totally incorrect." He had met Mrs. Shoecraft once before. At this second meeting in her home, he ate lunch with Billee and Willard. Mrs. Gardner was not there, although she had prepared the meal. Schroeter did sit on the edge of the bathtub to talk with Mrs. Shoecraft. "It certainly was a moving experience, I will concede that," he said. "But I don't cry very easily, and I did not cry then." Further, he did not say that he hoped the plaintiffs won the lawsuit. He said he believed, as did doctors, that Mrs. Shoecraft's cancer was "spontaneous" in origin and was not caused by exposure to herbicides.

The Shoecrafts owned some rugged land north of Phoenix in a place called Agua Fria, which is Spanish for "cold water." Mrs. Shoecraft decided she wanted to build a house in Agua Fria so she could get away from the hated house that made her feel so sick in Ice House Canyon. "Even then we knew that she couldn't," Willard said. But he and John drove up to Rock Springs, a small town near Agua Fria, and "bought a little house. We convinced her we would live in the house and then drive over to this site," he said, "and we could get a contractor and we would go over there every day and build the house. But there was no way this was ever going to happen. We got up to Rock Springs, and we moved all the dishes and everything out of the house, and we got up there, and she was able to be there about one day."

They came back to Globe. Mrs. Shoecraft said she knew a doctor in Salem, Oregon, who could give her Laetrile. "So that night we were heading for Oregon," Willard said. "You know, if she said she wanted to go to Europe, we would have gone to Europe."

It was just a few days before Christmas 1976. They rented a motel suite in Salem and headed directly for the doctor's office. He was accustomed to finding people like Mrs. Shoecraft on his doorstep. They came long distances, hoping that he "would have some kind of miracle for them, and most generally I don't." He looked at Mrs. Shoecraft's small, wasted arms and legs, her distended abdomen, her swollen thighs, and understood, even though she had "rambling conversations" that she was having "timetable pains," similar in force and frequency to labor contractions. The doctor drained two quarts of fluid from her abdomen to make her more comfortable. She said, "I was exposed to cancer-causing chemicals in 1969." He took her blood pressure; it was 90 over 60—very low. He counted her pulse; it raced at 110. Her skin was dehydrated and felt like parchment. He gave her three vials of Laetrile intravenously and a pain pill to help her sleep.

The doctor went to her motel room the next day and the day after that. He gave her something called "supplemented orally hydrolized protein," and on Christmas Eve Mrs. Shoecraft said she felt considerably better. "Somewhat sassy and argumentative" was the doctor's observation. He gave her more Laetrile and diet instructions, and on Christmas night Willard phoned him at home because she was "hurting very badly." The doctor rode his motorcycle sixty miles through the sleet to give her another shot of Laetrile.

Before dawn broke on December 28, an ambulance took Mrs. Shoecraft from her motel room to Salem Memorial Hospital. The doctor believed from the amount of fluid and the size of the mass in her abdomen that she had ovarian cancer. "In the worst way they didn't want to get into a hospital," the doctor remembered. "She was kind of a tough little gal, and she wanted to tough it out if she could."

The receiving surgeon at Salem Memorial's emergency room saw a woman brought in who was moribund, or approaching death. He gave her morphine and other sedatives. Willard stayed near as the Shoecrafts' twenty-ninth wedding anniversary passed with the eve of the new year. "I simply kept her comfortable for the week that she lived," the doctor at Salem Memorial said. "She died the sixth."

The death certificate listed cause of death as malignant tumor in the ovaries and carcinomatosis, or the spread of cancerous tumors throughout the body.

Her family and the Gardners flew over Agua Fria, dropping seven yellow roses and her ashes on the hard, rocky ground below. "You know, she didn't like flowers in her house," Mrs. Gardner said. "If you wanted to really aggravate her, you'd send her flowers. She said, 'I am not going to be put under a bunch of dead weeds. When you look up in the sky and see a sunset, that's where I am.' "

Letters of condolence came to Willard in Globe. Some were from what might be politely called her outer-fringe friends. "Got down to Arkansas and all hell broke loose," one of them wrote. "Gypsy and black witchcraft. Tried their best to eliminate me. All my mail is opened." A doctor from a Tijuana clinic asked, "So what do we do now, without her?" He enclosed a story from the *National Enquirer* headlined PLANTS ARE HELPING WIN FIGHT AGAINST CANCER, STROKES AND OTHER DISEASES. Dr. Knight sent an anguished note. "Billee was a rare blend of poetess, dreamer, lover of beauty and nature," he wrote, "coupled with a fierce determination to see justice done and to expose to the impersonal, blinding light of pure justice those whose little selfish minds would destroy mankind (if necessary) for a mess of potage." A woman fighting her own lawsuit against herbicides in California told Willard that "Billee catalyzed efforts all over the world to band citizens together against trespass by toxic chemicals."

In the years following Mrs. Shoecraft's death the temperature of the herbicide debate climbed upward. Pressure was applied from two directions: pregnant women in the Pacific Northwest who believed they were miscarrying in abnormally high numbers after the spring spraying season, and Vietnam veterans who wondered if their medical problems might not be related to herbicide exposure during the war. A local television station in Chicago aired a documentary, *Agent Orange: Vietnam's Deadly Fog,* in the spring of 1978 to tremendous viewer response. In Globe, Bob McCray described the "real heavy fog, just like a mist out of a spray nozzle" that descended on his family in 1969. The

McKusicks were interviewed among their menagerie of deformed animals and plants. "It will never again produce," Bob McKusick said as he pointed to a dead algerita. "It is sick, and it is like a lot of the animals and people and things in the area—they are in a living death in many cases." University of Wisconsin scientist James Allan described the increased number of tumors found in rats that were fed minute quantities of dioxin. "We found that low-level consumption even as low as five parts per trillion of dioxin in the diet was capable of causing an increased incidence in tumors in experimental animals," he said. Dr. Irving Selikoff of New York's Mount Sinai Hospital, one of the first to uncover the hazards of asbestos to workers, explained that whereas asbestos had a delayed effect, dioxin did not. "With dioxin," he said, "we do see things fairly early on. For example, our colleagues in the laboratory have shown us that quite rapidly we find effect on blood cells. We find effect on the liver, we find effects on the enzymes. You find effects on the immune system and these are unusual. At the same time, it is difficult to categorize these vague symptoms at the moment. When somebody gets an effect on memory, how do you quantitate that? What kind of statistics do you have when somebody can't remember where he has left his car? Or he has to stop work because he does not remember which exit to take on the highway, or when his sleep is now badly disturbed or where he can't concentrate or where he is now irritable all the time."

As the dioxin debate went national (ABC's *20/20* ran a two-part series that summer), industry and its support groups worked valiantly to get their side of the story out. The Council for Agricultural Science and Technology (CAST), an influential voice on Capitol Hill, proclaimed itself an impartial research organization, but it clearly had a proindustry bias. Among the corporations and trade groups paying up to $6,000 a year to bolster CAST's budget were Dow, American Cyanamid, Monsanto, and Union Carbide. "CAST operates in an unorthodox fashion," the magazine *Bioscience* reported, "routinely producing scientific consensus statements on controversial issues with a force and speed rarely seen in other scientific organizations." The magazine reported that during a round of media interviews with the authors of a pro-2,4,5-T CAST report, an Associated Press reporter determined that Dow was not only sponsoring the meeting but was paying for the services of the woman publicizing it. Biases aside, CAST filled a necessary place in an increasingly lopsided media dialogue that was too often dominated by antipesticide voices.

The contention of CAST, the industry, and a not unsizable number of scientists was that the hazards of dioxin were overstated. Apart from health problems in chemical plants, where workers were exposed to abnormally high levels of dioxin, there were no documented deaths and remarkably few human illnesses in the thirty-year history of herbicides. The health effects in the Missouri horse arena and in Seveso, Italy, were caused not by herbicides but by large amounts of dioxin in waste sludge and in an industrial release. Although dioxin appeared to be a carcinogen in rats, they said it was only a weak teratogen, and the scientific evidence did not support allegations of spontaneous abortions, miscarriages, and other reproductive problems. Numerous tests had proved that dioxin did not build up in the environment but was degraded by sunlight or bound up in soil particles, where it decomposed over a period of years. Herbicide use in the United States deposited no more than one ounce of dioxin across 5 million acres a year. A sheep with a guinea pig's sensitivity to dioxin would have to eat all the vegetation on 9 acres of land to get a lethal dose.

Dow threw a new wrench in the works when it issued a report declaring that dioxin was produced not by the chemical industry alone but by power plant stacks, car exhausts, home fireplaces, and charcoal grills. The company contended that dioxin was formed in trace amounts in all forms of combustion and had been in the environment "since the advent of fire." Coming on the heels of the discovery of dioxin in the Tittabawassee River, which runs through Dow's Midland plant, the report was viewed with some skepticism. "If you were standing outside Dow Chemical Company looking in at us, you'd say, 'Aha! You've got dioxins in the Tittabawassee River, and therefore they are coming from 2,4,5-T, trichlorophenol and pentachlorophenol. The way they are getting into the river is you are discharging them in the waste streams, they are going through your collection pond, and they are coming out into the river," Dow's Dr. Robert Bumb told *Chemical and Engineering News.* "And I've got to admit that makes good sense if you are standing outside and looking in. If you are inside Dow, you wouldn't say that. We monitor those plants very, very closely and we didn't think at the time the dioxins were coming from those plants." Dow theorized that the dioxin in the Tittabawassee River was from combustion sources at its plant, and not from discharges from the herbicide plant. Tests on soil and dust from rural and urban areas in Michigan showed dioxin was everywhere, although most of it was one of the less toxic members of the big seventy-five-member dioxin family.

Jerry Sullivan, a Phoenix lawyer with the hurly-burly desk-thumping air of a Brooklyn DA, took over the Globe lawsuit just before Mrs. Shoecraft died. He inherited a case that had practically nothing going for it. Few expert witnesses had been lined up to testify about the hazards of dioxin at the trial. There was only $10,000 left from various out-of-court settlements to spend on hiring witnesses and running the case. The plaintiffs and their complaints were widely thought to be sort of kooky. "The climate in Arizona then was that this case was ridiculous, the people were all crazies, and herbicides didn't hurt people," Sullivan recalled.

Sullivan examined the facts: Healthy people got sick, their animals died, and their plants dried up after they had been sprayed with herbicides. "I thought, at least I have a prima facie case here. So I said, 'Hell, let's go. Bring the people in.'" He met with the McKusicks, the McCrays, and the Satamas in Phoenix. "I'm not an environmentalist. I met these people, and I really wasn't convinced, but after listening to them, I thought what had happened was just atrocious," Sullivan said.

Mrs. Shoecraft called him up and "interviewed" him on the phone. She was tough on lawyers, but Sullivan passed muster. He knew that she was dying from cancer and probably wouldn't live through the trial. He asked her to film a video deposition. She refused. "She was an independent person," Sullivan said. "She was convinced she was not going to die." She told Mrs. Gardner's husband that she was too "indispensable" and had "too much unfinished business" to die. "The one flaw in Bill's program was that she had all the knowledge," agreed Mari Shoecraft. "She never thought that she would die." Mrs. Gardner was flabbergasted when she heard that their new lawyer wanted to put Mrs. Shoecraft on videotape. "I said, 'Are you insane? Do you want to pick her bones?' My hands were sore from the pain of her grip. I thought, 'What a vulture' and hung up on him. But he was absolutely right." Sullivan said, "When she died, it just broke everybody."

Sullivan got on the phone and started calling scientists. "I don't have dime one," he told them. "I'm just a lawyer out here in Arizona, trying to get someone on our side. I understand you have opinions about dioxin." He called a lot of people who wouldn't get involved in the case. But he slowly built up a group of experts who were willing to lend a hand. Dr. Arthur Galston, the Yale plant pathologist who had visited Globe previously, agreed to testify on effects of dioxin. Dr. Patrick O'Keefe, from Harvard University's chemistry department, would testify on dioxin's toxicity. Jacqueline Verrett, the Food and Drug Administration biochemist who had testified at the Hart Senate hearings in 1970, would talk about dioxin's fetus-deforming effects in chicken embryos. They were good, but were they good enough to go up against the six medical doctors, two veterinarians, eight herbicide experts, and twenty-two of their own scientists and medical people whom Dow planned to put on the stand in Phoenix? Sullivan knew he needed some respected science and medical people to talk about the human health effects of dioxin. He was leery of Dr. Knight, who "wasn't in mainstream medicine." He called up the scientist at the University of Wisconsin who had been running studies on the effects of dioxin on rats and monkeys.

Dr. James Allan and his colleagues had fed eight female rhesus monkeys a diet containing just 500 parts per trillion dioxin for nine months. After the first three months the monkeys developed dry, scaly skin and swollen eyes, and they lost their facial hair and eyelashes. By the sixth month most were anemic and underweight and had reproductive disorders. From the seventh month on five monkeys died. At autopsy, scientists found that a breakdown in the monkeys' blood-forming organs and a resulting decrease in the number of blood platelets had caused widespread hemorrhaging. There also was an unusual increase in the ductal tissues throughout the bodies, fewer than normal numbers of cells in lymph tissues, and inflamed stomachs. Moreover, the bile ducts and gallbladders were stretched beyond normal dimensions. "The data presented in this report indicate that profound cellular alterations are induced in many tissues following the ingestion of minute concentrations of [dioxin] by primates for 9 months," Dr. Allan wrote. The areas that warranted special attention in animals, including humans, exposed to dioxin were the blood-forming organs, the immune system, and the reproductive system. Abnormal cell change "suggests a carcinogenic action," he wrote. In similar experiments on rats, Dr. Allan and his associates found that all the animals exposed to dioxin levels of fifty parts per billion and above died within a month. Levels as low as five parts per trillion produced tumors. The liver was the primary site of dioxin localization. The scientists suggested the possibility that dioxin was a tumor promoter, rather than a direct carcinogen. In other words, it put stress on the cell in some way so as to make it susceptible to cancer.

With Dr. Allan on board and the rest of his experts locked in, Sullivan felt confident. "I thought an Arizona jury would sock it to 'em," he said. But in the middle of everything Sullivan ran out of money. He knew a firm in San Francisco that was big in products liability law. He called his clients. "You guys are broke," he said. "We need some dough. I'm gonna fly to San Francisco and talk to one of the big guns." He made a deal with Cartwright, Sucherman, Slobodin & Fowler. If the firm would finance the case and give Sullivan an office, it could have his fee. The firm agreed. "Sullivan was a male Billee Shoecraft," Mary Lou Gardner said with obvious affection. "He went to San Francisco and peddled our case." Some of the other plaintiffs preferred Jack Slobodin, a partner in the San Francisco law firm who came on the case with Sullivan. "Jerry was too up and down," Bob McCray said. "I think he gave us very poor advice." Like Sullivan, Slobodin said he was not an environmentalist. "In the beginning, I had doubts," he said. "What convinced me was going down to Globe and seeing those animals with the deformities and the birth defects. I knew that would convince a jury."

In his trial brief, Slobodin said twenty-one people had been exposed to a herbicide containing dioxin, "the most poisonous substance synthesized by man, the most poisonous small molecule known to man, and the second most poisonous substance known in the world." When silvex was registered with the federal government in 1954, Dow had "not done research on the long-term effects of exposure." In a 1964 factory accident Dow discovered that dioxin was a contaminant in 2,4,5-T. Dow tried to reduce the amount of dioxin in its herbicides but did not tell the government about it and continued to claim, as it had in a 1964 press release, that 2,4,5-T "was absolutely non toxic to humans or animals." Slobodin said Dow would assert in trial that the amount of dioxin in the herbicides sprayed over Globe was 0.15 part per million, "but the concentration has been estimated to be much higher and is not known." If dioxin was as toxic to humans as it was to guinea pigs, he said, "one gram [about the size of a pill] would be enough to kill more than 10,000 men, and a medicine drop could kill about 1,200 people." He said the label on the Dow silvex sprayed over Globe did not warn about dioxin. He said Dow did not warn the Forest Service about the dangers of spraying the herbicide in populated areas. He said the plaintiffs were suing for wrongful death, physical injuries, fear of cancer, and fear of birth defects.

Dow responded with a historical review of herbicides and their importance to agriculture. Rudolf Schroeter wrote in his brief:

> As an inevitable incident of manufacture, both 2,4,5-T and silvex contain trace amounts of a very toxic compound known as [dioxin]. Like other compounds—from aspirin to zinc—[dioxin] in sufficient quantities can produce . . . adverse health effects. However, there is so little [dioxin] in these herbicides that in an agricultural or forest management setting such as here, people can be and are exposed only to the tiniest fractions of the kind of exposure which even laboratory animals tolerate without effect. Similarly, the herbicides 2,4,5-T and silvex themselves have been well studied and found to present no more than the minor health hazards against which the customary and government-approved labels [such as Dow's in this case] warn.

Schroeter said the Dow workers exposed to dioxin in the 1964 factory accident developed a "dermatological phenomenon" known as chloracne, but follow-up studies showed they suffered no other medical problems.

In Washington, meanwhile, the Environmental Protection Agency was emerging from its long silence on dioxin. In order to activate the agency's Rebuttable Presumption Against Registration (RPAR) process—which Dow had once described as a witch hunt in a seven-page manual called *A Trojan Horse Named RPAR*—there needed to be a "trigger." An RPAR was the first step in taking a chemical off the market. Once a chemical was targeted for RPAR, the manufacturer put together a case showing how the product benefited society. The government and groups opposed to the chemical tallied up its risks. An administrative law judge weighed the information and decided whether the chemical's registration should be canceled. It was usually a long process, and the chemical could continue to be sold until the judge issued his opinion and the EPA acted on it. The process could be triggered if a chemical was proved to be acutely toxic, like mustard gas; or if it was shown to be harmful to people who were regularly exposed to the chemical for a certain period of time; or if it was proved carcinogenic. By late 1977 the EPA believed it had a case against 2,4,5-T on the basis of the third trigger, or carcinogenicity. There was evidence of tumors in animals exposed to dioxin-contaminated 2,4,5-T in three laboratory studies, including a two-year experiment run by Dow during which rats fed dioxin developed liver and lung cancer. In the spring of 1978 the EPA issued an RPAR against 2,4,5-T. The herbicide continued to be sold while both sides prepared their case for hearing. In the interim the agency received new information about abnormally high numbers of miscarriages that were occurring after forests around the small community of Alsea, Oregon, had been sprayed with 2,4,5-T. In a controversial report that was heavily disputed, scientists studying the area determined there was a statistically significant relationship between miscarriages and the spraying season.

On January 8, 1979, New York attorney Victor Yannacone filed a class action suit on behalf of all Vietnam veterans against Dow and four other chemical companies that manufactured dioxin-contaminated herbicides for use in Vietnam. The veterans were suing not only for their own health problems but for deformities in their children. Denying that the herbicide was responsible for any health effects in the veterans, Dow countersued the federal government and tried to have the class action suit dismissed on the ground of derivative sovereign immunity—which is to say, a government supplier enjoys the same immunity from lawsuits as does the government itself.

On March 1, 1979—with the start of the annual spring spraying season just weeks away—the EPA took the most drastic measure it could and ordered an immediate emergency suspension of all major uses of both 2,4,5-T and silvex. "Studies completed only days ago show a high miscarriage rate immediately following the spraying of 2,4,5-T in the forests around Alsea, Oregon," announced EPA Deputy Administrator Barbara Blum. "This alarming correlation comes at a time when seven million pounds of 2,4,5-T are about to be used to control weeds on power line rights of way and in pastures, and to manage forest lands across the nation." The only exceptions to the emergency suspension were applications to rangeland and rice crops.

In Midland, Dow was bitter. "We are shocked and disappointed . . . but not surprised," declared Dow's Dr. Etcyl Blair, the research chemist who had ascended to a vice-president's office, "as this is just another example of 'government by decree.' We can only conclude that the action is a political move to subvert the agency's own scientific review process. This is an example of government at its worst. . . ."

Dow failed to win an appeal of the emergency suspension, and cancellation hearings began in Washington. Environmental groups said Americans were being used as "Dow's guinea pigs." Dow called it Chemical McCarthyism. "Is scientific integrity going to triumph over witch-hunting, or is it going to fail?" Dow's Cleve Goring, who managed the company's global development of herbicides, declared in a television interview. "We think 2,4,5-T is a very important symbol. If we were to lose on this issue, it would mean that the American public has been really taken back a couple of hundred years to an era of witch-hunting, only this time the witches are chemicals, not people. So that's the importance of the issue. It certainly isn't the importance of the product or the size of the product to the Dow Chemical Company." He said the herbicide represented less than 1 percent of Dow's $7 billion annual sales.

In a letter he wrote to the Reverend Robert E. Roos, a disgruntled stockholder, Dow chairman of the board Earle B. Barnes said:

> There is no product that we manufacture that we have more toxicology and health data on than 2,4,5-T, and we consider it extremely safe. Unfortunately for the product, the U.S. government requisitioned its use in the Vietnam War to defoliate the jungles in the search for the Viet Cong. It, therefore, became a symbol of the Vietnam War that some people have become obsessed with destroying, along with anything else related to that unfortunate experience.
>
> The hostility raised toward this product by the aftermath of the Vietnam War gave it a high profile. This was picked up by a lot of the extreme activists among the environmentalists who are opposed to using any kind of chemicals for agricultural purposes. . . .
>
> These activists have learned the trick of the Hitler-type

propaganda in Germany; that is, if you tell a lie often enough people will begin to believe it. The news media have been willing channels for certain quacks who make claims not backed by any data, because they are not concerned with the truth and their only intent is in doing away with agricultural chemicals. Because 2,4,5-T is one of the oldest and safest ones known, if it can be destroyed, the newer ones for which there is less data can be more easily destroyed.

There is one other piece of information you need to know. There is a flourishing business in northern California and other northwestern states in growing marijuana in open spaces and in the forests. . . . Marijuana is very readily destroyed by 2,4,5-T and the U.S. Forest Service has been using it to kill underbrush, and the marijuana is also killed. So, Reverend Roos, there is a lot that doesn't always meet the eye. . . . There are some very strong forces combining to do away with our agricultural chemical business.

There was little evidence to link the herbicide protesters with marijuana growers. "The connection made between the two groups seems based primarily on generalizations over the 'kinds of people' who protest and those who grow marijuana," conceded Jay Heinrichs, the associate editor of *American Forests* magazine. But the chemical and forestry industries were not convinced. Forestry lobbyist Robert Matthews declared marijuana money "the major means by which the anti-herbicide movement is continually updated, motivated and financed, and probably to the complete ignorance of the typical anti-herbicide advocate." Timber companies were particularly distraught over the loss of 2,4,5-T and silvex, which they sprayed on scrub oak and other broad-leaved weeds that deprived pine seedlings of sun. Crown Zellerbach estimated its costs tripled when the herbicides were banned.

The bitter confrontations of the napalm days returned to Midland. At Dow's 1980 annual meeting in the Midland Center for the Arts a group of religious investors introduced a resolution requiring the Dow board of directors to review the safety of the herbicides and to justify their continued production economically. At the front door, stockholders were handed small Dow cards that asked, "What is a part per billion? How big is a part per trillion? Well, if you're 32 years old, the equivalent of a part per billion would be 1 second out of your 32 years. A part per trillion is 1 second out of 320 centuries!" Dr. Samuel Epstein, professor of occupational and environmental medicine at the University of Illinois, spoke in defense of the resolution. He told Dow's 2,400 shareholders that dioxin was present in 2,4,5-T and silvex in concentrations of about 100 parts per billion; that dioxin produces birth defects and reproductive abnormalities at the part per trillion level; and that it was the most potent known teratogen and the most potent known carcinogen. "As overwhelming scientific evidence on the hazards of 2,4,5-T and dioxin has accumulated in the last decade, Dow has attempted by every possible means to challenge and discredit them, and to insist on unattainable degrees of scientific precision," Dr. Epstein declared. "Dow has repeatedly challenged the relevance of experimental animal data to humans, although Dow continues to insist on the acceptance of negative toxicological data on its own products as evidence of their safety."

The resolution was defeated by a vote of 131,264,537 shares over 7,650,537.

Shoecraft et al. v. *Dow* was scheduled for trial in October 1980. Then it was delayed until March 1981. Jerry Sullivan was fired up and ready to go. He was not convinced that he could link Mrs. Shoecraft's ovarian cancer to dioxin exposure. But that was "just an added wrinkle." He believed this case had a *purpose*. He told the people from Globe, "No one is ever gonna have a better chance against Dow than you people." The experts were lined up, the exhibits ready, the voir dire questions typed out for the jury. "Does any juror believe that the state of nature must give way to industrial needs?" the plaintiffs wanted to know. "Have you ever felt critical of the chemical manufacturing industry?" Dow wondered. "Or have members of your family or community expressed privately or publicly criticisms of the use of chemicals in home agriculture and the environment?"

Jury selection was slated to begin on Monday, March 2, before a visiting federal judge from St. Louis. Dow had made a settlement offer, but the plaintiffs rejected it. "The weekend before trial Jack [Slobodin] and I went up to Globe and had conferences with every family," Sullivan said. "Dow had offered a lot of money, and it was still going up. Everyone wanted to settle. The Cartwright firm had eighty-five thousand dollars in the case. Slobodin asked me what I thought. I told him, 'You know damn good and well that Billee Shoecraft wouldn't ever talk settlement. This case has a purpose. We should go on with the case.'" The judge asked the lawyers from both sides to eat breakfast together the Saturday morning before the trial. Still, there was no agreement. "I said, 'Fine, you want to settle, give us three hundred million dollars,'" Sullivan said. "'Give us all the money you made on this herbicide from the beginning.'" Monday morning Rudolf Schroeter phoned the judge and told him not to call the jury. "I was ready to start trial," Sullivan said. "I thought we were really going." Dow offered some more money —"just a piddly amount more than they had offered before," Sullivan said. "But money talks. This country is run by big money. And here

Elements of Risk / 193

was a big firm from San Francisco saying, 'Settle!' " But Bob McCray claims that it was Sullivan, not Slobodin, who counseled settlement. "I was refusing to settle out of court," McCray said, "but Sullivan called me and said, 'You've got to sign.' He gave me a lot of bad advice." The people from Globe were called into the courtroom on Monday morning. The judge told them what the offer was. "Everyone wanted to settle," Mary Lou Gardner said. "What do you do if they make you an offer more than you asked for? Slobodin said, 'If you want a crusade, go find another attorney.' I thought the battle was lost. My friend was dead. She laid down her life for us."

Bob McKusick thought it was time to settle. The goals he and Mrs. Shoecraft had set after the 1969 spraying had been attained. The Pinals had never again been sprayed with herbicides. McKusick's health was deteriorating, as was his wife's. He felt strongly—and some of this was based on his psychic perceptions—that if he did not settle, he would die —"one way or the other." Either his health would give out, or someone would see to it that he could not carry forward the suit. "What's the price of a few bullets?" he asked. He felt as though he could anticipate certain things happening in the suit before they happened. "I told Sullivan to skip Dow's first offer and to take the second," he said.

Jerry Sullivan said he was horrified. "We get doctors to testify. We get epidemiologists to testify. I had people *calling* me to testify. And then they *settle!*" he declared, banging his fist on his desk. "You throw some money at people, and they cave. Billee wasn't there to egg them on anymore. Bob McKusick had had a heart attack. Their health was going. They had other interests. And Dow started throwing money at them. We would have had a jury finding of proximate cause, I know that. We would have had a finding that dioxin causes cancer. And then those people sold out.

"I don't really blame them," he said after a while.

Dow admitted no liability by settling the case out of court. The plaintiffs were not allowed to divulge the amount of money they received ("It was fairly large," said Bob McKusick's daughter, Cathy, "but it wasn't fair"). When the judge asked them if they found the settlement satisfactory, they all agreed, except for the Finn Mr. Satama. "I lost the money, I lost my health, I lost my wife, I lost my daughter, I lost everything," he told the judge. "What I have? No marriage. Nothing. That's all what I say."

"I think we let a lot of people down when we settled," Bob McCray said later, "people from different places around the country that had been sprayed with herbicides."

Arizona Republic columnist Tom Fitzpatrick wrote a story about the settlement the next day. "I wonder how Billee Shoecraft, the spunky woman who suffered from that poisonous spray," he wrote, "would have reacted to a deal like that?"

In the same month that *Shoecraft* v. *Dow* was settled out of court, Dow and the Environmental Protection Agency announced in Washington that they had reached agreement to suspend cancellation hearings on 2,4,5-T and silvex. Closed-door meetings were commenced between the two parties to pursue a private settlement.

chapter 8

The agricultural chemical division of the Occidental Chemical Company is an unprepossessing metal building that sits a short distance from a fence that separates the chemical plant from John Mendes's 400-acre, 300-cow dairy farm. The workers call it Ag Chem and the company Oxy. The town is Lathrop, California; the valley, the verdant San Joaquin. Here, as in the rest of the valley, agriculture and industry treat each other with the eggshell manners of two senior citizens whose future depends on the continued good health of the other. When pesticides turned up in farmer Mendes's well water, Oxy piped good water in from a mile away, and the farmer was satisfied. Across the street from the plant a hand-painted sign in front of a shambling house offers RED WORMS, 50 CENTS and SHEEP FOR SALE. A half mile away fieldworkers are harvesting alfalfa.

San Francisco is eighty miles to the west, and there's rarely a good enough reason to make the drive. The nearest big city is Stockton, a port and agricultural processing center. Stockton is where Lowell Berry originally wanted to build his new fertilizer factory, but he sensed that Stockton was city enough to take issue with the inevitable stink, so he moved his plant to the open reaches of Lathrop instead. Lathrop had two geographical advantages: It was in a central location in the market area, and it was in a favorable rate zone for railway shipments of raw materials from Idaho. The $1.7 million plant opened for business in December 1953 with sixty-eight employees, a quarter-million-dollar annual payroll, and a unique way of making pelleted plant foods. It was called Best Fertilizers.

Lowell Berry's fertilizer empire began inauspiciously with two men mixing fertilizers in a wheelbarrow inside a factory in Oakland, across the bay from San Francisco. He got into the business during the Depression because, as he later pointed out, "I had to do something if I wanted to eat." Berry believed in the redemptive powers of industrialism. He liked to say, "Only the strong are productive and only the productive are strong." He moved his Oakland operation up to Lathrop in 1958, and soon Best Fertilizers had sales of $12 million, and Lowell Berry was paying his 325 workers $2 million a year. In 1962 he added a new insecticide plant, the occasion of which inspired him to proclaim, "We are facing a godless, ruthless, highly organized, dedicated Communist movement, the goal of which is to gain control of the world. We can remain strong only through superior productivity."

No one worried about pollution from Mr. Berry's factory except when smoke from his incinerator stacks turned nearby Highway 50 into bumper cars. A man was killed in a twenty-car pileup on the smoke-shrouded highway in 1963, and more accidents followed. The company eventually invested $460,000 in the "latest, most modern" air pollution control equipment. It was the least that could be expected from Lowell Berry, whose second favorite pronouncement was "We want to be good neighbors."

In the spring of 1963 Lowell Berry sold his company, which had grown to be the largest independent manufacturer and distributor of fertilizers and chemicals on the Pacific coast, for $12 million to the Occidental Petroleum Corporation. Occidental was run by Armand Hammer, a retired physician who had bought into a failing oil company six years before and was in the process of turning it into a major multinational business by buying up smaller companies in oil-related fields. Five years after buying Best Fertilizers, Occidental acquired the Hooker Chemical Company, founded by Elon Huntington Hooker in 1905, which had grown into the eighth largest chemical company in the country. Occidental swapped $800 million in stock for Hooker in a deal that was touted as the largest ever corporate acquisition.

The transition from Best to Occidental was orderly. The oil company installed a subsidiary, Occidental Chemical, in Lathrop to run the plant. The chemical products still carried the Best name, which lingered also in a giant sign on the side of the plant. The new management smoothed out some minor labor problems and held a roast chicken lunch to announce improved working conditions. Neither the workers nor the townspeople were overeager to cause problems for the plant, which was tied like an umbilical cord to the local economy. Occidental played up the obvious in a 1968 stunt in which the weekly payroll was distributed in $2 bills. The bills were numbered and tracked throughout the community to see how much was spent, to note where it was spent, and to "increase local awareness of the importance of the company's payroll to the economy."

Occidental could afford to strut: the chemical industry was doing blockbuster business in the 1960's, and a big piece of that business came from the sale of agricultural chemicals. Situated as it was in the fecund

San Joaquin Valley—the state's and the nation's richest farming area —with access to growers up and down the length of California (which produces fully one-third of America's table foods) Occidental serviced customers that were greedy for chemical products to nourish and protect their crops. The company was marketing more than 200 chemical products to meet that need, and one of the biggest sellers was a straw-colored liquid known as 1,2-dibromo-3-chloropropane, or DBCP.

DBCP was the nematocide developed and tested during the 1950's by Shell and Dow Chemical and formally registered in 1964, despite concerns voiced internally in the two companies and externally by the federal government about its damaging effect on the testes of laboratory animals. Shell and Dow workers had been making it, and growers using it, since the mid-1950's and the companies pointed to their blemish-free safety records as *ipso facto* proof that DBCP was harmless to humans. Shell had surveyed the health of its DBCP workers in the early 1960's as part of a bargain struck with the U. S. Department of Agriculture, which found the animal tests worrisome. If Shell could produce evidence that its workers were healthy, then the department would register DBCP for sale. Shell's workers passed their health exams with flying colors, although the only test for testicular effects was a cursory hand check by the company doctor, who looked for lumps and varicose veins and who had no idea that DBCP had damaged the testes' internal sperm-manufacturing apparatus in animals. The Agriculture Department gave DBCP a Class II designation, its least hazardous category.

DBCP was made by Shell in Denver, Dow in Midland, Occidental in Lathrop, and a few smaller companies in other states. Some companies manufactured it from scratch by reacting allyl chloride with bromine, which produced a concentrated or technical-grade material. Workers had to formulate, or break down, the technical-grade DBCP into different concentrations for farm use by adding solvents and emulsifiers to dilute it. The finished product was canned, sealed, and stored on pallets in a warehouse until it was ready for shipment. Some companies like Occidental bought technical-grade DBCP from Shell or Dow, formulated it, and then sold it under their own labels or shipped it to farm supply stores under the manufacturer's label.

Occidental was known in the trade as a formulator or diluter. It got its DBCP in 50-gallon barrels from Shell during the late 1950's. Occidental had a falling-out with Shell and eventually struck a contract to buy DBCP from Dow. Beginning in 1964, rail tanker cars from Midland regularly delivered the concentrated liquid to Lathrop. In the 1960's Occidental was buying 50,000 pounds of DBCP a year, an amount that increased sixtyfold over the next decade. The chemical was emptied into huge holding tanks, then pumped into 1,000-gallon containers called batch tanks in the Ag Chem building. The batch tanks were affixed to weight scales so that workers—who stood on wooden platforms raised up over the tanks—could monitor the weight of the mixture as they added solvents and emulsifiers to dilute its strength. Afterward the finished DBCP was measured into cans by workers who manned a small assembly line off to the side of the batch tanks. The operation was very small; one worker might be formulating, and two or three canning. It was seasonal work, with 25 or so people working around the clock to get orders out during the growing season, and half that many working during slack times. The bulk of Occidental's 350 production and maintenance employees worked in the company's fertilizer operations.

"They'd bring in the DBCP in tanker cars," said Haskell Perry, who started working in Ag Chem in 1974, "and we'd have to hook up the tanker cars and put it into our holding tanks. Then that would transfer through lines into batch tanks when we were ready to mix it. The lines leaked, and we'd get some of it there. After that it was basically a closed system until it came time to mix it. Then you would stand above this open batch tank on a platform built over it, about ten feet high." The workers dumped barrels of chemicals into the tank, splashing DBCP on themselves.

"Once we started getting DBCP shipped in in thirty-gallon drums. Rumor had it there was going to be a tie-up in receiving DBCP, so we had a couple hundred of these drums just stacked on pallets out around the yard. That way we were just opening the containers and pouring it in."

For a time Occidental manufactured its own DBCP, and the workers would take hoses and spray down the nylon filtering screens when they became clogged with chemical residues. If it was hot, as it often was in the San Joaquin Valley, they sprayed bare-handed. When they were formulating or canning DBCP, workers were supposed to wear coveralls, rubber boots, and plastic safety glasses, and most of them did. There was no ventilation system drawing away fumes from the chemicals. The workers were supposed to wear rubber gloves but management didn't enforce the rule; in fact, one worker who insisted on gloves was branded a troublemaker by his bosses. Most didn't wear gloves except when they were running the organophosphate pesticides, like parathion, which the workers knew to be dangerous and which they called the poisons.

The organophosphates were first synthesized as pesticides in Germany. Later they were used by Nazi scientists who were searching for more effective nerve gases. After the war the Allies adapted the compounds into pesticides that were as exceptionally lethal as they were quick-acting. Parathion is much more acutely toxic than the chlorinated hydrocarbon pesticides like DDT, but it does not have DDT's persistence and so disappears from the environment more quickly. It works by short-circuiting the body's nervous system. The body manufactures an enzyme called acetylcholinesterase, which in effect enables one nerve to talk to another nerve. Acetylcholinesterase stops the transmission of impulses across synapses when a muscle action has been completed. If the impulse isn't stopped and the muscle continues to receive commands to act, the nervous system runs wild. Organophosphates destroy acetylcholinesterase. An insect sprayed with parathion develops uncontrollable spasms and dies. The human victim of parathion poisoning goes into convulsions, and if exposure is severe, the result is death, usually from respiratory failure. Nine drops of parathion swallowed or thirteen drops absorbed through the skin are fatal. More commonly, an overdose of parathion produces headache, blurred vision, nausea, and vomiting. The effects of mild parathion poisoning can be counteracted with an antidotal drug called atropine.

Organophosphates were a clear danger. Enzyme levels could be monitored in the blood, so Oxy set up a system of measuring a worker's susceptibility to the poisons by taking his blood count. If the count was too low, the worker was kept away from the poisons for a few days. Employment and safety rules issued to new employees explicitly described the organophosphates as poisons. A pesticide like TEPP could kill a 200-pound man if he accidentally swallowed just three or four drops. "It is essential that you know the nature of the material that you are working with," the rulebook said. At special indoctrination sessions on the handling of the so-called economic poisons, workers were warned that organosphosphates were designed to kill insects and could kill humans, too, if they weren't used properly. Workers were instructed to wear protective clothing, including coveralls, helmets, rubber gloves, and outer garments, boots, goggles, and respirators when they worked with the poisons and to shower after work and change into uncontaminated clothes. But until the mid-1970's shower facilities were primitive, and some men skipped them altogether.

The workers understood the potency of the poisons and generally followed the rules (some of them wouldn't follow any rules and did as they pleased), but they still got sick from the organophosphates—some of them several times a year. One windless July day with temperatures topping 100, the line was running badly, and Farnham Soto took his rubber glove off for a minute. His bare hand brushed against the conveyor belt. One of the poisons had been canned on the line the previous day, and some of it must have spilled on the belt. In fifteen minutes Soto was weak in the knees and vomiting. He was off work for six weeks. The sickness usually started with the sweats or a bad headache and then nausea. In bad cases the men slobbered, and the muscles in their eyelids and tongues twitched, and it was very hard to breathe. It ended with a trip to the company doctor for a dose of atropine and a day or two off from work until the blood count stabilized.

But no warnings were posted in the company rulebook about DBCP because it was classified as only moderately toxic. Workers breathed its fumes, soaked their coveralls with it, and dipped their bare hands into it. "All through the system, there was physical contact with it," recalled one worker. "You breathed it. You touched it. You got it all over your skin. The guys handled it like they would water, and there were vapors all over the place."

Haskell Perry was a "number one operator" in Ag Chem; that meant he did a little bit of everything, including reacting, mixing, and canning the chemicals. When he first started working around DBCP, Perry asked around the plant about it. One of the leadmen, who was a non-management supervisor, told him that DBCP was perfectly safe.

"It's like hand lotion," the leadman explained. "In fact, it ought to be good for your skin; it helps soften it up."

"Exact words," Perry remembered.

"Standard practice, when you came into the plant, was to take your street clothes off, put [your] work coveralls on, then at the end of the day take a shower, put your street clothes back on, and go home," Perry recalled. "We did have our couple of basic major poisons. They said, 'Oh boy, when you get near that, wear all your safety equipment.' But this stuff DBCP, they made no suggestions other than our normal attire —we changed into company coveralls and boots when we got to the job and wore safety helmets and glasses, and other than that there was no requirement.

"When I first transferred into Ag Chem in 1974, [the] safety code was very, very lax. They had a real small shower room at the end of the warehouse. It had dust all over it. It was a very unkempt place. It was from the time when only four or five people worked in the Ag Chem department, before everything got to be such a boom on pesticides. So it was just capable of bathing four or five people. And we used to have to stand out in the actual warehouse, where they kept the product, in a little tiny locker, where we changed standing on a towel. There were a couple of people who didn't even take showers. An old-timer there was very reluctant about taking showers. I think it was because it was a gang locker-type thing." The shower facilities were eventually upgraded.

The information sheets that Dow and Shell sent along to Occidental on DBCP didn't sound many alarms. Shell's original data sheet made DBCP sound like an alcohol rub. It was moderately toxic by inhalation or ingestion, slightly toxic by skin absorption, and no special hazard from skin contact was likely. After the animal tests in the 1950's had shown that DBCP had a potentially damaging effect on the lung, liver, kidneys, and testes, the data sheets were toughened somewhat. The companies warned that DBCP could cause injury to the eyes, was absorbed through the skin, and had highly toxic vapors when inhaled. Dow advised wearing protective gloves, shoes, and garments made of Compar rubber or other resistant materials; Shell did not. But neither Dow nor Shell warned about any testicular effect because the company's toxicologists did not think it was significant. They were convinced that problems would show up in the three vital organs long before the reproductive organs were affected. Workers were not monitored for reproductive damage because the scientists who wrote the 1961 report did not specifically recommend it and because it was not a common industry practice. But the scientists had warned in their report that: "Until further experience is obtained, close observation of the health of people exposed to this compound should be maintained."

Throughout the 1960's and 1970's demand for DBCP skyrocketed. It was the salvation of Georgia peach growers, who had been forced to move from the nematode-infested sandy coastal plain to the heavier soils of the northern part of the state, and of the California grape growers, who had watched their vineyards wither and their yields decline to one-third of normal because of the menacing root-knot nematode. The tiny worms had wreaked havoc, too, with production of pineapples, citrus, figs, bananas, strawberries, and a host of other crops. To meet grower needs, Oxy was making at least thirty-five products that contained DBCP as an ingredient and marketing them under the Best and Oxy brands and under Dow's trademark Fumazone. During the peak production years workers in Ag Chem were formulating nearly 3 million pounds of DBCP a year.

Mike Trout never planned to work in a chemical plant. Like most of the men who worked the pesticide line in Oxy's squat factory in the valley, he drifted into the job because it paid a good hourly wage. Mike had graduated from high school in 1970 and pumped gas for $1.75 an hour at a service station in Manteca, the town where he had lived since he was eight. After a time he took a $2-an-hour job at Weinstock's Tire Center in Stockton. His parents encouraged him to go to college, but he refused. "He had his mind made up," his mother said. Faced with their son's intractability on the college issue, Betty and Robert Trout settled for the next best thing. They called a friend who worked for Oxy and arranged a job interview for Mike. The pesticide business was booming in the early 1970's and Oxy was hiring. It was steady factory work paying some of the best money in the valley, and there was a management training program that Mike might eventually use to lift himself out of the plant. On January 8, 1971, Mike Trout drove from his parents' house in Manteca through the flat alfalfa fields to Lathrop, four miles away, and reported for his first day of work in Oxy's warehouse.

It seemed a golden, safe time to be a chemical worker in America. The Environmental Protection Agency had opened for business on December 2, 1970, with U.S. Assistant Attorney General William Ruckelshaus as its first administrator, and twenty-seven days later the Occupational Safety and Health Agency (OSHA) began operating under a charter to provide safe working conditions for the country's workingmen and women. The laws that OSHA enforced required employers to provide workplaces "free from recognized hazards that are causing or are likely to cause death or serious physical harm to employees." OSHA had the power to inspect workplaces, issue citations, and shut down factories when an "imminent danger" existed. An advisory agency called the National Institute for Occupational Safety and Health did health and safety research and recommended worker protection standards for OSHA to enforce. At its best, OSHA would give workers the right to know about hazards in the workplace and to participate in health and safety decisions. The employment and safety rulebook that Oxy handed out to new employees even quoted from OSHA's charter. "To assure so far as possible," it said, "every working man and woman in the nation safe and healthful working conditions and to preserve our human resources."

It seemed as if the entire resources of the federal government were grouped for battle against environmental poisons. OSHA was pacing its cage like a lion, promising to patrol and clean up the country's factories and to devise worker exposure standards for known carcinogenic chemicals. The EPA came off its mark like a pit bull at a dogfight, and industry felt the teeth marks. In its first month of operation the EPA forced the detergent industry to drop $100 million worth of plans for a phosphate substitute that had caused birth defects in laboratory animals; brought enforcement actions against three major cities for delays in building municipal sewage treatment plants; asked the Justice Department to bring suit against Armco Inc. for polluting the Houston Ship Channel and the Jones & Laughlin Steel Company for dumping wastes into the Cuyahoga River; and threatened Union Carbide with enforcement action unless the company cleaned up the air pollution at its Ohio River plant. A government scientific panel issued a report directly contradicting the chemical industry's threshold theory by declaring that no level of exposure to a chemical carcinogen should be considered "toxicologically insignificant" for humans. In the "psychobabble" of the time, the country had a raised consciousness about its personal and environmental health. Immediately after January 1, 1971, all cigarette advertisements were banned from television.

Industry was shocked by the severity of the EPA's opening moves. In trying to right the balance from no regulation to adequate regulation, the EPA overregulated in the beginning, and industry was furious. When the Justice Department sued Armco Inc. and won a court order prohibiting further waste discharges into the Houston Ship Channel, Armco president C. William Verity, Jr., an influential Republican, wrote a bitter letter to President Nixon. "It is inconceivable to me that your administration believes environmental problems can be solved by shutting down industry," Verity said, warning that the court order had effectively eliminated about 300 jobs "in one stroke of the pen." It turned out that the 300 employees had been laid off several weeks before, when part of the Armco plant had been shut down for maintenance repairs, but the White House, mindful of Armco's $12,000 in campaign contributions to committees supporting Nixon's 1968 candidacy, intervened in its behalf. The White House wanted the court order stayed while industry and government worked out a more suitable settlement, but when the story leaked to the press, and Ruckelshaus declared that he would quit "if environmental decisions are overruled because of political considerations," the White House backed off. Armco quickly agreed to a compromise pollution abatement program. But the stage was set for one of the decade's biggest battles—jobs versus environment.

Following in Armco's footsteps, Union Carbide and other companies threatened to close plants if they were forced to comply with environmental regulations. Sympathetic business and government leaders warned that the new environmental and occupational health protections would throw large numbers of people out of work during the 1970's. The Commerce Department predicted in 1971 that environmental regulations would cause such a severe economic dislocation that major new relief programs would be necessary. The chief executive officers of America's largest corporations formed the Business Roundtable in 1972 to help stem the tide of government encroachment, and the assistant labor secretary promised confidentially to delay controversial workplace standards so OSHA could be used as a fund raiser for Nixon's reelection campaign. OSHA became little more than a paper tiger, and worker health and safety laws were barely enforced. Job-related carcinogen standards adopted in 1972 were not enforced until a more sympathetic administration took over four years later, and the Oil, Chemical and Atomic Workers International Union called a five-month strike against the Shell Oil Company in order to breathe life into the once-heady promises of "safe and healthful working conditions" for America's workers.

When Mike Trout transferred into Ag Chem in the spring of 1974, it was one of the worst times to be a worker in America. Like a fake-fronted movie town, OSHA and the EPA looked fine until you walked through the door. The agencies were pressed to the wall by a retaliatory business lobby and by congressmen whose home districts were losing jobs. At the same time they were supposed to rectify decades of neglect and build a regulatory house almost from scratch. In 1972 Congress strengthened the nation's major pesticide law—the Federal Insecticide, Fungicide and Rodenticide Act—by requiring data on not only a chemical's acute but its chronic effects, including carcinogenic, mutagenic, and teratogenic. The EPA was ordered to go back and review every major chemical product registered since 1947. The agency could hardly keep up with its current work, much less the old, and quickly fell behind. The country itself was changing as the fervency of the environmental movement butted heads with the recession. When Russell Train replaced Ruckelshaus as EPA administrator in 1973, congressmen tried to win assurances at his confirmation hearing that Train would not be too aggressive an environmental taskmaster.

But Mike Trout and his fellow workers in Ag Chem were oblivious of the desultory protection coming out of Washington, just as they were ignorant of the true nature of most of the chemicals they worked with at Oxy. Russian studies in 1971 had confirmed what Shell and Dow knew in the 1950's about DBCP's deleterious effect on animal reproduction. Soviet scientists reported that male rats fed DBCP for two and a half months could not fertilize female rats. Two years later American scientists showed that DBCP caused cancer in animals. The 1973 studies were part of a National Cancer Institute program to test various environmental chemicals for chronic toxicity. Rats and mice dosed with DBCP through stomach tubes developed stomach and breast tumors. The stomach tumors grew with uncommon rapidity, some appearing as early as ten weeks after the experiment had begun. One of the government scientists wrote a courtesy letter to Shell Chemical in September, alerting it to the "rather startling" results of the study, especially the "unusual effects" that appeared "relatively early" in the experiments. Shell was told that "these results show that chronic exposure to DBCP could be a health hazard" and was advised to provide its workers with sturdy protective clothing. The entire study was published in December 1973 in the *Journal of the National Cancer Institute*. The study also showed that ethylene dibromide, or EDB, the fumigant developed by Dow back in the 1940's during the race with Shell to find a better nematocide, also caused cancer in animals. "These results show that chronic exposure to either DBCP or EDB could be a health hazard," the journal reported. "Anyone exposed to DBCP or EDB should . . . use protective clothing, masks and other means to avoid absorption of either material."

Neither company saw fit to pass information about the test results to its workers. But industry wasn't alone in ignoring warning signs about the chemical. In the same year that the cancer studies were reported, and with the 1961 report showing liver, lung, kidney, and testicular effects in hand, the EPA nevertheless reregistered Dow's DBCP product Fumazone. Dow had included the joint 1961 Dow-Shell study with its registration application, but the EPA registered Fumazone without comment on the study.

So it was that Mike Trout and his co-workers in Ag Chem formulated more than 10 million pounds of DBCP over the next four years protected by little more than rubber aprons and boots. The supervisors in Ag Chem (the workers called them supes or white hats) were more concerned about the length of Mike's and some of the other workers' hair, and they studiously enforced a hairnet rule on the line. Mike wore his hair in the shoulder-length style popular with men his age in the early 1970's, and he and his friend Haskell Perry, who also had long hair, and some of the other workers would get lippy with the bosses now and then. They got reputations in the plant as smart alecks or worse. "We copped an attitude," Haskell said. Haskell was married, with two small children, but he was only two years older than Mike, and they shared common interests. They played ball together and camped together; they drank beer and drove around in Mike's 280-Z together. They fitted in well with the men in Ag Chem, most of whom were young, some of whom were hell raisers, most of whom smoked cigarettes and some of whom smoked marijuana. They were tough guys, but when the surface cracked a little, there was usually a boy underneath who was nervous about girls and ate Sunday dinner every week with his mother.

Still, the tough guys butted heads at the plant with some of the supervisors, one in particular—who was a strong Mormon and whose convictions about how people should behave were often violated by his workers. He thought they were disrespectful of authority and had an attitude problem. They thought him too rigid and unforgiving, especially when he tried to apply his moral ideals to the men in the plant. To the workers, Ag Chem was a factory, where men sweated and swore, not a Sunday school. Once when Haskell Perry cursed him, the supervisor called the California Highway Patrol to the plant and tried to have Haskell arrested for public profanity. "I had a couple run-ins with him because I have a slip of the tongue when I'm talking to people who make me mad," Haskell said. "He'd just pull his authority on you. You had to talk to him like you were talking to a third-grade child, you know—polite and nice. After a while it got to the point where you weren't even allowed to talk to him, you had to go through his lower supervisors. He had his two or three guys that he didn't have a very high opinion of, and Mike and I were two of them."

Mike had a gentle side rarely seen in Ag Chem. He was shy around women, he loved children, and he lived at home until he was twenty-one. Home was the same house on Acacia Street—a carved wooden trout hanging over the front door announces whose house it is—where Mike's parents have lived for twenty-two years. The street is in a neighborhood of streets named after trees in Manteca, which rises up low and wide from the San Joaquin Valley floor. It is a town built on a vast alluvial plain and is best described by clichés. It is pleasant, secure, a good place to bring up a family. The 1950's-style Main Street has been usurped by a 1960's-style fast-food and convenience-store strip where boys cruise girls, and vice versa. In the autumn banners are strung from traffic lights to announce the annual Pumpkin Festival, bunches of purple grapes sell for 22 cents a pound and casaba melons are two for a dollar at roadside stands, and the priest at St. John's drives a pickup truck.

If there was a passion in Mike's life, it was baseball—fast pitch, slow pitch, any kind of pitch. He directed only casual attention to other sports, although as a golfer he could reach the green in two strokes on a par-five hole and once took the prize (a garden hose and one dozen balls) for the longest drive—317 yards—in an Oxy golf tournament. He was tall (six feet one) and solidly packed (165 pounds), and his powerful arms were displayed to best advantage on a baseball diamond, where relays fired in from center field earned him the nickname Rocket Arm. For a time he was playing ninety games a season for four different teams.

It would take a stronger and more enduring word than "passion" to describe what Mike felt for his child, a boy named Michael Lee Trout, Jr., who was born in August 1974, ten days before Michael Lee Trout, Sr., turned twenty-two. Mike had met Marta Tagliabue on a fix-up date in the spring of 1973. They had gone to high school together but were four grades apart and knew each other only slightly. On their first date he went to her high school graduation party. She took a telephone operator's job, and they dated through the year. They married the next winter in St. Anthony's Roman Catholic Church, honeymooned on the California coast, and rented a small apartment on Powers Street in Manteca when they returned. Marta had grown up in the country, surrounded by vineyards and walnut groves, and she felt hemmed in by the thin apartment walls, with neighbors on all sides. Apartment life was all the more intolerable because she was expecting a baby. Mike would not allow his wife to work during her pregnancy, and he preferred that she never work again. He was old-fashioned in the sense that he believed the man supported the family. He asked for a promotion, and in late March 1974—one month after he was married—Occidental assigned him a $4.19-an-hour job in Ag Chem as a helper,

Elements of Risk / 207

assisting the "number one" and "number two" operators, who mixed and canned the chemicals.

Mike was elated about becoming a father. He had always been fond of children, even as a boy, and he wanted a big family. As a five-year-old, perturbed by what he considered his mother's tardiness in producing another brother or sister, Mike would canvass his neighborhood, playing with the neighbor's babies. His mother cooperated by delivering a third boy, a younger brother, and a few years later, a baby sister, whom Mike would rock to sleep every night while his mother cooked dinner. "If there was a baby around," his mother says, "that's where you would find Mike."

After Mike, Jr., was born, the Trouts began looking for a house. They needed more room for the baby, and for the two or three more children whom they hoped to have and whom they began trying to have after the boy was born. They could afford a house now because Mike's job in Ag Chem paid substantially more than his warehouse work. In 1973 he'd made $6,032 in the warehouse. In 1974 he grossed more than $10,000, working in Ag Chem.

In the 1960's a modern subdivision of earth-toned, shingled ranch and split-level houses was built several miles outside Manteca's city limits. It was called Raymus Village, and its streets were named after Indian tribes. A faded billboard still stands at its entrance. A young couple, their faces unlined and serene, are frozen in peeling paint, watching their son frolic with his dog in the billboard's rendition of the good life to be had in TODAY'S PLANNED COMMUNITY. HOMES. SHOPPING. PARK. This is where Mike and Marta found a house, a cream-colored ranch style with brown shingles on Comanche Drive, in the spring of 1975. With a nine and a half percent bank loan and a $250 monthly mortgage payment, the house was theirs. Mike's supervisor from Oxy lived nine houses away in a two-story colonial on Navajo, and although the adults never socialized, the children from both families sometimes played together. Mike played ball and doted on his family, and life seemed as it should be, save for the fact that Marta could not get pregnant again, even though she was using no contraceptives. It was a puzzle to her, but a puzzle that paled in comparison with a new problem that emerged toward the end of their first year in the house, when Mike began feeling sick and out of sync.

At first it was as illusory as a fleeting headache in the morning, a wave of blurry vision driving to work, something that could be passed off as flu or fatigue. Sometimes after baseball games he would come home and vomit. Once near Christmas he woke up sick, and Marta thought it was a hangover. "You must have had too much to drink with the boys last night," she said, teasing him. But the headaches and nausea didn't pass with the new year, and his family made him promise to see a doctor.

"He'd had these headaches for quite a while," explained Mike's father, Robert, a trim, amiable man in his middle fifties with the tattooed arms of a sailor and the peaceful look of a survivor who lived through a war, a heart attack, and two bypass operations. In 1959, when Mike was eight, Robert Trout transferred from Sonoma State Hospital, where he worked as a laundry supervisor, to Deuel Vocational Institution, a medium-security prison in the valley. He moved his family from the northern California mission-vineyard town of Sonoma to Manteca, which had 7,000 people then and rivers for fishing and forests for camping. Mike and his brothers were infected by their father's love of the outdoors and especially of baseball. Robert Trout coached Little League and Babe Ruth teams in Manteca when they were growing up. Later the four Trout men played on a slow-pitch team together: Mike in center field, his brothers, Steve and Mark, playing first base and shortstop, and their father pitching. "Mike really smoked 'em in from center field," his father said, with a connoisseur's appreciation.

His mother remembered the day Mike and Marta went to Doctors Medical Center in Modesto, where Mike had X rays taken. "When I saw him later that day, I said, 'What happened?' And he said, 'I've got a tumor.' Just like that. 'I've got a tumor,'" Betty Trout said. Two years younger than her husband, she had short, curly hair and an enviable figure, wore glasses, and talked with disarming charm and directness. "Well, for goodness' sake, you'd have thought we were crazy. We came into the living room and had pizza and beer and acted like it was nothing to worry about."

It was Friday, February 13, 1976. Doctors had located a tumor in Mike's brain by injecting a radioactive isotope into his arm and then recording the radiation given off by the isotope. They knew that the tumor was taking up critical space in his brain, causing compression and pressure inside the skull. It was this pressure that was causing Mike's headaches, nausea, vision problems, and fatigue. Unlike other malignant tumors, brain tumors rarely mestastasize, or spread to other parts of the body. The more critical fear in Mike's case was that the tumor would invade the surrounding brain tissue. The skull is a closed system housing a brain that weighs roughly three pounds in a finite space. A tumor that weighs any more than one-fifth of a pound crowds the brain so severely that death is inevitable. But some tumors are enclosed in shell-like capsules and, if removed early enough, can be prevented from breaking through the capsule and spreading into surrounding areas of the brain. These are called astrocytomas, and they are a type of tumor in the larger category of gliomas, which account for nearly half the brain tumors in adults. Mike would have a fighting chance if his was an enclosed astrocytoma of a sufficiently low pathologic grade—Grades I and II grow slowly and enable people to survive for years; Grades III and IV are more deadly.

Mike was sent home for the weekend, and surgery scheduled for the following Tuesday. That same night he fell into convulsions and was rushed back to the hospital. In surgery on February 16 at Doctors Medical Center in Modesto, a Grade III astrocytoma was removed from Mike's brain. It was classified as having a moderate degree of malignancy, but Mike's doctors were optimistic because it looked as though the tumor had not broken through its shell case. They prescribed six weeks of radiation therapy once Mike had left the hospital to slow down tumor regrowth and gave him medicine to reduce swelling in the brain, convulsions, and mood swings.

"Mike wasn't upset about the operation," Betty Trout said. "As for us, well—you just think, everything is going to be all right. The only time I really got shook was when Marta and I talked to the doctor's assistant, and he said, "Mike may not come out of this. He may be paralyzed."

"They thought he was going to be paralyzed on the left side," Robert Trout said. "He had quite a large tumor."

"When we went in to see him before surgery—I've got a big family and there were twenty-five of us in that hospital room—they let Little Mike get in bed with him," Betty said. "Mike was laying there with his head all wrapped up in a surgical turban and . . . well, he came out of that operation with flying colors."

Brain tumors are as complicated and mysterious as the brain itself. Some 11,000 new brain tumors occur in the United States each year, yet the cause—and the mechanism of the effect—is only minimally understood. Mike Trout's doctors did not have a clue to why a healthy, athletic twenty-three-year-old man should develop a Grade III astrocytoma.

Mike took a six-month leave of absence from work. Every Wednesday he drove to Modesto for radiation treatments. Every day he pedaled his son downtown for lunch-hour visits with Marta, who had been compelled to return to work at the telephone company to supplement their income. Once he went hunting with his father near Angels Camp in the Sierra and shot a buck. If he felt sick or depressed, he didn't talk about it with his family. Whenever he could, he took off with Haskell Perry and drove up in the hills for the afternoon or camped overnight in the mountains. It was during those times with his friend that he opened up. Mike was bald from surgery, and a row of jagged stitches stood out in stark relief above the hairline on his skull, where doctors had cut a semicircular incision downward through the scalp and then drilled holes through the bone to expose a window into the brain. His scars startled people. "We told everyone he was the Bionic Man," Haskell said.

They both were growing to hate working at Oxy. Besides the pressure from their supervisor, they were starting to become slightly suspicious of the chemicals. Haskell had never thought much about it, but a doctor he'd seen on and off during the past years for minor health problems had told him once that working at Oxy might be hazardous to his health. Haskell had never been "poisoned," like so many of the others, but he would occasionally feel sick to his stomach when he would wipe his brow on the canning line or eat a candy bar on break without first washing his hands. Mike told Haskell that he felt stuck, that he wanted to leave Oxy but couldn't because of his illness, and that he couldn't quit because he couldn't support his family on unemployment and he needed the health insurance benefits. "No one in the valley would have hired someone with cancer," Haskell said.

It was Haskell who got out first. His marriage was breaking up, his life was in disarray, and he began taking to heart the doctor's warning about working at Oxy. In June 1976 Haskell quit work and signed up for unemployment. In August Mike Trout went back to work in Ag Chem.

Nothing much ever went right for Wesley Jones. His mother and father both had cancer (it killed his father), one of his sisters died in a mobile home fire, and he dropped out of high school in the eleventh grade. He married his girl friend, Deborah, when he was seventeen and she was fifteen and pregnant. He worked in a cannery, in a pump and welding factory, and as a bookkeeper for an interior design business to support his pregnant wife. He was also arrested for burglary shortly after they were married and put on probation. In May 1971 his daughter Kathleen was born.

Wesley and Deborah knew that their deck was stacked from the beginning, and they tried to beat the odds, but the rawness of their age, the heat of their tempers, and the dominance of in-laws wore the marriage down. When the child was eighteen months old, they separated briefly, and Wesley sought psychiatric help. He told the intake interviewer at the San Joaquin County Mental Health Services—where both his mother and his mother-in-law had previously been patients—that he was depressed about problems with his marriage and his job. His was a dilemma that many men have faced at one time or another. He couldn't live with her, and he couldn't live without her. Neither wanted a divorce, but neither knew how to make their marriage work. Wesley was diagnosed as having an anxiety neurosis and was given medication for nervousness.

The couple reunited and started to get marriage counseling. Deborah became pregnant again that winter, and Wesley was delighted. He told his friends that he had always wanted one wife and one son. He looked around for a better job and found one on the Ag Chem line at Oxy. Wesley was physically suited to the work. He carried 165 pounds on a well-muscled five-foot-seven frame. He had a full head of shoulder-length red hair, a pack-a-day cigarette habit, and an Irishman's cocky charm. The men on the line liked him.

Wesley started at Oxy in the summer of 1973 and almost immediately got into trouble. In early July he had an acute attack of organophosphate poisoning. He was nauseated, had vertigo, was shaking and confused, and was throwing up blood. The company doctor treated him with the antidote atropine, but his blood count showed extremely depressed enzyme activity, and Oxy took him out of Ag Chem and put him in a different department for two weeks. When he came back to Ag Chem, he continued to have minor bouts of poisoning, and the company nurse recommended his exposure be decreased. But the money was good, and Wesley didn't lobby hard for a transfer. In October Deborah Jones delivered a second daughter, Lesley.

Wesley's health—physical and mental—continued to deteriorate. Three months after his daughter's birth he had his appendix removed and was off work for several months. Next he had a tonsillectomy, then problems with his stomach (pain, nausea, dry heaves), and tests turned up some abnormal liver and spleen readings. As part of the examination Wesley's doctor took a sperm count. The count was flat zero, no sperm. The doctor tentatively diagnosed azoospermia; the term comes from Greek words, which mean "lifeless sperm," and is defined as an inability to produce sperm.

Deborah had been trying to get pregnant again, to give Wesley the son he wanted. Her failure could now be laid to Wesley's sudden and perplexing infertility. But the couple was preoccupied with more immediate problems. The marriage was foundering badly again, and Deborah blamed Wesley's job. She thought his personality had noticeably changed since he went to work for Oxy. He was irritable all the time, and he yelled at her for no reason. "He was jumping all over me with words," she said. His mood turned aggressive, menacing. Once he threw a soda bottle at her and punched her on the shoulder and left the house. The "last straw" was when he became violent with the girls, shouting at them until they shook with fear. She took the girls and moved out.

On April 29 Wesley swallowed 100 pills belonging to his mother and sister, including Noludar, Librium, and Valium, and a full bottle of Allerest. Doctors at San Joaquin Hospital's emergency room pumped out his stomach and sent him to the county's mental health outpatient clinic for observation and rest. Wesley said he was despondent because his wife and daughters had left him. He told doctors that his wife was in love with a hippie. He said he had been drinking too much. The diagnosis was depressive neurosis with episodic excessive drinking.

After he had recovered from his suicide attempt, Wesley went back to work, but his absences were frequent. His stomach hurt all the time, and his headaches persisted. A doctor he saw in February 1976—the same month that doctors found a brain tumor in his co-worker Mike Trout—suggested that his azoospermia might be related to toxic chemicals. Wesley was referred to the Palo Alto Medical Center for a thorough evaluation. Once again his sperm count was zero.

In the spring of 1976 the Environmental Protection Agency became concerned when continuing laboratory studies confirmed that DBCP was carcinogenic. The studies had been initiated three years earlier under the sponsorship of the National Cancer Institute. For seventy-eight weeks DBCP was injected via tubes into the stomachs of rats and mice. At the end of the experiment 60 percent of the rats and 90 percent of the mice developed stomach cancer. Female rats also had a high incidence of breast cancer.

Some scientists distrusted the gavage method used to expose the animals to DBCP, which is akin to pouring a chemical down an animal's throat. But the National Cancer Institute researchers thought the results suggested "a potential environmental hazard." In response, the EPA put DBCP on its preliminary list of candidates for the bureaucratic procedure called Rebuttable Presumption Against Registration.

Shell submitted a detailed package of information about DBCP to the EPA. The company vowed that it would "take every prudent step necessary to counter regulatory problems that might threaten our position" as the largest supplier of DBCP to the domestic market. Shell also told its employees about the results of the animal tests, as did Dow, although both companies downgraded the results because they didn't like the way the studies had been done. Neither company passed the information to Occidental, and the DBCP workers there never heard about the cancer tests. In Shell's Mobile, Alabama, plant, a notice on the employees' bulletin board said:

> Animal testing has indicated that another of the materials we handle, NEMAGON, can produce cancer in rats and mice. This testing was carried out by inserting tubes directly into the stomachs of the test animals and injecting liquid NEMAGON in fairly massive doses. While this sort of test is difficult to

relate realistically to an industrial atmosphere, and the exposure levels possible under the worst conditions, it does say that the potential exists. Inhalation tests on rats, conducted for six hours per day, 5 days per week for 90 days at 1 PPM [part per million] and 5 PPM, produced incidents of spotted balding and some severe eye and respiratory system irritation. No carcinogenic effects were observed.

A number of agencies and private research groups are involved in a great deal of testing activity related to carcinogenic capabilities and other health hazards of many compounds. While a lot of these tests could be called unrealistic, and exaggerated, the long-range results should be beneficial to all of us. And above all they tell us that our Plant philosophy of "no avoidable exposures" is a good, valid approach to our business.

Despite its hedging about the validity of animal tests, Shell at least tried to give its workers a worst-case scenario of the problem. Scientists like to see test results "replicated" by another test before they accept the results as valid. Further tests on DBCP seemed to confirm its carcinogenicity. Dow-sponsored studies produced levels of stomach cancer in rats and mice similar to the original findings, and rats that breathed DBCP in doses of 0.6 and 3 parts per million in EPA-backed chronic tests developed nasal cavity tumors. Shell urged its Mobile workers to wear full-face respirators and impervious clothing made out of polyethylene or neoprene to keep DBCP from penetrating their skin. No such warnings were handed out at Oxy. The workers in Ag Chem continued to think that DBCP was relatively harmless, and many of them not only worked without respirators but worked bare-handed.

In July Wesley Jones checked into Stanford University Hospital for a testicular biopsy. Doctors cut through the scrotum and made tiny L-shaped incisions in each testicle in order to extract tissue for lab analysis. Both tissue samples showed almost total absence of sperm. One of his doctors advised Wesley to quit his job.

While he was making plans to leave, Wesley pressed a workmen's compensation claim against Oxy based on his infertility. Rex Cook, secretary-treasurer of Local 1-5 of the Oil, Chemical and Atomic Workers (OCAW) International Union, knew of three other men who had worked at Oxy who had fertility problems. One of them was Richard Perez, who quit Oxy in 1974 after finding out that his sperm count was zero. Another was Ted Bricker, a longtime Oxy employee who had been treated for organophosphate poisoning five times in eleven years and who had just that year had three sperm counts come up zero. Rex Cook decided that fertility problems would be a priority issue in the upcoming contract negotiations. He told Wesley's doctor that "a significant problem existed at the Occidental Chemical Plant." In July the doctor wrote a letter to the state Department of Health's occupational branch chief outlining the Jones case and noting that chemical exposure seemed to be the cause of azoospermia. State authorities took no action; it's not certain who saw the letter since no one could even find it in the files when it was looked for at a later date.

Wesley Jones quit Oxy on September 3, 1976. Oxy contested his compensation claim, and he lived first on disability, then on welfare. Michael Trout had just returned to work as Wesley was leaving. The men in Ag Chem knew about their co-workers' fertility problems but only in a general, gossipy sort of way. It was a sensitive subject, and because these were men who were made uneasy by personal discussions, they talked about it only in the most jocular way. The standing rumor in the Oxy lunchroom was once you had started working in Ag Chem, you could forget about having children. Richard Perez told some of his colleagues that he was sterile before he quit. "It really got to be a conversation piece after he made it known down there that he was sterile," said longtime Oxy worker Jack Hodges. "There's been guys talking, you know, about trying to get their old ladies pregnant, and [it's] awful damn funny there ain't no kids," said Clifford Enos. And few of them talked about it at home. The problem might have emerged far sooner had the workers been women and more comfortable talking about the vagaries of their bodies.

Richard Perez had quit his job at a Kentucky Fried Chicken store in 1967 after getting an oil burn on his arm and gone to work in Ag Chem because it paid better. He was eighteen when he started, and soon he and his wife, Gloria, had a son. During the next seven years the couple tried to have a daughter. In the winter of 1974 Perez went to a doctor on his own to have his fertility tested. He was azoospermic, or sterile.

By the summer of 1976 Oxy management was certainly aware that something was terribly wrong in Ag Chem. They didn't know "why," but they knew "what": At least four of thirty-five workers had fertility problems. Oxy asked a specialist from the University of Utah Medical Center to tour the plant in September to assess the situation, and he suggested three options: review the medical literature to see what kinds of pesticides cause sterility, run some animal experiments, and get sperm counts on the men in Ag Chem. He thought the numbers reflected a "considerably higher rate of sterility than one would normally expect in a population of middle-aged males." At first Oxy fingered the obvious suspect. Could the troublesome organophosphates affect fertility? No, the Utah specialist reported back. There was no known link between sterility and that kind of pesticide, "although it has been seen with other pesticides."

Just the summer before, in fact, a pesticide factory in Virginia had been shut down after workers were hospitalized with a mélange of health problems, including sterility. The pesticide in question was kepone, a chlorinated hydrocarbon chemical sold as ant bait. The establishment of a link between kepone and male infertility was a turning point in the way doctors looked at hazards in the workplace. A handful of industrial chemicals had long been known to affect male reproduction—lead and methylmercury, for instance—but men were thought to be far more resistant than women to reproductive harm partially because the effects showed up in women in more dramatic ways. As far back as the turn of the century doctors were recording a 60 percent rate of stillbirths and spontaneous abortions among women working in British lead factories. They eventually began to understand that toxic chemicals could cross the placenta barrier in a pregnant woman and damage the unborn child. But the male sperm cells were thought to be protected by a biological barrier comparable to that which retards the penetration of chemicals to the brain. Consequently, women industrial workers were monitored more carefully and, in some cases, barred from working with certain chemicals.

In its confusion Oxy couldn't decide what to do and so stalled for time. There was an overriding reluctance to publicize the problem until it was nailed down. When the doctor from Utah inquired about Oxy's plans in November, the company seemed more concerned with "generation and dissemination of reports" about the problem than with devising a plan to combat it. As fall blended into winter and winter turned to spring, Oxy kept the Ag Chem line running and told the workers nothing about its suspicions. Finally, in May 1977—nearly three years after Richard Perez had quit because he was sterile—Oxy had two industrial hygienists survey the plant in order to "determine whether the implicated health effects are real."

It was a cool, windy morning in early May when Doctors Robert Spear and Stephen Rappaport from the University of California at Berkeley set up their measuring devices in Ag Chem. The wind kept a good air flow moving through the plant, providing a natural ventilation that helped make up for the lack of mechanical venting of the chemical fumes. The line was running DBCP that day; one man was formulating it and two were canning it. The industrial hygienists had scanned the medical literature on pesticides before driving up to Lathrop, and DBCP was high on the list of suspects. The most obvious clue was the joint Dow-Shell report on DBCP published in 1961, with its conclusions about reproductive effects in lab animals.

The hygienists hooked up their monitoring equipment to three workers on the line. When the amount of DBCP detected by the equipment was averaged over an eight-hour workday, the men were exposed to less than half a part per million of the chemical, well under the one part per million limit recommended by Dow's Dr. Torkelson and Shell's consultant Dr. Hine in the 1961 report. But the industrial hygienists were concerned about the short bursts of higher-concentration DBCP the workers were breathing, and they wondered if the plant might not be considerably different on a hot, windless day in the summer. "We believe that the concentrations of DBCP fluctuated widely and that peak concentrations in excess of 5 parts per million probably occurred," they told Oxy. "In view of the toxicity of DBCP, we feel that the present facility will be found to be marginal for handling this material safely under all temperature/ventilation conditions."

They were troubled by DBCP's "documented effects on the reproductive system of animals." They were blunt in their assessment of DBCP. It was "the most hazardous of those chemicals chosen for review." They advised Oxy immediately to improve ventilation in the plant and to get in touch with Dow and Shell to get some guidance on what to do. Another survey was scheduled for July.

Ten days after the plant survey, on May 15, Oxy sent Wesley Jones to a doctor in San Francisco for a physical examination. His workmen's compensation claim was pending, and Oxy wanted an opinion on whether or not his infertility was work-related. The examining physician was toxicologist Dr. Charles Hine, the occupational health specialist at the University of California at San Francisco who had studied DBCP extensively during the 1950's, who shared credit for the joint Dow-Shell report in 1961, and who had examined Billee Shoecraft for herbicide effects.

Wesley Jones told the doctor that fertility problems prevented him from having the son he wanted. He told him he was suspicious of the chemicals he had worked with in Ag Chem and brought a list of them along to the examination. Dr. Hine listed his patient's other health complaints: "He lost 20 pounds, has a feeling of weakness and getting played out easily. Decreased sexual desire (once or twice a week), headaches, blackouts at work, shakes, colds, nosebleed . . . , shortness of breath, etc." The doctor knew about Jones's background; the date he started at Oxy; that he had worked there with a number of agricultural chemicals; his organophosphate "intoxications"; his stomach problems; his sperm count and testicular biopsy results; his marriage problems and suicide attempt; his probation for robbery; the written opinions of other doctors who thought his sterility was related to chemical exposure (one doctor referred to a paper on kepone and its effects on testicular function); the diagnosis of a psychiatrist who found that

Wesley had a history of "slight emotionality" but no evidence of organic brain syndrome; the statement of Wesley's union chief that there was "a significant problem" at Oxy; and Wesley's own assertion that three and maybe four other men in Ag Chem were infertile.

Dr. Hine concluded that there was no relationship between Wesley Jones's infertility and his work at Occidental. He further concluded that Jones was not work-disabled. He made no reference to DBCP. "I have found nothing in the medical literature or in my files that would indicate aspermia to be induced by chemicals in the absence of profound intoxication," Dr. Hine wrote. "In fact, there is nothing to suggest that this is likely to occur except in rare instances in man, even with physically overt intoxication."

Dr. Hine told Oxy that he was going to search the medical literature further and that he would be interested in touring the plant first hand. But the dominoes were lined up and waiting to fall at Oxy. After three sperm counts had confirmed that he had no active sperm, Ted Bricker started talking with Rex Cook from the union. They both agreed that something had to be done. At about the same time a filmmaker from Los Angeles researching a documentary on industrial hazards had routinely called the Oil, Chemical and Atomic Workers Union for information and possible leads. Rex Cook told him there was a problem with fertility among some chemical workers in a factory in the San Joaquin Valley but there was no hard information on the numbers affected or the cause. The filmmaker offered to pay to have the workers' sperm counts tested. Rex Cook and Ted Bricker thought it might work; they began trying to convince the men in Ag Chem to volunteer for tests. After they had exerted considerable effort, seven men agreed to go over to Summit Labs in Manteca.

Mike Trout was one of the volunteers. He had recovered nicely from brain surgery, save for a loss of strength and agility that reduced his baseball prowess. Doctors said his long-term prognosis was reasonably good. But the disappointment of his life was Marta's inability to have another baby. Michael, Jr., was nearly three, and they had been trying since his birth to conceive again. He told Marta about the lunchroom conversations and about the men who knew for a fact they were infertile. They decided he should have the tests done.

Sperm counts are a quick and fairly inexpensive method of assessing a man's fertility. After two or three days' abstinence—to ensure peak semen volume—a man ejaculates into a jar in the doctor's office. If distaste or religion prevents masturbation, a couple can engage in coitus interruptus at home and bring the sperm into the doctor's office within two hours after ejaculation. A fertile man's ejaculate equals between two and five cubic centimeters of fluid, or about a teaspoonful. To gauge fertility, a doctor examines the ejaculate through a microscope. He looks for a minimum number of sperm in every cubic centimeter of ejaculate. He also examines the shape of the sperm and how energetically they move.

The medical profession engages in an ongoing debate over the number of sperm considered adequate for fertility. The consensus seems to be that anything below 20 million sperm per cubic centimeter of ejaculate constitutes oligospermia, or an abnormally low number of sperm. The commonsense theorem is that the closer you get to zero, or azoospermia, the higher the chances of infertility. The numbers have been adjusted down over the years in response to gradually decreasing average sperm counts in the United States. In 1929 the average sperm count was 90 million. In 1951, when John McLeod, the dean of American infertility doctors, surveyed the sperm of men whose wives were attending a prenatal clinic, he found that 5 percent had counts less than 20 million. In 1975 a study of male college students in Florida produced an average sperm count of 60 million, and studies later in the decade showed that between 8 and 9 percent of American men had sperm counts under 20 million.

Recent years have also seen doctors placing more emphasis on sperm quality. Traveling at a speed of three to seven inches an hour, the sperm —propelled by whiplike tails—swim twelve inches uphill against gravity and against acidic mucus currents in the woman's vagina and uterus. Only one sperm will survive to penetrate the egg. The rigors of this journey require sturdy, elongated sperm capable of moving forward at a good clip.

Besides sperm count, doctors can measure the follicle-stimulating hormone (FSH) in the blood. FSH is essential to sperm production. The hormone, which is secreted from the pituitary gland near the brain, tells the testes to make sperm cells. An elevated FSH means that something is out of kilter because the body is working harder than normal to stimulate the manufacture of sperm.

The seven sperm counts were analyzed at Summit Labs in late June. The lab would release the results only to a physician, and the union asked that they be sent to Dr. Donald Whorton, an occupational health specialist in Berkeley who had done some consulting work for the union in the past. Over a period of three days in early July, envelopes containing the seven lab reports made their way through the mail to Dr. Whorton's office. "When I examined the contents, the first slip had a name and a sperm count. 'Result: no sperm seen,' " recalled Dr. Whorton. "The following one read, 'Sperm count, result: no sperm seen.' The next one: 'Sperm count, result: no sperm seen'; likewise for about five or six of them." All seven had abnormal counts.

Dr. Whorton was puzzled. Why were the sperm counts so low? Had all these men had vasectomies? "What does this mean?" he asked Rex

Cook. Cook told him about the infertility rumors in Ag Chem and about the men who had independently had sperm counts taken. Dr. Whorton told Cook that he wanted to set up a face-to-face meeting with the seven workers to explain the results, rather than to send the lab slips to them through the mail. He wondered what would happen if he "went to this small town to talk to these men and told them it looked like they were all sterile, probably as a result of something to which they were exposed at the plant" and then left. He told Cook he was reluctant to "bet the ranch" on the results of one test. He had never heard of Summit Labs, and his first instincts were to have the results reconfirmed by tests in the laboratory he commonly used. He also suggested that Oxy be immediately apprised of the situation.

On Monday morning, July 18, Rex Cook called Gregory Vervais, Oxy's manager of employee relations, with the news about the abnormal sperm counts. Cook told him that the workers suspected that DBCP was the cause. Vervais immediately went to James Lindley's office with the news. Lindley was an Oxy vice-president and the man at the top. He says Rex Cook's phone call was the first time ever he'd heard about fertility problems in Ag Chem. Lindley and Vervais drew up a list of things to do: (1) Call Dow; (2) meet with Dr. Whorton; (3) call some regional formulators; (4) call the National Agricultural Chemicals Association (the industry's lobbying group); (5) call Dr. Mitchell Zavon (Shell's longtime medical director, who had recently been hired by Oxy's sister subsidiary, Hooker); and (6) notify Oxy's legal counsel in Houston.

The next morning Dr. Whorton drove up to Lathrop to meet with Oxy and the union. In the afternoon he met with six of the seven men (one worker refused to talk about the tests and didn't come to the meeting). Dr. Whorton explained to them what the numbers on the lab slips meant. He told them that he'd like to have another set of tests done for confirmation and asked five of the six to take the tests over again at his offices in Berkeley on Friday. (The sixth had a prior vasectomy.) Dr. Whorton wanted to give them three days to build the sperm supply back up again, and he asked them to abstain from sex during that time. At the end of the day Dr. Whorton reported his plans back to the union and Oxy. Oxy decided to cooperate with him after Dr. Zavon had confirmed Dr. Whorton's qualifications.

Although no one was talking about which chemical in Ag Chem might be the cause of infertility, DBCP must have been on everyone's mind because the union had passed the word that DBCP was the chemical the workers suspected and because Oxy placed a call to Dow the day after the meeting with Whorton. Dow immediately sent a copy of the joint 1961 Dow-Shell report on DBCP to Lathrop.

Oxy's troubles put Dow on the alert. The company no longer manufactured DBCP at its Midland complex but had transferred the bulk of its DBCP operation to Magnolia, Arkansas, the year before, in 1976. A Dow management committee began to lay plans on how best to determine if there was a problem with its own workers.

On Friday, July 22, the five Oxy workers made the hour's drive to Berkeley to Dr. Whorton's office for the sperm retests. Blood was drawn to test for abnormalities in kidney and liver function, in the male hormones, and in cholesterol and thyroid levels. While they were waiting for the results, each man was given a physical examination and asked to provide a medical history. Dr. Whorton could get clues to underlying causes of infertility just by looking at a man. He checked to see that there was normal hair distribution, on both the head and the face. He made sure there was no gynecomastia, or swelling of the breasts. He measured the testicles and felt for abnormalities in the scrotal sac. He listened to the heart, the lungs, felt the abdomen. He did a rectal examination. He asked about childhood and other diseases: Had any of the men had mumps, tuberculosis, venereal disease, or chronic prostate problems?

When he examined Mike Trout, they talked about his sperm count. The initial count had been 4.2 million—solidly in the subfertile, or oligospermic, range. Motility was less than normal at 40 percent, and doctors had noted 10 percent abnormally shaped sperm. Mike told the doctor that he and Marta had not been using any birth control methods and wanted to have more children. He told him, too, about his brain surgery and subsequent radiation therapy and that his last brain scan evaluation had been good. He'd had mumps as a child, but they had occurred prepuberty and so probably were not a factor in his infertility. When Dr. Whorton examined him, he felt a slight varicocele, or varicose vein, in his left testicle. It was something that might play a role, but the fact that he had previously fathered a child lessened its importance. There was nothing else out of the ordinary in his medical history. Mike smoked a pack and a half of cigarettes a day and had done so since his late teens, but he was not much of a drinker, and his only recurring health problem was an upset stomach that was relieved by antacid pills.

Late that afternoon the lab sent the results of the sperm analyses over to Dr. Whorton. The news was devastating. The first-round results had been accurate. There was no change the second time around. Four men were azoospermic, with no sperm at all. One—Mike Trout—was severely oligospermic. His sperm count was less than 1 million. There were too few sperm in his ejaculate even to evaluate the shape. All five workers were given the bad news before they drove back to the valley that evening. "Okay, it's real," Dr. Whorton told them. Then he telephoned the union and the company to set up a meeting the next day, Saturday, in Lathrop.

Elements of Risk / 221

It was decided that everyone who worked in Ag Chem must be tested and that the company should locate two former employees with known fertility problems. In the meantime, Oxy called a meeting Monday morning of all Ag Chem employees to fill them in. Everyone who had worked fewer than three months in Ag Chem, and the lone female production worker, were transferred to other jobs. One worker asked to leave, and he was allowed to go. Given no choice, the rest of the men kept the Ag Chem line running. They were in the middle of a DBCP production run.

In the midst of everything, the two industrial hygienists who had surveyed Ag Chem in May came back to take their July reading. They toured the plant on July 26. By noon the temperature had reached ninety-five degrees, and there was little wind. This time they found the workers on the line wearing rubber gloves and half-mask respirators. The concentration of DBCP in the air was less than half a part per million, but quick bursts of exposure on the canning line averaged 1.8 parts per million. The hygienists reiterated their criticism of the lack of a mechanical venting system in the plant but congratulated the "substantial effort" made by supervisory personnel "to provide healthful working conditions." Two days later Oxy notified the state's Occupational Safety and Health Agency (Cal-OSHA) about the fertility tests. The next day Lindley found out that the first round of expanded tests on other workers in Ag Chem was turning up more sterile men. He suspended all pesticide production in Ag Chem.

The news of the sterility scandal at Oxy could not be contained. It was leaked to one of the television stations in San Francisco and broadcast nationwide. On Friday, the day that the pesticide line in Ag Chem was ordered stopped, television crews from the ABC and CBS affiliates in San Francisco drove up to Lathrop. The local press was already there, and they beset the workers as they came out of the plant. "This has got me a little depressed," Mike Trout told the Associated Press. "My wife and I have been trying to have another child." He told the Stockton *Record* that he was planning to sue Oxy. But he said: "Money's not going to bring me any happiness. It might make me rich, but it won't bring me happiness." Jack Hodges, one of the older workers and a union steward in Ag Chem, told reporters that he had first suspected trouble from listening to "lunchroom talk." He said, "I started looking around and there weren't any children being born." Ted Bricker, who, although sterile at age thirty-one, considered himself lucky because he didn't want any more than the two children he already had, said the idea that one of the chemicals they worked with was causing infertility had been a "theory among the guys for at least four or five years."

It was a made-to-order front-page story. The AP story was sent to thousands of newspapers across the country and widely used. The lead said:

> LATHROP, Calif.—The men noticed it first, swapping stories over lunch at the chemical plant where they worked in this tiny central California town. That was a couple of years ago. Today, part of the Occidental Chemical Company plant is closed, and doctors are scrambling to figure out what made several young workers sterile.

Jim Lindley was interviewed by the television reporters, and he let the crews photograph the Ag Chem building. "At this point, we just don't know what the cause is," Lindley said, adding that he had never seen or heard of a report linking DBCP to sterility in lab animals until the week before. The story was broadcast Friday and Saturday nights and picked up by the national networks. Lindley thought the networks did a good job of reporting.

Lindley was well prepared for the reporters' questions by Dr. Zavon, Hooker and Oxy's new medical director, whose long years in the pesticide business taught him what to expect. In late July Dr. Zavon had typed up a list of anticipated questions and "suggested answers":

1. How long have you known about the problem of sterility among your employees?

We first learned of this on July 18.

2. What about the workmen's compensation claim presented by Mr. Wesley Jones?

He was examined a year ago and found to have a problem. We did not associate his difficulty with a work exposure. Only when a group with similar work experience were found to all have a similar problem was there any reason for concluding that Mr. Jones' problem might be work related.

3. Didn't you check for the effect of these pesticides on your employees?

Yes, we certainly do and have done so for a number of years [Dr. Zavon later substituted "periodically" for "certainly"]. The accepted practice in pesticide exposures in California is to check on the cholinesterase concentration in blood and plasma when cholinesterase inhibitors are involved. That is fairly specific. Other than that, a general medical examination is common but not necessarily customary. You must appreciate that checking an employee's sperm is just not normally done. That is why this finding is so utterly unexpected and raises all sorts of questions as to causation and future surveillance.

Elements of Risk / 223

4. What are you doing about the other employees on this job?

All who agree to it will have the same examination being given to the employees already examined. We also expect to have complete physical examinations conducted on employees who work in other parts of the Lathrop plant.

5. Who is doing the examinations? Is he your company physician?

No, the examiner, a specialist in occupational medicine, was selected by the union. We are working closely with him and he has conferred on this problem with our corporate medical director and various colleagues in different medical specialties.

6. What kind of examinations are being given?

A general medical examination and an examination of the sperm. There is also special attention being paid to liver and kidney function. If indicated a biopsy is done and special endocrine studies performed.

7. Do you have any idea of the cause?

No, I'll have to leave that to the medical men and the scientists to try to figure it out and tell me.

8. There are many formulating plants in California and none have ever reported this problem. Is your plant worse than the others? Why Oxychem? [The word "worse" was scratched out and "different" substituted.]

No, we don't think we're any worse than the other formulators. [That sentence was rubbed out and replaced by "Our plant is similar to many other formulation plants".] I haven't the data, but in terms of chemical exposure we may be better than some of the others. In fact, we will soon have specific data as we're just winding up an industrial hygiene survey which was started this spring.

9. Are you closing down the operation?

No, our medical advisors have advised us to continue operations in the hope that we could more readily find the cause. They have also advised that the damage seems to have been done, whatever the cause, and that it didn't seem to make much sense to throw people out of work when we didn't know whether the cause was still present or was gone. Also, since this doesn't appear to threaten the life of our employees we are, at least for the moment, continuing operations.

Dr. Zavon apparently had second thoughts about the last "suggested answer" and deleted it from his list. But some of the workers would have agreed with him. Stanley Mize, one of the original five workers who tested as sterile, was childless at thirty-five. "I'm already in the pot so I might as well stay and find out what it is," he mused. "What good's compensation going to be if they close the plant down on you?" One of Mize's twenty-three-year-old co-workers said that he would volunteer to work in Ag Chem when it reopened because he was on top of the layoff priority list.

Cal-OSHA wanted an emergency meeting with Oxy, and representatives of the state agency met with Oxy management that Friday evening. Also attending were officials from the federal OSHA and its advisory arm, the National Institute of Occupational Safety and Health (NIOSH), which conducts scientific research on workplace hazards and recommends worker exposure standards. Dr. Whorton was there, too, with his wife, also a doctor, who was assisting him in the case, as well as the two industrial hygienists who had surveyed Ag Chem and a hygienist for Cal-OSHA who had spent the previous three days in the plant. The Cal-OSHA hygienist thought that Ag Chem's equipment and procedures were "better than many, but not as good as some." NIOSH asked Dr. Whorton to conduct a health evaluation of the plant as part of its investigation, so he was now working for the federal government, the union, and the company—a splitting of allegiances that caused some of the workers to distrust him.

During the two weeks following the initial infertility findings Dr. Whorton examined another thirty-four workers from the Ag Chem area, including the supervisors, maintenance men, and lab workers. Three were women, but only one of the women actually worked in the production area, and she was taking birth control pills. Of the remaining thirty-one men, eleven were vasectomized. When Dr. Whorton added the original five to the twenty men who had not had vasectomies, he found that a total of eleven were azoospermic or severely oligospermic, with counts of 1 million or less, and three had counts under 30 million. The rest measured above 40 million, which was considered within the normal range. Thus, fourteen of the twenty-five male Ag Chem workers who were potentially fertile had below normal or zero sperm counts.

When Dr. Whorton analyzed the data, he found a striking correlation between the sperm count and the number of years spent in Ag Chem. The eleven workers who were azoospermic or severely oligospermic had worked in Ag Chem for at least three years. None of those with normal range counts had worked there for more than three months.

The eleven workers with depressed sperm counts also had significantly higher levels of the follicle-stimulating hormone (FSH) —a telltale sign of testicular damage.

Confronted with Dr. Whorton's report, Dr. Zavon was privately concerned about the incidence of infertility at Oxy and the "difficult problem" it posed in the face of "the inevitable media sensationalism, attempts to generate legal action, and the almost certain political opportunism." In a memo he wrote on August 1 to James Galvin, Hooker's Agricultural Products Groups president, Dr. Zavon laid out his concerns about "this unusual report, perhaps the first time an occupation has been associated with infertility without any more prominent signs of illness." He wrote:

> The incidence of infertility among males in the age range 17–55 is probably of the order of 5 to 10 percent at most. Therefore, it is worthy of note when a group of men within this age range show an incidence of infertility of approximately 67 percent without any history of vasectomy. It is even more noteworthy and raises a strong presumption of causality when such a group all work in the same work area and presumably have similar exposures to the workplace environment.

Also on August 1, a memo from Jim Lindley circulated through Hooker headquarters in Houston and Niagara Falls and Occidental Petroleum corporate offices in Los Angeles. Lindley briefed the executives on the Lathrop plant itself, what it made, how many people worked there, and "the medical problem." He reviewed Dr. Whorton's findings and suggested that the "one ray of hope" might be the worker with the highest sperm count, who had worked in Ag Chem for two years in the early 1970's and been out of the building ever since. "Maybe there is a chance for regeneration," Lindley wrote. He briefed his superiors on testicular biopsies scheduled on several workers on August 11, which "may shed some further light." He reported that Dr. Whorton had heard many complaints from the men in Ag Chem about loss of smell and a high divorce rate.

Under a section he titled "The Future, what OCC West plans to do about it," Lindley wrote:

A. Locate and medically test all former Ag Chem department employees.
B. Establish and medically test a control group of plant employees not connected with Ag Chem. Drs. Zavon and Whorton are coordinating this effort.
C. Establish and medically test a control group of local citizens not connected with OCC [Occidental Chemical Company].
D. Drs. Zavon and Whorton have approved restarting pesticide formulations operations utilizing as workers only those persons who have vasectomies or are already sterile.
E. Cooperate with the Steering Committee [Zavon, Whorton, and two Cal-OSHA representatives], Cal-OSHA, NIOSH, and EPA in de-

veloping programs to further investigate sterility in pesticide formulation activities.

F. Respond forthrightly to media requests for information.

There was, in fact, a brief, unauthorized start-up of the Ag Chem line on August 2 using vasectomied or already sterile workers, but Cal-OSHA called and ordered it stopped. In the first place, no one yet knew for certain which of the 200 or so pesticides formulated in Ag Chem had caused the fertility problems. All signs pointed to DBCP, but there were a few others worthy of investigation. And if DBCP were indeed the culprit, might it not have other health effects? A man might not want to have any more children, but what if you were exposing him to a carcinogen?

Indeed, cancer was on everyone's mind. When Rex Cook called a meeting at the union hall in Lathrop, the room was loaded with anxiety and rumor. Here was a group of big, boisterous men—more than a few of them were pushing 250 pounds and Dr. Whorton had joked that he felt as if he were examining the Oakland Raiders—trying to cope with two of the most frightening words in the English vocabulary: sterility and cancer. Dr. Whorton drove up again from Berkeley to answer their questions as best he could. Despite some of the workers' doubts about which side of the fence he was on, Dr. Whorton had been deeply involved with occupational health for the past eight years. He knew that Oxy's support was essential in getting the answers he needed to help the workers. He knew also that the company had committed some grievous errors in Ag Chem, but it did not completely fit the industrial bandit image that was being promoted in the press. Jim Lindley and Greg Vervais were genuinely concerned about the workers. And Dr. Whorton knew from his own work in the field that Oxy probably could not have headed the problem off any earlier by getting sperm counts from the men. That kind of male reproductive monitoring was rare in the chemical industry. Besides, the workers themselves probably would not have cooperated if the company had arbitrarily demanded sperm tests. "If you had asked me in June 1977 if I could go to a plant and talk to a hundred men and convince them to masturbate in a jar, I would have said you couldn't run fast enough to get out of there," Dr. Whorton said later.

But the doctor saw the other side just as clearly. He knew that there had been a strong warning about DBCP's reproductive effects in the 1961 Dow-Shell report and that someone should have paid more attention to it. He knew the workers had been given practically no information about the chemicals they worked with and that they were bitter because they felt as if they had never had a chance to make an educated choice about staying or leaving. He knew how devastated by their infertility some of the younger men were, especially those who were just starting families. One worker, Frank Arnett, not yet twenty-one, had just married his seventeen-year-old sweetheart shortly before the scandal broke. Tests showed that his sperm count was severely depressed, and he and his wife were heartbroken. Dr. Whorton knew that some of the men had become impotent after learning that they were infertile —clearly a psychological reaction because infertility did not affect the mechanics of an erection. He could imagine the toll that would be extracted from many marriages. And he knew that almost to a man, they were worried about cancer. If chemicals could destroy their fertility, they reasoned, couldn't they just as easily invade other parts of the body and cause disease?

The men asked a lot of questions at the meeting in the union hall. If they were azoospermic, what were the chances of recovering fertility? Was there more reason for hope if you had a few sperm instead of none? What about the future? Cancer? Birth defects in their children, if they managed to have any? Dr. Whorton told them that DBCP was an animal carcinogen. If a biological effect had been exerted on the reproductive system by DBCP, there was probably an increased risk of developing cancer. Just because it caused cancer in animals, though, didn't mean it would cause cancer in humans. As a precautionary measure he advised the men to get annual physical examinations with special attention given to the cancer warning signs.

Dr. Whorton wanted to know how much damage had been done to the sperm-manufacturing equipment in the testes. Had the primary germ cells themselves been destroyed by the action of the chemical, or was there just some sort of physical blockage gumming up the works? Ten men underwent testicular biopsies. The majority had no sperm-making cells in their testes. As with the sperm counts, the longer the exposure to DBCP, the more severe the damage.

The evidence mounted against DBCP. Not only did it have documented reproductive effects in animal experiments, but it was by far the largest volume product in Ag Chem. Supporting evidence came from tests on Dow's and Shell's DBCP workers.

Through a complicated series of patents and licensing arrangements, Dow and Shell essentially controlled the DBCP market in the United States. Each produced about 45 percent of the 25 million to 30 million pounds of DBCP marketed annually. Shell had made DBCP at its Colorado plant for twenty years and then moved production to Mobile, Alabama, in the spring of 1976. Dow had done much the same, making DBCP at Midland, Michigan, for eighteen years and then relocating the operation to its Magnolia, Arkansas, plant in January 1976.

As soon as they heard about Oxy's troubles, Shell and Dow began looking at their own DBCP workers. Shell tested twenty-one workers from Colorado and Alabama. Sixteen of them had depressed sperm counts. In Arkansas, twelve of fourteen Dow workers had abnormal counts. Dow had not been convinced that DBCP was at fault and had sent around a low-key "Dear Customer" letter in mid-July advising formulators to adhere to the one part per million exposure limit. But faced with the intractable reports from its own plant in Arkansas, where DBCP workers had allegedly been protected by Dow's tight exposure rules, the company shut down production and sales in mid-August. Shell did the same, and both companies announced national product recalls. Dow's Dr. Verald K. Rowe, who had conducted some of the early research on DBCP, told reporters, "It came as a total surprise to us." When reporters asked him why Dow hadn't told the workers about the possibility that DBCP could affect sperm counts and why the company hadn't tested them periodically to see if that was happening, Rowe conceded: "I think our medical people would have to agree, in retrospect, this would have been a nice thing, a good thing to do."

The scientists at Dow were distraught when they heard the news about DBCP. "I was shook," recalled Dr. Ted Torkelson, who had also done some of the early research work on the nematocide. "It hit me like a ton of bricks," said Dow vice-president Dr. Etcyl Blair. "It was kind of like infidelity with your wife or something."

On August 11, the day that Dow suspended production of DBCP in Arkansas, James Heacock, chief of the California Department of Health's occupational health division, received an ominous report from one of his investigators:

> The effect of DBCP in producing sterility in human males has been firmly established from studies done by physicians at UC [the University of California] and at the Dow Chemical plant. Animal studies have shown that DBCP is also a very potent carcinogen, producing not only primary cancer, but rapidly spreading metastatic lesions as well. In my opinion, it is imperative that this material should no longer be manufactured, formulated or applied for agricultural use, effective immediately.

The next day, August 12, 1977, the California Department of Food and Agriculture issued an emergency order suspending the use of pesticides containing DBCP.

chapter 9

News of the DBCP sterility scandal spread around the world. It was the first documented case of male infertility in the workplace, and it caught the attention of scientists as well as the news media. Carl Djerassi, a chemistry professor at Stanford University who did much of the development work on the female birth control pill, read about Oxy's troubles in the newspaper. Where others saw a sow's ear, he saw a silk purse. On August 18, hardly more than a month since the workers first learned of their plight, Professor Djerassi wrote a letter to Dr. Alexander Kessler, chief of the World Health Organization's human reproduction unit in Geneva:

> What I am writing you about is what I consider one of the most exciting recent leads in the male contraceptive area and one that ought to be taken up right away by your male contraceptive task force.
> You may have read in the paper that a group of employees at a subsidiary of Occidental Petroleum Corporation, who [sic] has been formulating for years the soil fumigant dibromochloropropane, has been found to be completely sterile. Manufacture of this material is now discontinued and these workers are being examined carefully in terms of their health record and, of course, will be followed up. If it should turn out that they exhibit no other deleterious clinical syndromes, then this would clearly be an extraordinarily important lead in the male contraceptive area. If it should also turn out that their sperm count returns to normal during the coming year after they have stopped working with this material, then this could be sensational.
> What is really important is that inadvertently a major clini-

cal experiment on a male contraceptive may have been carried out prior to the necessary animal toxicology.

The professor discussed his plan with Donald Baeder, president of the Hooker Chemicals Company, which was Oxy's sister subsidiary. Baeder liked the idea. He asked Jim Lindley at Oxy to file a patent on the use of DBCP as a contraceptive. In a memo "From the Desk of Donald L. Baeder" on September 6, the president of Hooker told Lindley and Dr. Zavon to cooperate with the World Health Organization if it reacted "positively" to Professor Djerassi's idea. "Carl Djerassi believes DBCP could be a very exciting lead in what is a world need in male contraception," Baeder told them. "At my concurrence, he has taken the initiative to get some free R & D." Before signing off, Baeder asked Lindley, "Jim, did you file a patent as we discussed several weeks ago?"

Dr. Zavon liked the idea, too. In a memo to Baeder on September 14 he wrote: "One of the first things that came to mind when this problem arose at Lathrop was just the possibility that Djerassi raised in his letter to Kessler." But cooler heads prevailed, and despite the support of Hooker's top business and medical people, the patent was never filed.

While Oxy and Hooker were privately putting their feet in their mouths, Dow and the agricultural industry were doing a more public job of it. Growers and their support groups were worried that DBCP might be banned for use on certain crops. It was a particular nightmare for the peach industry, which argued that the loss of DBCP would reduce America's fresh peach supplies by nearly a quarter. Like the chemistry professor at Stanford, the National Peach Council spotted the silver lining in the thunderclouds. "While involuntary sterility caused by a manufactured chemical may be bad," Robert K. Phillips, the council's executive secretary, wrote to OSHA on September 12, "it is not necessarily so."

The peach growers envisioned DBCP factories staffed by older workers past their childbearing years and by young men desirous of free chemical vasectomies. They reasoned that workers who did not want to have any more children but who had not gotten around to having a vasectomy or who wanted to circumvent religious bans on birth control could volunteer to man DBCP production lines. "After all," Phillips wrote, "there are many people now paying to have themselves sterilized to assure they will no longer be able to become parents." The peach council's letter was leaked to the press to predictable derision, but the idea of putting older and already sterilized workers on the line was not so new. Oxy had done exactly that until the government shut it down.

In serious and joking ways, the outside world inexplicably belittled the workers' anguish over their loss of fertility. One of the first lawyers contacted by several Oxy workers jocularly suggested that working in Ag Chem was a cheap way of getting a vasectomy. And a Dow vice-president sent to monitor sperm tests at its Magnolia, Arkansas, DBCP plant rather cavalierly told the *Arkansas Gazette* that "we have actually had situations where some guys have had vasectomies and want to check and find out how good their vasectomies were."

In fact, as sperm test results were tabulated at Magnolia, it became clear that Dow had a problem at least as bad as Oxy's, and probably worse. Twelve of the first fourteen workers examined had depressed sperm counts. When the results were added up, more than half of Dow's Magnolia workers—fifty of eighty-six—were sterile or had low counts.

Magnolia was a tree-shaded town of 12,000 people in the far southwestern corner of Arkansas. Dow had moved its DBCP operations there from Midland in January 1976. The money was good; a man could make $12,000 a year in steady work and avoid the vagaries of the region's only other major employments in timber and oil.

The workers weren't talking; the local newspapers speculated they trusted Dow and did not want to rock the proverbial boat for fear of losing their jobs. But the picture painted by Dow supervisors was of a reasonably well-informed and very well-protected work force. The one part per million exposure limit was enforced, although not continuously monitored, and workers wore neoprene gloves, jackets, and boots. The toxicology data on DBCP were reviewed with every employee on an annual basis, although testicular effects were not mentioned because Dow medical people believed that problems would develop first in the lung, liver, or kidney. Dow said it also told its DBCP workers that the chemical was a suspected carcinogen.

Dow officials had been reiterating from the beginning that DBCP could be safely used if air concentrations were held below one part per million. But if safe exposure levels had been maintained at Magnolia and the workers practically wrapped in neoprene, why, then, were half of them infertile? "We don't have enough data to say," Dow vice-president Dr. Etcyl Blair told the Arkansas newspapers. "We do not know every minute how much exposure [the workers] received. Someone may have spilled some. This is what we have to sort out."

The newspaper reporters were still puzzled. Since Dow was the touted repository of scientific information on DBCP in the chemical industry—not only had its scientists been the lead researchers in the 1961 DBCP study, but the company had developed its own exposure standard in the absence of government regulations—why had it not included the reproductive warnings from its 1961 study on its own safety information sheets about DBCP? "I guess at that time we weren't all that concerned about it," Dr. Blair said. "That's hindsight. Our concern was the effect on these other organs more than it was that." Blair also tried to minimize the workers' concerns about cancer. DBCP "probably is not going to have an effect," he said. "The things that are really going to cause the cancer are going to be: do they smoke cigarettes, do they eat a lot of fatty substances, these kinds of issues from a physician's point of view or from a scientist's point of view are really the things that are going to cause cancer." That same week the industry trade journal *Chemical Week* ran a story saying that DBCP produced the earliest cancer tumors ever seen in laboratory animals.

In early September some of the Magnolia workers broke their silence. The stories they told about working with DBCP were far different from those circulated by Dow management. The workers who talked to the local newspapers were cautious: They wouldn't let their names be used, and they insisted that they did not want to hurt the company. But they were bitter. Four of the workers who tested out sterile had just gotten married that summer. "We don't think they know how it is up yonder," they said. "Up yonder" meant Dow headquarters in Midland.

The workers said that when DBCP production began in early 1976, they got the chemical "all over our hands, all over our boots, and you'd splash it on your pants legs some catching samples." They said the unmistakable smell of DBCP shrouded the plant, despite the cautions of the 1961 report which said, "The warning odor MUST not be ignored." They admitted avoiding the neoprene slicker suit and gas masks during hot weather, but the newspaper pointed out that Dow itself had studies showing that neoprene did not provide adequate skin protection against DBCP. The workers conceded they had been told about the results of animal experiments implicating DBCP as a carcinogen, but they said the company downplayed the findings.

Dow, meanwhile, was concerned about its Midland employees who had worked with DBCP for eighteen years before the operation was moved down to Arkansas. "There doesn't appear to be any lack of children being born in the Michigan division," Etcyl Blair said. In early September Dow began broadcasting special hourly messages on its extensive closed-circuit television system at the Midland plant explaining the DBCP problem and asking workers to volunteer for testing by company doctors. Dow said it did not want to make a "sideshow" of the fertility tests and so was allowing workers to volunteer, rather than have them searched out and contacted by management.

With DBCP the common factor in cases of sterility at three factories in three states, it seemed the inescapable culprit. California had put a temporary lock on DBCP use by suspending registration of products, but there was no movement at the federal level. Nor was there an official workplace exposure standard. Dow and Shell had shut down DBCP production, but there were in the country eighty or so smaller companies at which workers dealt directly with the nematocide. Finally, on September 9, the federal OSHA issued an emergency temporary exposure standard for DBCP and prohibited eye and skin contact with the chemical. The exposure limit was ten parts per billion averaged over an eight-hour workday—a limit which was subsequently lowered to one part per billion, or 1,000 times less than the one part per million standard traditionally recommended by Dow. The standard was so tough to meet—too tough, some critics said—that it would have been impossible to manufacture DBCP without violating it.

OSHA's action was just one prong of a trio of regulatory moves announced on September 9 at a press conference in Washington. The EPA took the first step toward a national ban on DBCP when it announced its intention to suspend registered uses on nineteen food crops—including squash, lettuce, tomatoes, and several root crops—and to impose restrictions on other uses. And the Food and Drug Administration announced that it was examining food crops to determine whether or not they retained harmful levels of the chemical. Regulators had long thought DBCP degraded into relatively harmless by-products, but recent tests had shown some of the chemical's residues on food.

Like actors in their scripted roles, the stars of the DBCP drama moved through the traditional acts: the scandal, the public outcry, the federal regulatory response, the company denials, and the attempts to head off government clampdowns and product liability lawsuits. But in this last act the DBCP companies were less surefooted than they had been when caught in the same performance with other products. The industry had fought for years to keep DDT on the market and believed, even now, that its banning in 1972 had been a political, not a scientific, decision. Operating on the same premise—that politics, not science, was at the base of regulatory action—Dow had rebuffed the initial attempts of the federal government to restrict the herbicide 2,4,5-T in 1969 and had managed to keep it on the market throughout the 1970's by manipulating a weak EPA and using innumerable regulatory loopholes. But the single-minded obstinacy exhibited by the industry with these other chemicals was not as evident with DBCP largely because company officials were privately admitting to themselves that the nematocide was not only a messy public relations problem and a potential litigation disaster but also quite possibly a dangerous product. The DBCP scandal may have struck a sensitive psychological chord. "This is a chemical that gets you in the balls," observed one lawyer, who thought that corporate reaction to DBCP was more swift than it might have been if another chemical had been involved. "This was a male corporate reaction to a chemical that gets men."

In internal conversations at Shell, Dow, and Oxy, executives did little equivocating about the connection between DBCP and male infertility. They were less willing to concede the link between DBCP and cancer. And there was no uniformity of opinion about how to handle the situation. Privately all three companies acknowledged the chemical's danger. Publicly all three companies blamed a combination of ignorance, oversight, government bungling, each other, political manipulation, public hysteria, and even the workers themselves.

At Shell, 55 percent of workers tested had depressed sperm counts. A brief summary report on DBCP compiled for Shell Oil Company executive vice-president J. P. van Reeven in advance of a September board of directors meeting was blunt in its assessment of DBCP. "There is a very strong likelihood that exposure to dibromochloropropane may have a negative effect on sperm counts of chemical plant workers," the report said. The whole history was laid out: the 1961 study and its warnings about reproductive effects, the lack of follow-up animal reproduction tests after that study, the failure to tell workers about possible fertility effects, the cancer studies, and the grisly aftermath. At its September 29 meeting at One Shell Plaza, in New Orleans, the board approved a quarterly stock dividend of 40 cents a share, okayed a life insurance program for senior staff, approved higher salaries for certain executives, discussed whether or not the company should exercise its option on a 146-acre tract of land in a "fast developing sector" of suburban Houston, and took care of some other business. Yet the board steadfastly ignored the whole DBCP issue, and Shell itself adopted a low public profile and a defensive private posture. When a Shell executive gave something of a pep talk to his dispirited sales and management staff during the height of the DBCP controversy, he emphasized the public's growing distrust of the chemical industry. He quoted from a speech that Shell Chemical president Jack St. Clair had given on "Ethical Challenges for the Chemical Industry." St. Clair had said, "We are now seeing unprecedented concern about chemicals and their potential health hazards. Surveys show that there is a feeling among the general public and leadership groups that there is a problem which is not under control—the problem being that neither industry nor government knows the full effects of the chemicals that surround us—or if they do know, they are not telling the American people the whole story." Shell's president said that he had seen "a great increase in the feeling that the chemical industry is too secretive, that we aren't providing enough information about the products we make, and particularly, that we aren't warning people about the dangers of the products they come in contact with daily." St. Clair believed that "despite all the testing, and taking all the known precautions, we really don't know everything there is to know about many of the chemicals we produce including, in some cases, harmful effects we did not imagine at any stage of product development."

The Shell executive admonished his troops to take to heart St. Clair's speech and to do their part in the larger battle by paying attention to safety in the workplace. " . . . You may wonder what good it does to start worrying about protective clothing now," he conceded, "after we've been handling Nemagon without them for all these years."

He ended on a philosophical note. "The real significance," he said, "may be that we are entering a whole new era in pesticide use. It's the era of cancer and learning to live with it."

At Dow the company's efficient public relations staff produced a brochure called *Dow's Reaction to DBCP*. The brochure reviewed the chemical's history, including the fact that scientists who wrote the 1961 study had tested down to a five part per million level but had recommended a one part per million exposure standard on the basis not of hard evidence but of the crap shoot belief that "odds of a problem at 1 part per million" were small. Dow continued to hold to the theory that it took more than a one part per million exposure level to DBCP to cause depressed sperm counts, even though no animal tests had been done to prove it. In the company's quarterly report to stockholders in October, Dow president Zoltan Merszei reported that sales for the first nine months were up 12 percent to $4.7 billion. He also wrote a letter to the stockholders reviewing Dow's "new company objectives." Merszei said, "They put us on record about our values and our attitudes toward business and society." Among them: "To seek maximum long term profit growth" and to be "scrupulously ethical in the means to our ends and in the ends themselves"; to be "responsible citizens." Merszei wrote: "Some people might prefer to see us downplay our concern for profit in favor of doing great and good things for society. Somehow, they misunderstand the role of business. We use profit as a tool, as the MEANS to an end."

As the frontline formulator Oxy was theoretically in the tightest legal bind, and its internal discussions reflected such pragmatic concerns as whether DBCP caused cancer in man, and could a man rendered infertile ever recover a viable sperm count. Furthermore, Oxy was mulling over the chances of getting back into the DBCP business again. In October Dr. Zavon, the corporate medical director, told Jim Lindley, the top man at Lathrop, that DBCP was most likely the chemical that caused the workers' infertility. He waffled on the cancer issue but was not nearly so sanguine as Dow and Shell. He didn't like the stomach-feeding method used in the National Cancer Institute studies that showed DBCP causing cancer in rats, but he conceded "the finding in rodents always raises a presumption of risk in man from the same substance." He wanted some answers: What is the no-effect level in animals for reduction in sperm? Is there a "point of no return" or will all those affected eventually again produce sperm? Is DBCP a carcinogen for human beings? Dr. Zavon wrote:

> If an answer to the cancer question is needed to help make a business decision regarding DBCP, a good scientific answer is unlikely to be available soon. A scientific political answer is available now. That is: no evidence is presently available to demonstrate that DBCP causes cancer in man.
> A politically acceptable way can probably be found to continue using DBCP without significant risk to manufacturing, formulating or applicating employees. But more toxicology will have to be performed. If you wish to remain in the DBCP business, I would suggest a toxicological program be initiated. . . .

In October the California Department of Industrial Relations, the state agency charged with protecting the health and safety of workers, opened four days of hearings on DBCP. Department director Don Vial called it an inquiry the purpose of which was "to find solutions to immense occupational health problems which we now face." Critics suggested that the hearing was really just a dog-and-pony show staged by state bureaucrats to fix the blame on a corporate cover-up. It was an after-the-fact hearing with no regulatory clout, and although it provided a clearinghouse for information about DBCP, it became something of a free-for-all as the government blamed the corporations and the scientists, and the scientists and the corporations blamed the government, for what one writer called the "agrichemical Watergate of the 1970s." On another level, some insiders hoped the hearing would make it easier for injured workers to sue Dow and Shell by showing that the two giant suppliers were more culpable than Occidental—a guilt not of commission but of omission of warnings about the hazardous nature of DBCP. "It's very easy to hide and withhold information and indeed information was withheld," Don Vial observed. "The price of that is 90 workers."

The centerpiece of the hearing was a lengthy grilling of Dr. Charles Hine, the University of California toxicologist who had collaborated with Dow scientists on the 1961 DBCP study. In the days following the sterility revelations, questions had been raised about his role in the 1961 study as well as about his university-industry ties. At the hearing Dr. Hine was served up as the fatted calf, and he conveniently provided the cooking oil by appearing furtive on the stand. The interrogator was Peter Hart Weiner, the department's young and ambitious chief counsel who had been preparing for the hearing "night and day" for a month. At the hearing Weiner bore down hard on Dr. Hine's failure to test to a no-effect level in the early DBCP studies.

Dr. Hine was a big man with a booming voice. He explained that he was surprised when liver, kidney, and testicular damage occurred at the five parts per million level because his early projections put that as the no-effect dose. Since the effects were not as severe as they were at higher doses, he predicted that reducing the exposure fivefold, to one part per million, would hit the no-effect level. There wasn't another chemical in DBCP's wide family of chemical relatives that had a lower threshold value than one part per million, and it seemed a safe estimate.

"In fact you didn't test for a no-effect level. Correct?" Weiner asked Dr. Hine.

"No, I say we did not," Dr. Hine replied. "I think . . . that it's quite proper to criticize this study. That we did not go down to a no-effect level. Our explanation was, there are other things to do, and we felt that a fivefold reduction would have given the no-effect level."

Don Vial, the division director, fed soft punches. " . . . I'm a little bit confused," he said to Dr. Hine. "I think you were reported in *The Wall Street Journal* and *The Washington Post* [to have] indicated that Dow and Shell didn't follow your recommendations adequately, and then I think most recently in the San Francisco *Examiner,* you said you blew it."

"Counsel," Dr. Hine said, " . . . I don't know what appeared in either of those papers. I certainly did not say that. I said we had given clearcut recommendations, and I don't believe that I . . . had made the statement that they did not follow them."

"I agree," Vial said, backing off. "And, I . . . no. And I really don't wish to pursue that."

At that point, Weiner broke in. "Dr. Hine, I don't know what you told the San Francisco *Examiner* which quoted you as saying 'I blew it.' I do know what you told me, and I want to quote you from my notes of our conversation. You said, and we have a corroborating witness here as well. . . . You said, well, by today's standards, well, even by those days' standards, it was kind of stupid not to go to a zero effect level. Unquote.

"Have you changed your mind since then?" Weiner asked.

"Counsel, I don't think that I said it was stupid," the scientist replied.

"Well, those are . . . the exact words you used, Dr. Hine."

"Excuse me," Dr. Hine said. "It was probably ill advised . . . let's say stupid, not to have gone to it."

"I can understand that," Weiner said.

"In retrospect, you can say, okay, it was stupid, by those standards," Dr. Hine said. ". . . and I think that the majority of investigations would have gone to a no-effect level. And I think we should have gone to a no-effect level, and I admit an error in this thing. And perhaps, looked at today with the seriousness, perhaps the term stupid is applicable."

But Dr. Hine lashed out at the state for its own ineptness in regulating DBCP. He pointed out that the state had been sent the 1961 study, which had been filed and forgotten. And neither the state nor the federal government had ever set a worker exposure standard for the nematocide. In fact, at the time of the hearing acceptable exposure standards had been established for little more than 1 percent of the 60,000 chemicals in common use in this country, and neither the state nor the federal governments had any kind of central system for baseline information about those chemicals.

The state investigators were only mildly repentant. Don Vial suggested that the system of worker protection was handicapped from the start because it was predicated on information generated by "private corporations that have an economic interest in marketing lucrative poisons." What was to prevent companies from playing down unfavorable reports on chemical products? What kinds of pressures were put on an industry consultant not to go public with damaging information about a chemical's health effect? "Here you've had some 30 years of dual relationship with the university and private interests such as Shell," Vial told Dr. Hine, "and the real problem that's bothering me is just how do we know when your advocacy on behalf of private interests ends and your responsibilities as an objective researcher begin?"

"Certainly this relationship I've had as a consultant with industry has never at any time been secretive," Dr. Hine countered. "If you were to ask me at any time about this, I am certain that the information could have been released to you. Not one government official ever asked me or asked Shell what should be the threshold limit value for this compound."

The rest of the hearing was taken up with industry's defense of its safety standards and scientists' warnings about the potential dangers of DBCP. Dow vice-president Dr. Etcyl Blair delivered five pages of assurances about the company's long history of chemical testing and industrial hygiene programs and about Dow's "paramount" philosophy of "product stewardship." Simply stated, "product stewardship" meant that Dow was committed to the revolutionary notion of making and selling safe products. Dow had a full-time director of product stewardship, among whose duties was to "help translate our policy" to a "grass roots awareness and positive attitudes." Further, Dow had an Industrial Health Board, which watchdogged the workplace and set internal company exposure standards for chemicals not regulated by the government. Ten percent of Dow's annual $165 million research budget went toward health and environmental studies. Dow's own scientific papers —between 10,000 and 15,000 had been published in recent years— demonstrated "a long and consistent tradition among the people of Dow to exercise great care and concern for the human and environmental effects of chemical products we make for society's use."

If Dow is "the chemist for the chemical industry," as Don Vial put it, and if its commitment to product stewardship was so strong, why had its own workers not been warned about the hazards of DBCP? Replied Blair: "It's rather like teaching your children or something, don't get in front of a car in the street. You don't spend your time talking all about all the various things that can happen to you if you get hit by a car, you might break a femur bone, you might do a lot of different things, but you are going to have problems if you get in front of a car. Our contention is that if you have exposures greater than 1 part per million you are going to have a problem, so our efforts have been directed in that line."

No tests had yet confirmed that one part per million was the no-effect level for DBCP, but Dow held firm. Dr. Perry Gehring, Dow's director of toxicology, testified at the hearing that he would work in a DBCP factory if air concentrations were less than one part per million and if his skin were protected. Dr. Hine respectfully disagreed and said he probably would not join Dr. Gehring in that factory under those exposure conditions until further testing was done. And Dr. Zavon, Hooker and Oxy's medical director, testified that he had known about the 1961 study since it was originally published. But he said he never related the study's testicular effects to DBCP workers because the research had been done on weanling, or young, animals and the workers were not exposed until they were adults. "It's a quirk," he said. "You look back and wonder why we didn't know, but we didn't." The industry witnesses argued that even if companies had tried to monitor workers' sperm counts, they would have met with failure. "If employers had requested a mass masturbation among their employees—and that is what sperm counts are—I don't think that based on the 1961 study, there would have been any response at all," said Howard Kusnetz, Shell's manager of industrial hygiene.

Dr. Zavon did not think DBCP was a potent carcinogen, and he branded people "alarmist" who thought it was. Like Zavon, Dow's scientists criticized the high dosages and the stomach gavage method used by the National Cancer Institute's cancer researchers to expose animals to DBCP. They said dumping high doses of DBCP into a rat's stomach could not be compared with the workers' exposure from inhaling the chemical in the air. But "potent" was the word used by a University of California researcher to describe DBCP's cancer-causing action in animals. On a cancer scale rated by potency, Arlene Blum, a research associate in biochemistry at the University of California at Berkeley, who had done work on DBCP, said the chemical would fall midway between aflatoxin—a powerful fungal toxin—at the top and saccharin—the artificial sweetener—at the bottom. She calculated that DBCP workers had been exposed to a daily dose of just under half the amount that gave half the rats cancer in laboratory experiments. "... So this suggests a really increased cancer incidence among workers that have been exposed to DBCP at those levels," she said.

Dr. Bruce Ames, a University of California biochemistry professor, talked about DBCP's mutagenicity. His "Ames test" was then being hailed as a simple and effective early-warning system for mutagenic and carcinogenic chemicals. Most scientists looked at the effects of a chemical on animals or, in the case of epidemiologists, on humans. But Dr. Ames was trained as a geneticist, and it made sense to him to look at DNA, the molecule that contains the cell's genetic, or hereditary, information. And since DNA operates by the same principles in all living things, be they bacteria or people, Dr. Ames worked out a simple test with bacteria in vitro, or in a glass dish. In a relatively quick and inexpensive fashion, he could incubate a chemical with bacteria in a petri dish and watch to see if it caused mutations. The theory was that mutagens attack the DNA in germ cells, destroying the hereditary information and causing birth defects. And since cancer is the uncontrolled growth of cells, Dr. Ames theorized also that chemicals, acting as mutagens, damaged the DNA in other cells, allowing them to grow uncontrollably and eventually become tumor cells. "So really when you get a chemical into you it can do two things," Dr. Ames said. "If it gets to the germ line, the sperm or the egg, then it can cause mutations, and these show up in your children and your grandchildren; and if it gets just to any other cell of the body, it can result in somatic mutation, and then 20 years later you start seeing tumors."

Dr. Ames came to believe that most chemicals that caused cancer were also capable of causing mutations, or birth defects. If the chemical caused mutations in bacteria, then it was likely a carcinogen, too. Like animal tests, the Ames bacteria test couldn't say for sure what kind of damage a chemical might cause in people, but it was an early-warning system that suggested the possibility of danger, especially if it was used to complement knowledge gained from animal experiments.

The problem was that a number of the chemicals already known to be powerful carcinogens weren't showing up as mutagens in Dr. Ames's glass dishes. The big break came when scientists realized that a chemical entering the body is sometimes converted or metabolized into another, more potent chemical inside. That converted chemical is the "ultimate" carcinogen.

Dr. Ames worked out a way to mimic the metabolic process in his petri dishes. Since enzymes in the liver control much of the metabolic activity, Dr. Ames put ground-up rat liver in his dishes along with bacteria. When a chemical was added to the concoction, the liver converted it into its "active," or ultimate, form. "Once we started using the liver," Dr. Ames said, "then we found that a very high percentage of the carcinogens we tested worked as mutagens in the system. . . . About 90 percent of the chemical carcinogens we've looked at came out as mutagens in the system." DBCP was one of them.

Dr. Ames was an outspoken scientist whose beliefs did not always dovetail with the prevailing industry view. He felt it was more prudent not to assume, as industry did, that a threshold of safety exists for every chemical, below which no harm occurs. He also believed that the "modern chemical world"—by which he meant environmental chemicals and food additives—could possibly cause future cancer and increased mutations in the "human gene pool." Because of the twenty-plus-year time lag between chemical exposure and development of most cancers, Dr. Ames predicted at the hearing that "much of the modern chemical world we just haven't seen yet in terms of its effect on cancer." That effect, he said, "we won't see until 1980, 1990."

Dr. Ames's talk of the "modern chemical world" and its effect was anathema to the industry scientists, who believed that the bulk of cancer was caused by cigarettes and bad diet. Dr. Ames believed this, too. But it was his contention that man-made chemicals as well as natural chemicals present in the diet—along with cigarettes and ultraviolet light—were capable of causing mutations and cancer. Dow's Dr. Gehring told the hearing that fewer than 3 percent of all cancers were caused by synthetic chemicals, and he suggested that attempts to estimate the risk of cancer from DBCP and other pesticides were "an exercise in futility." He said, "It is high time that the public recognizes there are numerous compounds capable of producing cancer when given in high amounts. Nature has contributed as many of them as has the ingenuity of man." The Dow scientists also disputed the idea that there is no known safe dose of a chemical. They took issue with the theory put forth by Dr. Ames and others that every dose, no matter how tiny, might have a cumulative effect. They said that would be like saying that if a 100-mile-per-hour wind could knock down your house, then a 10-mile-per-hour wind could do the same.

But the university scientists disagreed. In DBCP's case, they said, it wasn't just a matter of dosages; it was a matter of the body's enzymes being overwhelmed by synthetic compounds they could not detoxify and excrete. "These enzymes see these particular sorts of chemicals which they have never seen before," Arlene Blum said, "and they make them into an active form where they are carcinogens. . . . The DBCP incident points out that in addition to somatic cells being affected, the germinal cells—the hereditary cells—are also being affected. At high doses, we get effects like sterility. At low doses, there is the potential for genetic birth defects. I think this is really one of the most frightening things about all these chemicals."

"It's like a balloon payment for agriculture," said Don Vial, ". . . for all the things they didn't pay for."

Some of the DBCP workers from Oxy drove down to San Francisco to listen to the testimony at the hearings. Those who stayed in the valley kept informed by reading accounts of each day's revelations in the local newspapers or watching the television news. The hearings attracted wide media coverage not only in the local press but in the *Los Angeles Times, The New York Times, The Wall Street Journal,* and the two big wire services. The headlines were sensational; the stories, brief and blunt. The Associated Press reported from San Francisco on October 14 that "a pesticide produced by Dow Chemical Company and by Shell Oil Company was described by researchers here as a potent cancer-causing agent."

Wesley Jones was back in Stockton, trying to get all the information he could about the hearings in San Francisco. Ever since leaving Oxy, Wesley had been consumed with his case. He gathered pieces of information from everywhere he could on chemicals. He asked some of the men from the union who were driving down to the hearings to keep in touch with him and arranged to get transcripts of the hearings after they were completed. His was a more pressing need than normal. Wesley's wife, Deborah, was two and a half months pregnant, and they both were frightened about birth defects.

At first, when Deborah told him she was pregnant that summer, Wesley didn't believe it was his child. His sperm tests had shown that he was azoospermic. There didn't seem to be much room for argument. He and Deborah had reconciled after his suicide attempt, but the marriage was still on the shakiest of ground. Wesley acted like a different man; he was violent sometimes—he blamed it on his terrible stomach pains and the new worries his life had acquired—and the police were called to the house on many occasions. Wesley veered in and out of the family, sometimes staying away for a time and then returning and frightening them with his strange moods. When Deborah told him she was pregnant, he says he begged her to say that it was someone else's baby, so she could have it without worrying. But Deborah was steadfast: It was Wesley's, and they had to make a decision. She and Wesley both had read the doctor's sperm analysis and testicular biopsy reports, and they were afraid that any child conceived of such deformed sperm would have serious problems. Wesley took it upon himself to call a number of doctors in the valley; he said he asked them the odds on a sterile man's becoming fertile again and producing a healthy baby. He got mostly negative replies.

Deborah went to see a doctor in Lodi on October 6. She told him about the pregnancy and its potential complications, but he could give her no reassurances. The Joneses spent the next two weeks trying to decide what to do. Deborah believed that abortion was morally wrong, but she did not know if she could survive in a world where she would be caring for a deformed child as well as for a disturbed husband. The state hearings on DBCP began on October 12, and Wesley was reading the scientists' testimony on cancer and birth defects every day in the newspapers. He told Deborah she could go ahead and have the child, and if something was wrong with it, the baby could go to a doctor all the time, just like its father.

On October 19 Deborah had a saline infusion abortion at the Lodi Medical Center. She was seventeen weeks pregnant.

On October 23 a $20 million lawsuit was filed by twenty Oxy workers against Occidental, Dow, Shell, the University of California, and Dr. Charles Hine.

On October 25 a pathology report disclosed that Wesley and Deborah Jones's aborted fetus was a perfectly formed male child.

On October 27 the EPA ordered a widespread ban on the sale and use of DBCP. Its registered uses on nineteen food crops, mostly root vegetables, were suspended, and its application to lawns and golf courses was severely restricted. "DBCP poses an imminent hazard to the public and to farmers and other persons that apply it," EPA Administrator Douglas Costle said in announcing the ban. "The DBCP calamity again dramatizes the need for vigilant, responsible regulation of chemical production and use." The EPA predicted 21 cancer cases for every 1 million people who ate average amounts of foods contaminated with DBCP over a two-year period. Growers predicted agricultural catastrophe. In California alone, growers said the loss of DBCP would cost them $386 million a year.

On November 21 Mike Trout received a letter from Dr. Whorton reporting the results of a repeat sperm analysis. His count had dropped even lower than before, to below 1 million.

chapter 10

It was as swift and ineluctable as a biblical vengeance. Some of the older men had the fatalistic grace that builds up over a life of diminished expectations, but the younger workers were swamped by a kind of bitter rage that was as debilitating to them and their families as the physical impotence that overtook the hardest-hit. In the course of a few dazing weeks during which he turned twenty-five, found himself infertile, and was bumped down to a janitor's job at Oxy, Mike Trout lost forever the fecklessness of youth. Fertility was at the core of what he considered the central reason for life; he felt worthless without the capacity to reproduce. He smudged the broad line between fertility and virility and began to think less of himself as a man. His wife, Marta, intuited most of her husband's unspoken concerns, but she was incapable of sharing his distress. She sat helplessly in the house on Comanche Drive amid Mike's baseball trophies, their wedding pictures, and the baby's toys and watched their marriage disintegrate.

Through the long winter of 1977–78 the Trouts dealt with what would become the two constants of their lives: medicine and the law. Dr. Whorton wrote to tell them in November 1977 that a new sperm analysis showed that Mike's count had fallen below 1 million. In January 1978 they learned that a consulting pathologist who was sent tissue samples from Mike's testicular biopsy could only confirm the original diagnosis. The diagnosis was "marked hypospermatogenesis." Dr. Whorton was usually very careful to explain medical terms in common language, but this report didn't come from his office. Marta had to look up the long word in the dictionary. It meant "abnormally low sperm production."

In between medical reports they filled out the legal questionnaires called interrogatories. The interrogatories were the first maneuverings of the attorneys for the workers and the chemical companies as they worked to position themselves in the months after the initial lawsuits had been filed. Besides the basics (age, past and present addresses, date and place of birth, schools attended, date of marriage, age of wife, denomination of religion, and names of wife, children, parents, brothers and sisters, and wife's brothers and sisters), Dow wanted to know whether Mike had any previous work-related injuries, the condition of his family's health, whether he and Marta had been using contraception, what kind and when, and whether he had any erection or ejaculation difficulties. The first interrogatories were only a few pages long; they got longer and more complicated as time went on. Before the case would finally come to trial, some fifty separate lawsuits would be filed in four counties, more than twenty attorneys would be hired (and some fired), five judges would be assigned to hear the suit, and five years would pass.

Marta believed that Mike's infertility bothered him more than any lingering fears about his brain cancer. "The infertility just affected him terribly," Marta said. "He was full of rage. His younger brother tried to talk to him, but nothing made him feel better. He was so angry. The whole thing just about destroyed our marriage. Mike was very determined to see the suit through. He wanted Oxy to pay for what they did. Not pay money—just pay. He really hated that company."

"I hate what he had to go through, but at least Mike had the guts to go ahead," his mother said. "He filed the first lawsuit, and he did it by himself. He was also the first worker to be interviewed on television."

His father nodded in affirmation. "As soon as Mike started his suit, that's when they really got on him at Oxy," he said. "He was trying to get another job. He put in applications at other places. But if you have a malignant brain tumor, no one's going to touch you."

And so it was that Marta believed she had a literal miracle on her hands when she found herself pregnant in February 1978. But Mike was rigid with disbelief when she told him the news. He thought it was a hysterical pregnancy. "He said, 'Why are you torturing me?' He thought we should get a divorce. He thought I was going to leave him," Marta said. "It just destroyed him, and I couldn't do anything about it. He was uncontrollable."

Hardly more than a few weeks passed before another missive from Dr. Whorton came in the mail, and this one convinced Mike that Marta's pregnancy was real. A new sperm analysis showed that Mike's count had soared to just over 11 million. "I thought, Wow! This is it. He's all better," Marta recalled. The sperm count still fell in the range of oligospermia, but the proof that numbers are fallible showed in Marta's burgeoning abdomen. Dr. Whorton knew nothing about Marta's pregnancy, but the numbers told him that Mike was improving. "Since it has been nine months since your last exposure to DBCP, I would attribute this to recovery from that exposure," he told Mike. "I would also anticipate that further recovery will be seen with more time."

Dr. Whorton was finding out that the effect of DBCP was reversible in some men. One man who had left Ag Chem several years before with an extremely low sperm count had since regenerated to 28 million. Several others who had worked around DBCP for a period of years and then were removed from exposure had normal counts. The recovery rate seemed tied to the duration of exposure and the degree of injury. The men with the lowest chances of recovery were those who either had worked around DBCP for the longest period of time or were azoospermic, or both.

Marta Trout's pregnancy heartened Dr. Whorton. It meant that Mike had regained fertility from an almost azoospermic low, raising hope for the other oligospermics. The doctor had finished his tests at Oxy and at Shell's two plants in Colorado and Alabama, and the results, when combined with Dow's data from Arkansas, showed that roughly one-third of the total number of DBCP workers at those four plants had abnormally low sperm counts. Of 335 workers exposed to DBCP, 111 had counts lower than 20 million.

Marta's pregnancy also heartened her husband. "It made him a new man," she said. But they were worried about the child's health. Given Mike's fertility problems and the questions raised by doctors about DBCP's ability to cause mutations in the hereditary DNA, they clearly understood that Marta might be carrying a defective child. But they decided they had no choice. "They always planned on having four kids," Mike's mother recalled. "When he found out Marta was pregnant, they came over and discussed the possibility of deformed sperm and should they go ahead and have the child. And they decided that no matter what happened, they could cope with it." Marta was apprehensive; she was cramping more than she had with Michael, Jr., and one of her girl friends who was also married to an Oxy worker claimed she'd been told to use birth control pills to prevent pregnancy because of her husband's sperm's having been "damaged." But Marta had a sonogram, and the fetus appeared normal.

Mike was still doing nights and weekend shift work as a janitor at Oxy, a job that paid substantially less than Ag Chem. Some of his friends had quit directly after the sterility revelations, but a few others were back in Ag Chem. Oxy was making pesticides again and maneuvering to get back into the DBCP market. Oxy's sister subsidiary, Hooker, had decided to defend DBCP before the EPA in an attempt to keep its registration intact and the chemical on the market. In February, Jim Lindley had suggested to Dr. Zavon that he get quotations on what it would cost for animal studies to determine if there was a concentration of DBCP that produced no testicular effect. Then, during the early spring, Hooker officials approached Dow and Shell. Hooker needed toxicological data to prove that DBCP's registration should not be revoked. Dow and Shell had that data, and Hooker wanted to buy the information from them. Dow had already decided not to expend the time or money necessary to defend DBCP in the arduous bureaucratic process called Rebuttable Presumption Against Registration. Dow was already in the middle of RPARs on several chemicals, including the herbicide 2,4,5-T, and it did not want to add another to its list. So Dow sold its DBCP data to Hooker, which began preparing the case on DBCP to take before the EPA.

At the time the EPA suspended DBCP's registration on nineteen food crops the previous October, the agency had promised to have a report on the risks and benefits of the chemical completed by January 1978. The report arrived eight months late. There was nothing radical about it. The report said DBCP should continue to be kept off those nineteen food crops and allowed on others — provided workers were made to wear protective clothing and face masks. The report carried no regulatory weight; it was merely a position document, the keystone in the opening round of arguments about DBCP. Like a trial, at which both sides argued their cases, it would be followed by debates, hearings, examinations, rulings, and appeals.

Despite the state ban, California officials were in fact trying to conceive a safe way that DBCP could be used on crops again. The benefit side of the risk-benefit equation was substantial in terms of the agricultural bounty provided by DBCP. State officials knew also that farm workers were exposed to much less DBCP than production workers, and they suspected that a way could be found to apply the chemical safely to crops. Test plots were cultivated, but considerable amounts of DBCP residue were found on the crops, raising the specter of human contamination. Through the year tests continued on different application methods as the state searched for a way to reduce residues. The search was all the more urgent because growers were beginning to have DBCP withdrawal symptoms, and a black market in the illicit chemical was operating.

DBCP is applied in three-year cycles, and the state ban had come in 1977, just as farmers were preparing to administer their next application. Within a year orchards began to wither, and growers began to feel like drug addicts without their next fixes. A thriving underground market in DBCP developed in California. Farmers passed telephone numbers of secret sources to each other, and prices of DBCP increased by 300 percent. Sources were scarce. No one was manufacturing the chemical anymore save for a small southern California firm called AMVAC Chemical, which imported it from Mexico for a time and then began manufacturing it for use on unrestricted crops outside the state.

It was not the best of times for marshaling opposition to a pesticide. The pendulum that had swung so urgently in the direction of environmental stewardship during the early 1970's had been creeping imperceptibly back toward the center in the ensuing years. Something of a backlash was in the works by the middle of the decade, although few were sharp-eyed enough to spot it. Even the media, which had long been big government's staunchest friend, began to talk in more conservative tones. The backlash was fueled by a stalled economy—the old jobs-versus-environment bugaboo was kicking up again—but more important perhaps was a kind of exhaustion mixed with helplessness, which permeated the nation after nearly a decade of relentless doomsaying. "Does *everything* cause cancer?" people began to ask each other after reading the morning newspaper. Network newscaster David Brinkley captured the mood of the country in 1978—just as he had in 1970 with his "Nobody's for pollution. Everybody's against it" speech—when he reacted to the news that scientists had found suspected mutagenic material in charcoal-broiled hamburgers. "If all the foods and drinks we've been swallowing for centuries were as dangerous as they say," Brinkley observed, "we'd all be dead, but we're not." William Tucker writing that summer in *Harper's* magazine put it more stridently: "There seems to be no limit to what the federal government is willing to do to indulge the fanatical concerns about what we drink, eat and breathe."

This was the sermon that Dow had been preaching for four decades. There was a large dose of "I told you so" in Dow president Paul Oreffice's voice when he told the Detroit Economic Club in 1978 that the deficit-ridden United States government was so badly on the decline that it could become a "banana republic," except for the fact that "the EPA wouldn't let us produce the pesticides necessary to grow bananas successfully." Oreffice's sardonic speech drew laughter and applause from an overflow crowd of businessmen who identified with the Dow president's sarcastic humor about overregulation as viscerally as would a black audience confronted by Richard Pryor's acidic monologues about racism. Oreffice's speech was rooted in the phenomenon that Murray Weidenbaum had first recognized when he realized his office walls were covered with comic strips that all said, "Government is overregulating." The former Nixon administration assistant treasury secretary had said, "When it begins . . . to be the topic of humor, then you know this thing is getting big."

Besides humor, the fifty-year-old Oreffice projected a certain bold strength, a fearlessness, that was like a salve to the businessmen's wounds. From trade deficits to Jane Fonda, nothing escaped his wide radar. "I was the second choice," Bendix chairman William Agee quipped when introducing the Dow chairman. "Jane Fonda couldn't make it." The crowd loved it. Agee pressed on. "Oreffice claims that if [Ralph] Nader lived in his hometown of Venice, he'd be demanding airbags on the gondolas."

The consumer zealotry of Ralph Nader was, of course, anathema to Oreffice. And just the previous fall Oreffice had withdrawn $73,000 in Dow grants from tiny Central Michigan University because the school had paid the activist actress Jane Fonda $3,500 for a campus speech in which she blasted Dow and other big corporations for making free enterprise "virtually obsolete." Fonda's contention was that the nation's economy was becoming monopolized by a few giant corporations the powerful influence of which on the country was largely unrecognized. "It is your prerogative to have an avowed communist sympathizer like Jane Fonda or anyone else to speak at your university, and you can pay them [sic] whatever you please," a livid Oreffice wrote the university president after the speech. "I consider it our prerogative and obligation to make certain our funds are never used to support people intent on destruction of freedom." Oreffice subsequently went after Michigan State University after it had voted as a protest against apartheid to sell its stock in firms (including Dow) doing business in South Africa. "Do you wish to continue to receive, and are you willing to accept ... grants from our company, knowing that some portion of the funds will be derived from our operations in South Africa?" Dow wrote the university in 1979. The letter was a perfect example of the lengths to which Dow would travel to make a point: The company had in South Africa only a small sales/warehouse operation, which was partially supervised by black employees.

Oreffice had left his native Italy for the United States at the age of twelve, graduated from Purdue University with a degree in chemical engineering, fought against the North Koreans, and worked his way up the proverbial ladder at Dow from company sales trainee in 1953 to the job at the top in 1978 and an annual salary-bonus-benefit-deferred-stock package worth more than $684,000 by 1982. Everything about him shouted self-made American. At six feet and 175 pounds, his trim tennis player's body was usually displayed in well-tailored business suits. He was tired of hearing that the rich never paid taxes—Oreffice said he paid 51 percent of his gross income in state and federal income taxes and felt like "the only sucker in town on April 15." But he was especially tired of listening to what he called "threats to do away with the free enterprise system." He was the tireless "public defender of corporate America," and he did not limit his preaching to the already converted. Although he served at one time or another on countless industry boards—chairman of the board of the Chemical Manufacturers Association, trustee of the American Enterprise Institute, a member of the policy committee of the Business Roundtable—Oreffice also made twenty trips to Washington and fifteen to twenty speaking engagements around the country every year to spread his message about government bureaucracy to frequently hostile audiences.

The main thrust of his message was that government regulation had turned into "pure harassment" of American business by increasing production costs, fueling inflation, and dampening innovation. Oreffice reserved most of his vitriol for the Environmental Protection Agency, with which Dow had been fighting something of a holy war. The agency had ordered Dow to clean up emissions at its partly coal-fired Midland facility and had sent planes up over the huge plant to photograph pollution. Dow went to court and won an order blocking release of the photographs on the ground that they could help its competitors. "We can't get EPA to listen to reason," Oreffice said at the time. "We're at an impasse—sometimes I think that's the best way to be with EPA."

In a manner reminiscent of the future President Ronald Reagan, the Dow executive was masterful at manipulating the small, telling fact. Squeezing a grip exerciser at his Midland desk to relieve tennis elbow, the highly ranked local tennis player told a reporter that he was worried that regulation might even intrude on his tennis game by banning his prized, specially strung racket from tournament play. "As one who came to this country as an Italian immigrant escaping a fascist government," he declared, "I saw what happens when government bureaucrats become veritable tyrants."

The philosophical underpinnings of his message derived from the technological hubris that Dow leaders seemed to pass from one to the other like a genetic trait. There was no such thing as reward without risk, and the idea of living in a zero-risk society was ludicrous. "Even God did not attempt that," Oreffice observed.

But Dow officials contend they did not launch their battle in earnest until they saw regulation begin to be driven by politics, not by science. Dow chairman of the board Robert Lundeen remembered a time in the 1950's when he was chief process engineer for Dow's western division and the company was designing a program to meet Los Angeles's new pollution control standards. "Management took it as a pretty reasonable thing to do," he said. "It was a technical problem. We were quite easy working in that environment. The issues weren't resolved on the front pages of the *Los Angeles Times* or on the morning television shows."

But as time went on, "this thing got into the political arena, and this is when Dow began to buck," Lundeen said. "We began to realize that we were going to have to do some things that would be imprudent economically—not a wise expenditure of the company's resources—in response to emotional pressures badly founded. And that really bothered us. Unnecessary regulation probably got our backs up more than it might have others since we felt we'd been doing a pretty credible job for many years on handling these problems."

Two of Dow's biggest crusades were against attempts by OSHA to rank cancer-causing chemicals in the workplace and against the movement in Congress to pass a law requiring premarketing health and safety testing of industrial chemicals. In the first case, OSHA proposed a system of categorizing workplace carcinogens as well as establishing no-effect levels for chemicals and measuring their risks and benefits before setting worker exposure standards. The plan would have put "human" carcinogens, as proved by human medical evidence or two independent animal studies, into a top-priority category that allowed production only under the most rigorously controlled conditions; a second group of "animal" carcinogens, as proved by one animal test or suggestive epidemiological evidence, could be produced under less stringent but still limited factory exposures. The agency estimated that about 260 chemicals would fall in the first category, and just under 200 in the second.

"They had a scoring system that said cancer in so many organisms would make it a human cancer, with no attention being given to the credibility of the [testing] work itself," said Dow's Etcyl Blair, who began building a health and environmental science division in Dow in the mid-1970's and was on the front lines of most of the company's regulatory and legal skirmishes. Dow also believed that OSHA's system was skewed by an inaccurate perception of workplace cancer, so it helped create a special organization to lobby against the OSHA policy. The American Industrial Health Council, as it was called, operated as an arm of the Chemical Manufacturers Association. It represented more than sixty companies and forty trade organizations with a $1 million-plus budget. Dow president Paul Oreffice was its first chairman. "The apparent intent is to preclude effective regulation of occupational carcinogens and to play down the public health impact of carcinogenic industrial chemicals in the workplace and general environment," Dr. Samuel Epstein wrote in his book *The Politics of Cancer*.

The council argued that cancer was on the rise in America largely because of cigarettes and diet and because medicine had lengthened the average life-span (the incidence of cancer increases with age). Its contention that industrial chemicals were responsible for no more than 5 percent of cancer deaths in this country stood in direct opposition to the prediction of an OSHA advisory panel that workplace chemicals could play a role in up to 38 percent of all future cancer deaths. It also contradicted the theory of John Higginson, a prominent international cancer expert, that most cancers are caused by environmental factors, which became the prevailing view of American science in the 1970's. But Higginson later claimed that his theory had been misinterpreted and that he included cigarettes, diet, exposure to the sun, and alcohol in his environmental factors. Dr. Epstein and other scientists suggested that the industry and its supporters were ignoring the growing evidence that chemicals released into the air and water in factory towns were causing cancer in the local communities and glossing over "the much greater costs to society of failure to regulate industrial chemicals in the workplace, let alone in the general environment."

But as more and more scientists began downplaying the contribution of industrial chemicals to cancer in America—over the arguments of others who said even if the 5 percent figure was accurate, that still meant 20,000 needless cancer deaths a year—Dow and the industry redoubled its efforts to stymie wrongheaded regulation and legislation. In one of its most concerted battles, the company and its industry allies almost single-handedly kept the Toxic Substances Control Act locked up in Congress for five years, until President Ford reluctantly signed it into law in 1976. TOSCA, as it was called, required industry to run health and safety tests on its chemicals before they were put into production. The EPA had the power to reject a hazardous chemical before it was marketed and keep it from ever being sold. The company could challenge the ban in court, but the burden of proof of showing the chemical was safe was on the manufacturer. The law also set up an inventory of all chemicals then in existence with an eye toward creating an information bank on the world's chemical population, along with safety data on each compound. The initial inventory numbered 30,000. "I can't underscore adequately how important I think this one is going to be over time," Douglas Costle, EPA administrator under President Jimmy Carter, told Congress after the law had been passed. "It is unlikely in a chemical and industrial society like our own that we will ever live in a risk-free environment, but we have got to get a handle on how to minimize and manage that risk."

Pesticides were exempt from TOSCA because they were covered by other laws. But a Senate staff investigation revealed in 1976 that the nation's pesticide laws were "in a state of chaos." The report by the Senate Judiciary Committee accused the EPA of relying heavily on safety test data submitted by manufacturers as long as twenty-five years before and of failing to take corrective action in the face of warnings that the data were faulty and incomplete. "The American people cannot be reasonably assured that the federal government is protecting them from pesticides that pose a serious threat to their health," the report said. The country's major pesticide law—the Federal Insecticide, Fungicide and Rodenticide Act—had been updated in 1972 to require more extensive safety testing. Five years later the EPA told investigators it would take it another ten to fifteen years to finish testing more than 30,000 compounds.

Despite concerns over the country's inability to keep track of its pesticides, industry argued that a new law to monitor the rest of its industrial chemicals was unnecessary. It contended that TOSCA inflexibly mandated extensive, expensive tests for chemicals that few humans would be exposed to, soaking up so much money and time that the industry would be left with fewer dollars for innovative research for new products. Dow was spending 10 percent of its annual research budget on product safety testing, and executives had watched the time and cost devoted to developing a new pesticide climb from three years and $3 million in the 1950's and 1960's to nine years and $13 million in the 1970's. "... There is a better way to protect the public than in passing another law," Dow's Dr. Etcyl Blair wrote after TOSCA had been passed. "It involves relying on industry's strong commitment to risk management." He recalled later, "At one time they had a proposal that you had to fill this booklet out for all new chemicals; it was forty-five pages long. Now the amount of manpower just to fill those forms out was a staggering kind of thing. You had to anticipate all the risks. And who is the person who is going to sit down and anticipate what all the risks are going to be?"

The question goes to the heart of one of the most controversial public policy questions facing America: Who gets to interpret the scientific data? As one union official put it: "The political decision is WHO decides what's an acceptable risk. The company which has to worry about profits, or the workers and the public who face these risks every day?"

Dow's full-court press against TOSCA was widely remembered in Washington, and it strengthened the company's image as a tough, unbending group of people who considered themselves smarter than everyone else, especially government scientists. "Dow stands out even among the nation's other chemical companies in its fierce unwillingness to compromise on the chemical regulation issue," declared Environmental Action, an environmental lobbying group in Washington.

Dow board chairman Robert Lundeen began to see a "conflict of cultures" develop in America. "We stuck to our scientific and engineering guns as the proper basis for solving these problems," he said. "We would not accept the idea of the single molecule theory. We never accepted the premise that we had to strive for zero risk because it doesn't exist in life. We believed there were amongst people of goodwill who look at the thing objectively, reasonable tradeoffs that need to be made between the demands of society at large and any individual enterprise. And something else happened ... we developed the capability several orders of magnitude greater [than before] to detect what is in the environment. It has been the industry and university effort by and large that has been able to make us understand much more clearly the contaminants floating around the public domain."

During the late 1970's—with the Carter administration EPA a virtual hothouse for environmental advocates who operated on the theory that, as EPA Assistant Administrator William Drayton put it, society had made a fundamental decision "that the government will control the toxic side effects of the chemical revolution"—Dow began compiling sophisticated reports on the annual costs of complying with government regulations. Costs were separated into three categories: excessive, questionable, and necessary. In 1977, with after-tax profits of $556 million, Dow said it spent $186 million to comply with federal regulations, a leap of 44 percent from the year before and up 82 percent from two years previous. Nearly half the total was spent for compliance with "excessive or questionable" regulations. The cost of filling out government reports and forms alone was $20 million. "These costs will eventually translate down to the consumer, who ends up paying what is in effect a mandated, hidden tax as a result," declared Dow president Oreffice, who complained that the company had to comply with rules imposed by eighty federal agencies. Dow had in fact canceled plans for a $500 million petrochemical complex near San Francisco in 1977 because after two years the company had received only four of sixty-five permits required from state agencies.

What Oreffice didn't say in his speeches about overregulation was that Dow was also losing money because it was overcapitalized and debt-heavy in a recession. Profits slipped two years in a row until, in the first quarter of 1978, Dow lost the coveted position it had held for three years as profit leader of the chemical industry. The company had borrowed heavily and expanded massively overseas. With annual capital expenditures in excess of $1 billion and a long-term debt load of more than twice that, profits were squeezed. The Arab oil embargo had driven up the costs of raw materials, and fierce competition kept Dow from covering costs by raising the price of its commodity chemicals. But the company eventually rallied, partially because it could pay back its big interest load with dollars made cheap by inflation. "It's a more risky way to run your business," one of Dow's financial executives observed, "but we have felt that rewards come with risk."

As the end of the decade approached—with Murray Weidenbaum asserting that the cost of regulation was topping $100 billion a year, with the Business Roundtable calling regulatory compliance "mostly wasteful," and with the American Enterprise Institute bankrolling a magazine devoted entirely to criticism of regulation—Dow was no longer in the vanguard but just a voice in the crowd. One of those voices belonged to James Watt, a brash attorney from Wyoming who founded the Mountain States Legal Foundation, a conservative group of western businessmen who opposed restrictions on development in their states. These antidevelopment westerners came to be known as the Sagebrush Rebels, but their core purpose—"to fight in the courts those bureaucrats and no-growth advocates who create a challenge to individual liberty and economic freedoms"—was in sympathetic alliance with the aims of Dow and the rest of the business community. When West met East in the next decade, the resulting alliance was so powerful that it helped overthrow the proenvironment Carter regime in Washington and set up its philosophical leader, Ronald Reagan, in the White House.

On November 2, 1978, Mike Trout received one of his regular sperm count updates from Dr. Whorton. His count had improved dramatically to 24 million with nearly perfect motility. Dr. Whorton was not surprised. Six of the nine Oxy workers with abnormally low sperm counts had fully recovered within a year of the time their exposure to DBCP ended. (None of the twelve who suffered total sterilization had recovered.) The results suggested that if DBCP exposure were stopped before it caused complete degeneration of sperm cells, the damaged cells might recover.

On November 8, Marta Trout gave birth to a healthy eight-pound-one-ounce boy. She named him Matthew, which means "gift of the Lord." Marta considered Matthew her "miracle baby." The child was the "spark" of Mike's life. Dr. Whorton sent a congratulatory card. "When Matthew was born," Mike's mother said, "we took it as a sign that Mike was going to be well. You just couldn't think about that dumb thing coming back again." The family seemed whole and safe again, and for a while everything was golden. Marta did not know— and no one told her—that just twenty-three days before Matthew's birth, Dorothy Arnett, the wife of one of Mike's co-workers, had given birth to a deformed boy.

Frank Arnett had started working at Oxy in the autumn of 1974, when he was just eighteen years old. He worked mostly in the warehouse, handling the big wooden pallets that came over from Ag Chem stacked with packaged pesticides. Shortly before the sterility scandal occurred, Arnett married his high school sweetheart, Dorothy, who was seventeen. In Dr. Whorton's tests Arnett's sperm count was 2.5 million, and more than half the individual sperm had abnormal shapes. Like most of the other workers with low counts, his eventually regenerated, and Dorothy became pregnant. Their child was born with the urethra in his penis slightly misplaced, a pouch in his weakened abdominal wall, and the inability to flex the third finger on his left hand. The Arnetts were frightened, and Dorothy immediately began taking birth control pills to prevent another conception.

Frank Arnett had quit Oxy after the stir over DBCP and lost touch with his former co-workers, so the Trouts were oblivious of his troubles. Nor did they know about Steve and Sandra Sarras's baby boy, who was born in April 1979 with an underdeveloped left leg. Steve Sarras had worked in Ag Chem longer than Mike Trout, and he and his wife had been trying since their first child was born ten years earlier to have another. He had been one of the first seven Ag Chem men to be tested, and he had been sterile. When he found out his wife was pregnant, he reacted just as some of the other men had done. He questioned her fidelity because he thought himself sterile. But follow-up sperm tests showed that he had made a recovery.

Marta Trout was more concerned with troubles she saw developing closer to home. She first noticed it the evening of their fifth wedding anniversary. Mike didn't like to eat out—he was a homebody and rarely even went to the movies—but that night he took Marta out for a celebratory dinner in Manteca. "It was the first time I noticed that weirdness again," Marta recalled. "It was nothing I could pinpoint—just something a wife could tell."

Mike's doctors had always been hopeful about his recovery from brain surgery three years before. He was doing well, they said; the prognosis was good. His family and friends were optimistic, too, although they could readily see that he was a different man. He even looked different. His head had been shaved for surgery and his hair never grew back right. What had once been thick, long, dark hair was now wispy, and it started so far back on his forehead that it looked as if he had a receding hairline. "It grew back like an old man's hair," Marta said. His strength and coordination were never very good again; he weakened easily, and his reputation as a power hitter with a rocket arm was quietly put to rest. He moved to an infield position. On the nights he did play baseball, he slept through half the next day. Marta had ample compassion, but when Mike slept through Easter because he'd played baseball the night before, she was angry. The game had been one of his chief passions, but he gradually gave it up.

During the months after surgery, as his strength deteriorated, so did his coordination, his memory, and his moods. "He couldn't put a screw on something, for instance," his father said. His memory had started to get bad before surgery, and it worsened afterward. "I'd send him uptown for something," Marta remembered, "and he'd never come back." Or she'd find him sitting in front of the television in the living room, with half a sandwich on the table in the dining room, a jar of mayonnaise on the counter in the kitchen, and the refrigerator door hanging wide open. "He'd forgotten he made the sandwich," Marta said. She was frightened but did not know what to do. As he got more forgetful, she became his memory.

Mike wasn't a complainer, and by the same token he did not reveal much about what ailed him. The only way Marta could tell that his bad headaches were coming back again was when he came home from work in the afternoons and quietly went to bed. He was taking various post-surgery medications to reduce swelling and to regulate his moods, and sometimes Marta found him just sitting in a daze in the bedroom. "Being sick was very hard for him," she said. "He had been so strong and so athletic and so set on being the family provider." It was hard for her, too, although she didn't talk about it. Personality changes caused by brain tumors are so subtle and can mimic the bad moods of ordinary life so well that irritability and moodiness are easily misinterpreted by a spouse as a foul temper, and lack of concentration and forgetfulness can be branded as poor job performance. Mike's supervisors at Oxy thought he had a bad attitude, and they began to lean on him.

Mike had been lobbying to get his job back in Ag Chem because he needed the higher wage. Dr. Whorton told Oxy just before Christmas 1978 that he saw no medical reason why Mike couldn't be reassigned to the pesticides line, although "for reasons of prudency" he asked they not put him to work "with agents shown to adversely affect the testes." The men weren't running DBCP anymore, but Oxy's sister subsidiary, Hooker, was still working on a plan to start selling it again. On December 11 Jerry Wilkenfeld of Hooker's staff had written a memo titled "Re-entry to DBCP Market" and sent it to David Guthrie, the company's acting vice-president for environmental health and safety. "Should there still be a desire to continue consideration of manufacturing DBCP, there is a need for inclusion of an estimate of potential liability in the economic evaluation of the project," Wilkenfeld said. The memo went on to say:

> Assume that 50 percent of the normal rate [of] those people exposed [to DBCP] may file claims of effects from the exposure, [then] determine the number of potential claims for sterility and cancer. Based on the Insurance Department's (or legal department's) estimate of the probable average judgment or settlement which would result from such a claim, calculate the potential liability including 50 percent for legal fees and other contingencies. Should this product still show an adequate profit meeting corporate investment criteria, the project should be considered further. This of course assumes that we do not feel there is in fact any significant risk due to exposure at the planned levels.

Oxy's plans were thwarted when a scientific advisory panel to the EPA recommended cancellation and emergency suspension of all uses of DBCP across the nation. The recommendation came on the heels of another startling discovery about the nematocide. In the spring of 1979 tests showed that DBCP had invaded groundwater in the San Joaquin Valley. Several hundred wells in the valley showed significant DBCP contamination. Groundwater was contaminated not only near the Oxy plant in Lathrop but in places so remote from the plant that the chemical must have seeped from the farmers' fields down through the soil and into the water table.

In Lathrop the contamination was traced directly to Oxy, which had been silently dumping its pesticide wastes in an unlined storage pond behind the plant and throwing half-empty pesticide containers and other wastes in a ditch the workers nicknamed the Boneyard. Despite the best "hindsight" pleadings of its executives, the dumping clearly exposed Oxy's moral flaws. It wasn't so much that Oxy dumped its wastes out the back door. Most chemical companies had done the same thing at one time or another because they didn't know better or didn't care; toxic waste had not penetrated the American psyche in a big way until the summer before, when a neighborhood was evacuated from around a leaking chemical dump called the Love Canal, where Oxy's sister subsidiary, Hooker, had disposed of forty tons of hazardous wastes, including dioxin, years before. It was that so many people in the company, including top executives, seemed to know about the dumping and the threat it posed to neighboring drinking wells yet condoned it and allowed it to continue by their silence. "At its worst," state attorneys would argue later, "Occidental's conduct . . . amounted to willful and deliberate concealment of the company's clandestine disposal of toxic wastes. At best, it was maintenance of a no less cynical silence in the face of an obligation to speak, engaged in with the hope of escaping unnoticed."

Some of the things Oxy did—like detoxifying pesticide containers and manufacturing vessels with caustic soda and then dumping the contaminated washdown water behind the plant—were actually approved by the state, which enforced vague and contradictory pollution control laws during the 1960's and early 1970's. If the case had gone to trial, the state might well have lost on technical grounds—except for one thing: The federal government came into possession of hundreds of internal Oxy documents that painstakingly chronicled the private thoughts and plans of the company's top officials as they were confronted over the years by increasingly complex and costly environmental rules. Although any company whose files were laid bare would have suffered similar embarrassments, the Oxy memos were useful in separating rhetoric from fact. The memos were written primarily by Robert Edson, a former navy man with a degree in chemical engineering who became Oxy's environmental coordinator at Lathrop in 1973. He was described by company officials as sincere and genuinely concerned but perhaps too overzealous and sometimes not even aware that his requests for pollution control funds, for instance, had already been approved. Several of his memos were directed to corporate headquarters in Niagara Falls and Houston, and they indicate that a number of Hooker and Oxy officials were aware of the problems at Lathrop for several years before taking action to correct them.

On April 29, 1975, Edson wrote:

> After more than two years of study of our pollution-control problems versus the local county, state and federal laws, I have come to the conclusion that we must move promptly to stop all discharges of chemicals to the groundwater. . . . The laws are extremely stringent about pesticides. We percolate . . . our pesticide wastes and one percent to three percent of our product to the ground in the form of production losses. To date, the [state] water quality control people do not know about our pesticide waste percolation.
>
> Recently, water from our waste pond percolated into our neighbor's field. His dog got in it, licked himself and died. Our laboratory records indicate that we are slowly contaminating all wells in our area, and two of our own wells are contaminated to the point of being toxic to animals or humans. THIS IS A TIME BOMB THAT WE MUST DEFUSE.

Little more than a year later Edson again described the continuing dumping of pesticide wastes in a pond behind the plant that was located "less than 500 feet" from "our closest neighbor's drinking well." Edson declared that he personally would not drink from the well. He continued:

> . . . Most other organizations involved in pesticide handling have spent millions to solve their problems. No outsiders actually know what we do and there has been no government pressure on us, so we have held back trying to find out what to do within funds we have available. No one likes this project, most probably because very few people understand the total problem. Other companies' solutions are so expensive we haven't had anyone with enough nerve to even suggest that we follow their examples.
>
> . . . If we stop percolation now, I believe we may escape unnoticed. The next drop of pesticide that percolates to the ground is a management decision which I don't feel we can afford. To date, we have been discharging more than 10,000

tons of wastewater containing about five tons of pesticide per year to the ground. I believe that we have fooled around long enough and already overpressed our luck.

The Oxy memos showed the dark side of the antiregulatory group's "get government off our backs and we'll police ourselves" message. Without a policeman, would the industry voluntarily keep itself clean? Industry leaders liked to say it didn't make good business sense to pollute because the cleanup costs would outweigh any money saved. But Robert Edson's words belied the industry's assurances. "Do we correct the situation before we have a problem or do we hold off until action is taken against us?" he asked in 1977. "I recommend cleaning up our problems as much as possible prior to bringing state people into the problem," he wrote in the summer of 1978. "When the job we do looks good to them compared to other problems they have, I have found that they ignore us unless they receive complaints." His last memo was written in September 1978, just six months before the state found the Lathrop wells and groundwater contaminated with DBCP. "We are continuously contaminating the groundwater around our plant," Edson wrote. "We are continually disposing of process water by percolation to the ground. We do not comply with California's regulations for operating a disposal site."

Once the groundwater had been found to be spiked with DBCP up and down the length of the valley, the EPA had little choice but to keep it off the market until a way could be found to stop it from contaminating drinking water. The EPA's advisory panel had been prepared to say that DBCP could be used in certain areas as long as proper protective measures were taken. But the panel could not ignore the chemical's penetration of the water table or the concerns of California health officials, who believed that concentrations exceeding one part per billion in drinking water could present a significant cancer hazard.

The EPA held a nearly two and a half month hearing in Washington that fall to take testimony (7,300 pages and 90 exhibits) on the risks and benefits of continued use of DBCP. Among those lining up to argue the chemical's benefits were the National Peach Council, the state of Hawaii, the United States Agriculture Department, and AMVAC Chemical. They knew that the EPA had two choices. If the risks were substantial, it could declare the chemical an "imminent hazard" and immediately suspend all registrations while cancellation proceedings were undertaken. But if they could convince the hearing judge that the environmental, social, and economic costs of suspending DBCP during the year or so it would take to complete cancellation outweighed the risks, they could buy some time to keep the chemical on the market.

The peach growers testified that a DBCP ban would be financially ruinous. Alternative pesticides applied to the soil before peach trees were planted would cost an additional $49 an acre. Since DBCP was the only nematocide that living roots could tolerate, the loss of postplanting treatments would cause lower yields and cost them an additional $51 an acre. But Gerald Harwood, the EPA's administrative law judge, autopsied their argument and found it lacking. He knew that DBCP was applied as a postplanting treatment in three-year cycles. He thought it would be fairly easy to test the premise that trees without DBCP had lower yields by looking at the California peach crop. The peach output should be starting to sag by now since it had been more than two years since the state ban. But when he asked for production projections for 1979 from the California peach growers, he found that yields were up, not down. He also pointed to the testimony of a grower from South Carolina, where little DBCP had been used since 1977. The grower's crop was larger in 1979 than it had been during the years he used DBCP. "The imminent consequence here, to human beings," the judge said, "is the possibility of cancer or of sterility, or of abnormalities to children because a parent has been exposed to DBCP. There has been no showing that there will be irreversible effects in the peach industry, following a year's suspension or even if the suspension should last two years."

The judge found similar flaws in the arguments of the lawyers for Castle & Cooke, Del Monte, and Maui Land & Pineapple, Hawaii's principal pineapple growers. The projected $1.8 million loss spread out over two or three crop years was "not a significant loss for these pineapple growers," he said. As for citrus, he again pointed to California, where there was no evidence that growers were experiencing serious economic hardship as a result of not being able to use DBCP.

On the risk side of the equation, Judge Harwood heard evidence that DBCP might be a human carcinogen and mutagen and could have adverse effects on male fertility. There was also testimony on the likelihood that DBCP would contaminate food products and drinking water and would be inhaled by people using it. But before anyone could testify about health effects, the dispute over the applicability of animal tests to humans had to be ironed out. Dr. Roy E. Albert of New York University's Institute of Environmental Medicine argued that animal tests were good identification tools for human carcinogens. He explained that of the twenty-six substances known to cause cancer in humans, only two—benzene and arsenic—did not also cause cancer in animals and that even benzene was now looking like an animal carcinogen.

Dr. Albert reviewed the cancer studies on DBCP—the National Cancer Institute gavage tests that produced high incidences of stomach cancer in rats and mice and of breast cancer in rats and its inhalation studies that caused nasal and brain tumors in rats that breathed three parts per million DBCP for 721 days; a New York University Medical Center test in which DBCP applied to the skin of mice caused tumors to grow in their stomachs and lungs; and Dow tests in which rats fed Purina Chow spiked with DBCP developed stomach, kidney, and liver tumors. Dr. Albert noted that the skin exposure tests and the Dow feeding tests duplicated the findings of stomach cancer by the National Cancer Institute. He concluded that high incidences of tumors in both sexes of two species of animals added further confirmation to the conclusion that DBCP was carcinogenic.

The EPA had recently begun constructing what were called risk assessment models to try to provide a rough quantitative estimate of the risk presented to humans by eating and drinking chemically contaminated foods and water or breathing chemicals in the air. The technique was new and controversial. It basically involved putting data about dietary habits and the level of chemical residues on foods into a computer along with information from animal tests in order to calculate the increased risk of cancer. The model was based on the "one-hit" or "one-molecule" theory, which holds that a tumor can be induced after a single exposure to a toxic substance if conditions are right. In the case of DBCP, the EPA calculated that a person eating food contaminated with 10 parts per billion DBCP for one year would have a 1 in 10 million chance of getting cancer. A person who drank water contaminated with 10 parts per billion DBCP for one year faced a 2 in 1 million chance of getting cancer. A person who inhaled 10 parts per billion DBCP in the air for 1,000 hours had a 2 in 100,000 chance of contracting cancer. Other testimony had shown that fruit treated with DBCP contained sizable chemical residues. In one case grapes had a residue of 120 parts per billion DBCP 31 days after treatment. Lemons had 5.6 parts per billion DBCP in the fruit and 6.4 parts per billion in the peels. In another study the fruit of Valencia oranges treated with DBCP 343 days before harvest contained a residue of 8 parts per billion when picked. "The probabilities of cancer," the judge observed, "appear sufficiently great, particularly if a person is exposed to DBCP from all sources, to make DBCP an extremely hazardous pesticide."

The opposition criticized the "one-molecule" theory on the ground that small doses of a chemical may not be toxic at all because they are eliminated from the body through breathing or through excretion in the urine. The judge branded that theory "simply speculation."

Dr. Whorton flew from California to testify on the testicular toxicity of DBCP. He described his initial surprise at finding seven Oxy workers with abnormally low sperm counts. He noted the striking relationship between the duration of exposure and sperm count: None of the workers with sperm counts above 40 million had been exposed to DBCP for more than three months. The doctor testified that the median sperm count for a group of 35 men who were never exposed to DBCP at Oxy was 78 million, while the median for those 107 who had been exposed to the chemical was 45 million.

At Shell's DBCP plant in Alabama, workers never exposed to DBCP had a median sperm count of 88 million, compared to 46 million for those who were exposed. Only 1.4 percent of the Shell work force in Alabama was completely sterile—a fact that Dr. Whorton attributed to the brief fifteen-month period DBCP had been produced there. More than 6 percent of the Shell workers were sterile in Colorado, where DBCP had been manufactured for twenty years. The number of workers with abnormally low sperm counts, or oligospermia, was roughly 15.5 percent in both plants. (Fewer than half the exposed workers in Colorado participated in the tests. More than 80 percent participated in Alabama.) "In summary, DBCP has been clearly shown to be a testicular toxin affecting the primary spermatogonia," Dr. Whorton testified. "Our initial Occidental Chemical study has been replicated twice by us at other plants. We have yet to test a DBCP exposed population without observing an effect."

Testimony was also given on fertility studies at other DBCP plants. Forty-five percent of Dow's Magnolia workers had sperm counts below 20 million. Dow said its studies at Midland showed no significant differences between exposed and nonexposed workers. At Velsicol Chemical Corporation, twelve of twenty-four workers had sperm counts below 20 million.

The opposition criticized the reliability of the animal and epidemiological tests performed on DBCP, but the judge noted that the testimony of one of the principal critics was "so full of inaccuracies and unsupported conclusions as to make it of little value." Only one study was presented to demonstrate the nontoxicity of DBCP. It was conducted by a college student for his senior thesis, using a specially purified DBCP supplied by AMVAC. The study claimed that rats that drank DBCP-spiked water for seventy-five days had no fertility problems even when dosages were quadruple those used in the Dow-Shell experiments back in 1961. Judge Harwood took issue with the "lower professional level" of the college student's study. He also pointed out that the student had incorrectly converted dosages. "Little weight can be accorded this study," the judge said. He also discredited the opposition's assertion that the toxic agent was not DBCP but an impurity present in DBCP, although he acknowledged that as in life, the combination of DBCP and another chemical might be producing a synergistic effect. As Dr. Albert noted, however, "anything that is susceptible to being potentiated by materials in the environment has got to be a nasty actor in its own right."

Dr. Dante Picciano, professor of genetics at George Washington University and a former Dow consultant, testified that DBCP was a mutagen. He said mutagens interfere with the body's DNA or with the movement of chromosomes during cell division. He said mutagens were believed to be responsible for a large portion of the 5 percent of babies born with birth defects and that they were also suspected of playing a role in cancer and heart disease—again, because of their interference with DNA. He reviewed the tests run by Dr. Bruce Ames and Arlene Blum at the University of California that showed that DBCP caused bacteria to mutate.

He also testified about the controversial findings of Dr. Robert Kapp, who had examined the sperm cells of Dow's Magnolia workers as part of his doctoral research at George Washington University. Dr. Kapp reported evidence that DBCP caused mutations in human chromosomes and thus greatly increased the chances of birth defects. Chromosomes contain DNA, or the hereditary information. Each of a person's somatic, or body, cells has forty-six chromosomes grouped in pairs. But the sex cells—the sperm and the egg—have only half the normal number of chromosomes. The sperm, bearing twenty-three chromosomes, unites during fertilization with the egg, also bearing twenty-three chromosomes, to form a new cell, which has forty-six chromosomes. This cell develops into the embryo, which carries a unique combination of half-maternal and half-paternal genetic information.

Each human sperm cell contains one sex chromosome and twenty-two nonsex chromosomes. The sex chromosomes are either X or Y; half the sperm cells contain X, and half contain Y chromosomes. If an X-bearing sperm cell fertilizes an egg, a female is produced. If a Y-bearing sperm cell fertilizes an egg, a male is produced.

But when Dr. Kapp looked at sperm cells from Dow's Magnolia workers who had been exposed to DBCP for up to eighteen months, he found that 3.8 percent of them had sperm cells containing two Y chromosomes instead of just one—a frequency three times greater than that found in men not exposed to DBCP. This meant that DBCP not only depressed sperm counts but could interfere with cell division and mutate the chromosome-bearing sperm cells themselves. Sperm with two Y chromosomes was still very capable of fertilizing an egg. The result was a man with forty-seven instead of forty-six chromosomes. If not aborted spontaneously, chromosomal errors such as the XYY man —which occur in about 1 in every 1,000 births—are usually born with physical and/or mental defects.

Dr. Kapp ran further experiments on rats. He exposed them to DBCP for five days and then injected a chemical to arrest their sperm and bone marrow cells during a particular step in cell division. Under

Elements of Risk / 265

a high-powered microscope, Dr. Kapp saw significant numbers of abnormal and damaged cells and cells with chromosome breaks. He concluded that DBCP increases the incidence of chromosome mutations in both somatic and sperm cells.

At the EPA hearings Dr. Picciano explained the significance of Dr. Kapp's findings. "Mutations to somatic or body cells are believed to be involved in the cause of cancer," he said. "So that if we are [damaged by] the mutagen, for example, DBCP, we would have a potential to develop cancer. Mutations to somatic cells are also believed to be involved in aging and heart disease." What was worse, Dr. Picciano said, was the passing of sex cell mutations "to future generations ad infinitum, and there can be a geometrical progression of the mutation for future generations, ranging from small, minor malformations to very severe mental and physical defects in children."

In a twenty-five-page decision summarizing the arguments of both sides, Judge Harwood concluded that the predicted $42 million loss to growers deprived of DBCP was "on the high side" and that there was no evidence their market would be seriously "dislocated" as a result. "Costs to growers and consumers in the form of decreased yield or higher prices may be capable of some rough dollar and cents figure estimate," he wrote. "The costs to society of the damage to human health caused by exposure to DBCP may well be incalculable.

"I recommend accordingly," he said, "that the [EPA] Administrator immediately suspend all registrations of pesticides containing DBCP."

Nine days later, on October 29, 1979, EPA Administrator Douglas Costle issued the suspension order and a notice of intent to cancel all registrations of DBCP. "DBCP has caused sterility among workers producing it and has shown to be a suspect cancer agent and possible cause of chromosome damage," Costle said. He made one exception to his sweeping order by allowing Hawaiian pineapple growers to continue using the chemical. The pineapple growers were ostensibly exempted because DBCP left virtually no residues on the fruit, but no one could deny that the huge pressures brought by the pineapple industry on Washington had contributed to the decision.

The nationwide ban on DBCP was reported in the valley newspapers on October 30, the day before Halloween. Marta Trout had a special affection for Halloween. She loved to make costumes for her children and take them to the annual parade in Manteca and then out trick-or-treating later that night. But Halloween 1979 wasn't much fun. Mike had been feeling sick. During his six-month checkup in September his surgeon in Modesto had run a routine brain scan and noticed a slight shadow. It didn't look like anything to worry about, but he asked Mike to come back in three months instead of six for another look.

Mike was back working where he wanted to be—in Ag Chem—but he was so sick with nausea and headaches that he was off work most of October. The headaches were spectacular, throbbing affairs that tended to be worst first thing in the morning and sometimes even woke him up. They continued on and off during the day in the kind of "touchy" way that afflicts brain tumor patients. Because of the pressure on the brain, headaches can be activated by a sudden turn of the head or a slight cough. They usually go hand in hand with nausea and vomiting. His moodiness and depressions were worsening, too, and Marta didn't know what to expect from him from one minute to the next. But she was his fiercest ally, and when the company doctor thoughtlessly told her that he thought Mike's troubles were "all mental" and caused by the fact that "he didn't like his job," she blew up. "I remember it was the day of the Halloween parade that I got into a fight with him," she said. "I told him that what he said about Michael was the most ridiculous thing I'd ever heard."

Mike had, in fact, been having problems at work, but he rarely told his wife about them. Part of the problem was of his own making. He'd had a cocky attitude ever since the early days, when he and his friend Haskell Perry got into trouble with the supervisor for swearing and wearing their hair long. He believed that the union grievance process and workmen's compensation laws were there for a purpose, and he used them. Over the years he had filed disability claims for organophosphate poisoning and several back injuries. He was the first Ag Chem worker to file a civil lawsuit for his fertility injuries. As he got sicker, his resolve that somebody should pay grew stronger, and his hard edges became even tougher. He didn't like to talk about his personal life, so his supervisor probably didn't know as much as he should have about Mike's illness. But brain cancer is not the kind of illness a supervisor should be able to forget quickly. If Oxy had been a better organized and more compassionate company, his sickness probably would have been more carefully watched. "Mike had a crappy attitude. That was just Mike," said Farnham Soto, a co-worker and good friend. "The bad reputation that he had just stayed with him, even after he got sick. Nobody really knew what was taking place. The supervisor just figured Mike was goofing off. I don't think Mike told them how sick he was."

"Mike wasn't someone who could function well under someone who was consistently down on him," his mother explained. "It would get his back up."

"Mike was getting these bad headaches," Soto said. "He had a bottle of Excedrin in his locker, and he'd take eight or nine of them at a time, but they wouldn't do anything for his headaches. His reaction time had slowed down considerably, and the supervisors were harassing the crap out of him." Mike's father remembered that "his supe told him: 'If you take time off to go to that doctor again, I'm going to dock you.' Mike told me that out there, towards the end, they were really on him. He wanted to quit. But I said, 'That's what they want you to do, Mike, so just keep smiling and do what they say.' " Nobody told Marta that Mike couldn't count to 100 anymore on the canning lines, and that was one of the inhumanities she never forgave them.

Matthew's first birthday was in November. Mike spent another week in bed. He was vomiting; his headaches were getting worse; his memory was almost nonoperative. He was scheduled to go back into the hospital on December 18 for another brain scan. "I had guessed that he had another brain tumor because of the way he was acting," Marta said. "I was hoping to wait until after Christmas to find out that he had it." His mother remembered the shadow on the October scan and thinking to herself, " . . . I just wonder." She said, "I guess I had a suspicion that his tumor had come back. We never talked much about it. It was just motherly intuition. I went to the library in Manteca and hunted everything I could find on brain tumors. But you can't find out that much about them—it's such a new field.

"There was a Christmas parade downtown," she remembered, "and he had a real bad headache. But he didn't complain."

"Except that he came over to my house after the parade, Betty," Mike's grandmother, Mildred, reminded her. "And I remember him sitting down in the chair and saying, 'Oh, Grandma, I've got such a headache.' "

On Thursday, December 13, in Houston, members of Shell's DBCP Task Force met in a conference room at company headquarters. The task force had been assembled to deal with DBCP lawsuits or, as the director put it in his notes, "to coordinate Shell's response to the DBCP problems arising in California and potentially other areas of DBCP use. . . . To date, Shell has not received any significant public attention, although we are the target of a number of lawsuits by formulation workers at Occidental's Lathrop plant." The task force included fourteen Shell people from almost every division, who were divided into six teams—human health, groundwater contamination, water treatment, public affairs, data management, and legal. Their job was to fill medical and toxicological "data gaps," to find out how DBCP infiltrated groundwater and what could be done to get rid of it, to set up a communications system and monitor press releases on DBCP, and to "structure data and information needs for defense in litigation; [to] advise . . . on methods and procedures to afford maximum protection of Shell's position." The "public affairs" team prepared a "general backgrounder" for Shell people receiving inquiries from the press, complete with "wording to use in referring the inquirer to appropriate . . . representatives," along with an information package "of greater depth," which included "approved holding statements and specific answers." The "human health" team reported a need for expanded toxicological studies on DBCP at a projected cost of more than $500,000. "Water treatment" revealed that carbon filters were effective in removing DBCP from groundwater, as was "simple boiling." The "legal" team scheduled a meeting in Houston with Shell's insurance company, and the group as a whole planned to meet with Dow "to discuss potential cooperation" on DBCP research programs.

On Thursday, December 13, in Lathrop, Farnham Soto, who was the union representative in Ag Chem, heard that the supervisors were going to give Mike Trout a reprimand because of his work problems. Soto went to talk to Mike's bosses. "Hey," he told them, "Mike is sick." Mike finished out his shift that afternoon and came home about four to fetch his older boy. They drove into Manteca to pick up his father so the three of them could go Christmas tree shopping. They stopped back to see his mother on the way home. She'd been singing in the Stockton Chorale and was in good spirits. When she saw the size of the tree Mike bought, she broke out laughing. "They bought this little teeny tree—so little that Little Mike could carry it—but they said they got a deal!" she remembered. "I remember Michael saying, 'Boy, am I gonna get heck when I get home with this.' "

"We had a good time that night," his father said.

Marta went to bed before Mike did that night. When she woke up later and went out to the living room, she found an empty beer can on the table. It worried her because Mike didn't drink at all anymore. She thought that something might be wrong, but Mike was sleeping and did not seem troubled.

It was about 3:00 A.M. when he woke up, vomiting. He had been sick like this before. Marta stayed up with him through the night. Sometime before dawn he lost coherence. Marta called the doctor, who thought it sounded like a bad case of flu. Marta called Betty Trout and asked her to come get the children, and she called Mike's supervisor, who lived down the street, to tell him that Mike was ill. "Good morning, Mom," Mike said to his mother when she walked into the bedroom. That was the last coherent thing he said. His thoughts and words became scrambled. He called his wife Mom. "He just wasn't there," Marta says. Then he had a convulsion.

While they waited for the ambulance, Mike pointed to his head and screamed, "Get me out of this pain!" He was still yelling when the ambulance attendants put him on a stretcher and drove him to the hospital in Modesto. He was kicking and screaming so much that the doctors had to tie him down in his bed "like an animal." His family tried to talk to him. He was blank, senseless. The surgeons cut a hole through his skull shortly after 11:30 P.M. on Friday and found a new tumor inside the brain. "There was so much pressure from the tumor that when they opened Michael up, it just kind of shot out," Marta recalled. "I remember the doctor standing there with his mouth open, saying, 'Oh, my God, I just can't believe this.' " The tumor was between a Grade III and IV astrocytoma—one grade larger than the encapsulated tumor doctors had removed nearly four years before.

"The doctor said, 'This is just devastating,' " Mike's father recalled. "He said that in a few months' time it had grown from a shadow to the size of an orange."

After surgery Mike was breathing on his own, and his heart was strong, but doctors could not control the hemorrhaging in his brain. In the middle of the morning of the next day, a Saturday, he died. He was twenty-seven.

Sunday, the family gathered together in their grief at the house on Comanche Drive. The console color TV in the living room with Mike's baseball trophies on top was on, but no one was paying much attention to it. Then somebody heard the familiar ticking-clock introduction of *60 Minutes,* and suddenly the Oxy plant was on the screen and reporter Mike Wallace was saying, "Tonight, a look at confidential memos from inside this chemical plant in California, memos that show the company knew and did nothing about the fact that chemical waste from the plant was contaminating all the wells in the area." While Dan Rather and Harry Reasoner previewed the show's other pieces—a look at snake venom and a controversial plan to sentence criminals to jobs rather than to prison—the Trouts gathered around the television set. They'd had no idea that *60 Minutes* had planned a story on Oxy and did not know what to expect.

Mike Wallace told how Oxy had dumped pesticide wastes into the ground. Don Baeder, the president of Hooker, Oxy's sister subsidiary, maintained that California officials knew about and approved Oxy's discharge of pesticide wastes. "No one has been hurt at Lathrop. No one," Baeder said. "From our discharges, no one has been hurt and no one will be. No one will be."

The Trouts were astonished. They couldn't believe their ears. The Hooker president was talking about the pesticide wastes in the groundwater, not the workers' exposure to DBCP, but ordinary people listening to the show wouldn't have known to make the distinction. All Marta and Mike's mother could think about was Mike and how much he had suffered. The words "no one has been hurt at Lathrop" struck at them like a slap in the face.

"We all sat there and watched this guy say big as life that no one had ever been injured by their operations," Marta said. "It was the biggest lie."

"I could have thrown a tomato at that guy," Mike's mother said. She began composing a letter to *60 Minutes*. She wanted them to do "further research" because "we had a son who had been in contact with DBCP who just died from a malignant brain tumor."

"As soon as we found out it caused sterility," Don Baeder was saying on the television screen, "as soon as we found out it caused cancer, those operations were shut down."

"And you don't want to go back into the business of making DBCP?" Wallace asked.

"We are not going back into the business of making DBCP," Baeder replied.

"Well, then how come I have a memo here, dated December 11th, 1978, to D. A. Guthrie from J. Wilkenfeld, subject: 'Re-entry to DBCP Market'?" Wallace asked.

It was the memo outlining the company's plan to play off the cost of defending claims for sterility and cancer against the profits of selling DBCP. Oxy had never implemented the plan, and Baeder explained that the company would not have gone back into the DBCP business unless it had found a way to manufacture it safely. "Is this the way America does business?" Wallace wanted to know.

"America looks at many options in doing business," Baeder answered. He said there were risks attached to almost every product. He said Hooker had spent more than $130 million over the past three years on environmental health and safety.

"I wonder why," Wallace said acerbically.

"Because environment is important . . . " Baeder answered.

"Fine," Wallace said.

"We have a concern for our neighbors," Baeder said, "and that's why we're spending this money."

Baeder's public relations staff was furious that Wallace had dropped the last line from the memo, the one that said, "This of course assumes that we do not feel there is in fact any significant risk due to exposure at the planned levels." The staff members had made their own videotape of Baeder's interview with Wallace, which ran considerably longer than the piece shown on television. In the basement of Hooker's corporate headquarters in Houston, they reran the videotape to show people their side of the story. Getting out its side of the story had become an obsession with Hooker after Love Canal. In the summer of 1978 a middle-class neighborhood built around an old Hooker waste dump in Niagara Falls, New York, was evacuated after the leaking dump had been declared a health hazard. Twenty-five years earlier the company had buried 20,000 tons of chemical residues in the canal bed and then covered it with clay. The land had been deeded over to the local school board. The deed had warned that chemical wastes were buried there, but not what kind or how much. Hooker maintained that its legal liabilities for the dump had ended with the deeding away of the land. In the aftermath of the evacuation Hooker had adamantly refused to accept legal responsibility. But try as it might, the company could not shake the shadow of Love Canal. In a complete-this-sentence quiz hardly anyone in America would have failed to attach the words "Hooker Chemical" to those of "Love Canal." And in the same manner that "Watergate" entered the English vocabulary as a generic word for "government corruption," 'Love Canal" became the catchphrase to express everything from environmental disaster to corporate ruthlessness. "Press reporting of Love Canal has been so scandalous," Hooker chairman (and former Dow president) Zoltan Merszei declared. "This hate campaign in which Hooker has been viciously attacked is based on distortions of fact, the manipulation of medical data, and the foolishness of some government agencies which have to justify their own existence." In fighting back, Hooker spent more than $250,000 to place ads in major newspapers refuting critical editorials in *The New York Times* and *Business Week*. A deluge of special Hooker publicity pamphlets called *Factlines*—10,000 in all—were sent to newspapers all over the country with the purpose of setting the record straight on Love Canal. Reprints of a collection of speeches given by Occidental Petroleum chairman Armand Hammer and Hooker chairman Zoltan Merszei to representatives of the East Coast financial community were also packaged and mailed out. Its title: *The Other Side of Love Canal.*

Mike Trout left behind a twenty-four-year-old widow and two sons, one aged five years and the other thirteen months old. When they buried Mike next to Marta's brother in St. John's Cemetery, not far from an almond grove off Highway 120, his family inserted a color photograph of him in a special place on the headstone. He had on a baseball cap, and he was on a fishing trip, squinting into the sun. "Forever in our Hearts" was cut into the stone. "Matthew has no idea who his dad is," Marta said. "To him, his dad is just a picture on a headstone." After Mike's death Oxy sent Marta his accumulated vacation pay. She also received the proceeds of a $10,000 life insurance policy, plus Social Security benefits. The union sent flowers. Oxy sent nothing in the way of condolences. Marta was treated rudely when she called to get information about his insurance. "What I'd like to say about the company is unprintable," she said. One of Mike's supervisors called his mother. "[He] had the nerve to call me after Mike died," she said. "To say he was sorry. I told him, 'I hope God can forgive you, being as right now I can't. The way you treated Mike, there's no other way I can feel.'"

In January Cal-OSHA received two reports from its medical people on Mike Trout's death. Both suggested there was an association between brain cancer and his exposure to DBCP, although neither was able to put forth a strong case. "Excessive brain cancer does occur in some industries, particularly chemical, but the causative agents have not been identified to date with one exception, vinyl chloride," wrote Dr. Ira Monosson. "Although conclusive evidence that DBCP causes brain cancer is lacking, there also is no evidence to the contrary. Further, the possibility that DBCP indeed can cause neoplasms of the brain is far from remote and certainly does exist." Dr. Lawrence Rose described the "well documented" carcinogenicity of DBCP but said, "In this particular case, one would have to go over all of the chemical exposures, work history, health history, personal habits, etc. to develop a list of possible substances."

In February Marta Trout changed her husband's workmen's compensation claim to a death case and refiled their lawsuit as a wrongful death action. She felt compelled to continue fighting his battle because she knew how important it had been to him. In truth, the battle had become even more important to her. Even though they were named as defendants in her suit, she knew little about Dow and Shell. Her focus was on Oxy. "I don't think he was as bitter toward the company as I was," she said. In a declaration filed with the court, she described her husband's work as "loading, unloading, packaging, and preparation of the dangerous chemical known as DBCP." She said, "He was not the least bit apprehensive of the exposure by reason of the fact that no one had ever warned him of any danger or hazards in connection with the product, which he believed to be 100 percent safe."

Ronald Reagan rode the rails of his antiregulatory railroad into the President's office in November 1980. Reagan believed that too much government and too many laws were bad for America, so he corrected the course. "Now you in the steel and coal industry know what it's like to deal with EPA administrators who neither care about nor understand your industries," candidate Reagan had declared in a campaign speech in Steubenville, Ohio. "Those are the people that if they had their way, you and I would have to live in rabbit holes or bird nests." As President, Reagan chose an ideological compatriot to run the EPA. Anne McGill Gorsuch was a bedrock conservative Republican who had worked to repeal taxes, cap budget increases, and stem regulatory excess as a Colorado state legislator. While in Colorado, she forged ties with James Watt, who as director of the Mountain States Legal Foundation was titular head of the Sagebrush Rebellion. The political alliance endured, and in 1981 they both came to Washington to work for President Reagan—Mrs. Gorsuch as administrator of the EPA and Watt as secretary of the Interior Department. Their rise to power paralleled the shift in the political winds that began whistling across

America in the late 1970's, as the big-spending, big-government Democrats lost their consensus under the weight of inflation, energy shortages, and economic decline.

Mrs. Gorsuch began running the agency under the premise that regulation for the sake of regulation, and confrontation for the sake of confrontation, were counterproductive. She said her predecessors at the EPA had badly mismanaged the agency by measuring progress in terms of how much money was spent, how many scientists and administrators were hired, how many regulations were issued—measurements that she dismissed as "beancounting." What she intended to do with the EPA was delegate authority back to the states, improve management, reduce backlogs, improve the quality of the agency's science research, control hazardous waste dumps through the new Superfund law that paid for the cleanup of abandoned waste sites—all of which, she said, would lead to "improved results with fewer resources."

The new EPA director surrounded herself with so many advisers from industry that one newspaper editorialized, "The Reagan Administration is naming regulators who by virtue of attitude or inexperience are more likely to be non-regulators." Her deputy administrator was a college dean of engineering who banished the word "hazard" from EPA press releases because it was a "trigger word" that needlessly excited people. Her chief of staff was a Washington lobbyist for the American Paper Institute and for Johns-Manville, the company that was pressed with so many worker suits charging health effects from asbestos that it filed for bankruptcy in 1983. Most of her other aides were former business lobbyists and industry lawyers. They were, by nature, amenable to sitting down and working problems out with industry. The trouble came when they began holding private meetings with industry to the exclusion of public-interest groups. Six private "science courts" closed to both the public and environmental groups and attended chiefly by industry representatives preceded the decision by the EPA to take no action against the suspected carcinogen formaldehyde. Over at OSHA, where Florida construction executive Thorne Auchter had taken charge, the agency suspended its policy of regulating workplace chemicals according to the theory that no safe level of exposure exists for carcinogens and abolished its list of suspected workplace carcinogens, saying its distribution might unnecessarily frighten workers.

But the chemical industry was heartened by the changing of the guard in Washington. Despite former Dow president Ted Doan's contention that "we are country boys in this government business," Dow had a sophisticated and effective Washington office and a chief executive officer who knew his way around town. After the election newspapers reported that Dow president Paul Oreffice flew to Washington to discuss EPA appointments with top Reagan advisers, and once the staff was in place, Dow's Washington lobbyists were frequent lunch partners of Rita Lavelle, chief of the agency's toxic waste cleanup program. In a speech in 1981 Oreffice talked about Washington before Reagan:

> I happen to feel very strongly that government and regulation in this country were well on the way to going unchecked. And I am not talking about some deep, dark conspiracy. I am talking about a momentum that was building and sweeping us in that direction. One result was that the regulated . . . were being shut out of the process—while an elitist group, many of them technically illiterate, were front and center with their arguments. Now I have no objection with their being front and center as long as others, including the chemical industry, can also have a day in court. . . .
> It had gotten so that I dreaded going to Washington. Not only were we from industry not being listened to . . . we were literally being frozen out. That's changed now. The bridges are back in place, and we're now seeing the beginning of a healthy exchange of information and ideas. . . . But the critics and the doomsayers are out in full force. It has been astounding to me to watch the hue and cry raised over Mr. Watt at Interior and Mrs. Gorsuch at EPA.
> One of the most remarkable features of all this criticism is the complaint that agencies are talking to and reasoning with industry instead of confronting, contesting, and punishing. They seem to believe people in government should be ashamed to talk to and reason with scientists and engineers from industry and academia. Well, I say nuts to that.

During the first year the Gorsuch EPA was in power, the agency delayed seventy-eight air, water, and toxic substances regulations and canceled forty others. Pieces of the premarketing chemical testing law called TOSCA were targeted for change or elimination under the President's Task Force on Regulatory Relief. Hazardous waste cases sent to the Justice Department for prosecution fell from 200 a year during the Carter administration to 10 in the first year of the Reagan administration. A special EPA "SWAT" team, which moved against dangerous pesticides on the market, was marked for elimination, and the entire Office of Enforcement was abolished, scattering its ninety lawyers among other departments and diluting the enforcement effort. Budget decisions that would reduce agency staff by half and would cripple research programs were made.

More subtly the agency began to fine-tune policy in an attempt to bring it more in line with the Reagan philosophy. Policy memos from agency files showed an apparent effort by top officials to develop a new policy on carcinogens—a policy described by academic scientists as a "covert" shift toward tolerating higher public health risks. The plan involved classifying carcinogens into greater- and lesser-risk categories on the basis of their genotoxicity, or ability to cause direct chemical changes in the body's genetically coded DNA. But under this theory, epigenetic carcinogens—which affect other cell mechanisms but likely do not induce cancerous changes at low levels of exposure—would have a threshold below which the chemical was considered safe. Most scientists contend, however, that research does not support the concept of a safe dose of any cancer-causing agent.

Beyond policy, TOSCA—the nation's most important law for protecting the public against cancer-causing chemicals by screening them before they are marketed—was proving to be ineffective. Studies showed that only about half the forms sent to the EPA by manufacturers of new chemicals contained the required toxicity information, and only 17 percent had data on cancer, birth defects, and mutagenic effects. " . . . It is clear," one environmental lawyer told Congress, "that we are still allowing most new chemicals to enter commerce with little or no toxicity testing."

chapter 11

In October 1982—the month before trial was scheduled to begin—Occidental settled out of court for $425,000. Duane Miller called it "peanuts," but the plaintiffs needed the money to pay bills. The settlement ended twenty-five claims against Oxy, released the company from all pending and future workmen's compensation claims, and prohibited future lawsuits from being brought for DBCP injuries. Oxy said it was settling not because the workers had a good case against the company but because it recognized the "vagaries of litigation" and wanted to avoid "future litigation costs." The workers signed their prorated $17,000 shares of the pot over to their lawyers, who were deep in debt after fronting money out of their own pockets for five years.

A thirty-four-year-old attorney from a two-man Sacramento firm, Miller was coordinating the rash of lawsuits filed by DBCP workers against Oxy (until it settled), Dow, Shell, University of California researcher Dr. Charles Hine, and the university itself. Miller was intensely devoted to the case and had given up nearly five years of his life to it. The defense team considered Miller a "cause" lawyer and subtracted points for his hard edges, but Miller shrugged off criticism. "I've been over law school for some time," he said.

In all, five attorneys represented fifty-seven plaintiffs whose cases had been consolidated for a single trial in San Francisco Superior Court. Most were suing for fertility and genetic damages and fear of cancer and birth defects. There were three cancer claims, including Mike Trout's for wrongful death.

Both sides were doing a lot of bet hedging and calculating in the last weeks before trial. The plaintiffs were betting they could win more on punitive damages from a jury than they could ever get in a pretrial settlement, but they were also fighting a rearguard action in their own territory from workers who wanted to "take the money and run." The defendants were worried about the "fear of cancer" issue and about the three cancer cases themselves. Evidence linking DBCP to the three cancer claims was shaky at best and would probably fall far short of the "preponderance of evidence" test in court. But cancer was such an emotional issue that it was difficult to predict how a jury would react. A third factor—the fear of publicity—ran so strong among chemical companies that these kinds of cases usually did not go to trial. Regardless of innocence or guilt, companies believed that a secret settlement and a release from liability were far preferable to a month of screaming headlines and speculation in the media. Dow and Shell were also looking over their shoulders at a mammoth class action lawsuit brought by San Joaquin Valley residents over DBCP pollution of their groundwater and additional claims yet to be filed by some Oxy workers.

The nine-lawyer defense team set up a chalkboard and calculated how much they would offer to settle each claim out of court. "Having never dealt with fertility cases before, it was very difficult to figure out what our values would be," said H. Christian L'Orange, a thirty-six-year-old San Francisco attorney hired by Dow as its trial lawyer. "We spent untold hours out of the courtroom running these guys on a chalkboard, trying to figure out what we could hang damages on, what we couldn't hang damages on, what was the high point, what was the low point . . . and what really is a permanently sterile person worth in California."

Like the rest of the defense team, L'Orange evaded easy labels. He was an ex-marine captain who had served in Vietnam and who had been posted to the White House during the Nixon administration, who counseled Vietnam vets once a week in a Mission District storefront—he'd been exposed to Agent Orange himself in Vietnam—and who declared himself a "card-carrying member of the Sierra Club." Another defense attorney's father had died of cancer, and his wife had miscarried during recent months, and he intimately understood the emotional terror of cancer and birth defects but regarded them as a fact of life, not as a chemical industry problem. Another, retained by a company that was irrevocably linked to napalm and the Vietnam War, had been an antiwar protester in Washington, D.C. If character judgments were being made on financial position alone, Douglas Brown, one of the plaintiff's attorneys, would have won hands down. The drawling "country lawyer" from the Florida Panhandle had made wise real estate investments over the years.

Brown was a defense attorney by trade who had been recruited by the plaintiffs because, as he put it, "they didn't have a punitive damage case until I got here." In fighting and winning a case in Florida for a young family whose house had to be cut into pieces and incinerated after a one-gallon jug of DBCP had spilled in the garage and infiltrated the house through the heat ducts, Brown had acquired more than

100,000 pages of internal Shell documents about DBCP. The plaintiffs needed the "smoking gun" documents that Brown had wrestled from Shell for their own case in California, so they hired him as associate counsel. Brown's Florida victory against Shell was subsequently reversed on appeal (a $490,000 judgment against Kerr-McGee Chemical Corporation was left standing), but he still had the fourteen boxes of documents.

In a pretrial conference in San Francisco the lawyers began settling the cases against Dow and Shell. A zero sperm count, or azoospermia, was fetching between $180,000 and $200,000. "Everybody, including the plaintiff's lawyers, thought that was fair," L'Orange said. The price went up for men who had been sterilized at a young age and for the cancer cases. After more than a week of negotiating, the three cancer cases were settled for what Brown called "a huge amount of money." Settlements were eventually made on all but nine claims. The plaintiffs' attorneys wanted nearly $2 million to close out the case. The defense declined.

"At that point, we had a lot of other litigation out there, and the question was: Do we pay this and let the word get around and be confronted with substantial amounts of litigation?" Dow's Chris L'Orange explained. "The question was: Do we pay what we thought was a king's ransom or else do we make a decision to go on and let a jury decide? It was funny, the concomitant opinion of the defense lawyers and, ironically enough, of the judge was that the verdict was going to be somewhere between a million and a million five. Which again, we thought was realistic. So that's the reason we went to trial."

On October 31, 1982—the night before the settlement conference started in San Francisco—Marta Trout was getting her boys ready to go out trick-or-treating in Manteca. She had sewn an astronaut costume for Michael, Jr., and a Superman outfit for Matthew. Just as they were ready to leave, Marta realized that her child was about to be born. Her third son was delivered at 3:00 A.M. the next day.

Marta had been twenty-four when Mike died, and after the money had dwindled, she took a part-time job as a bookkeeper at her parish church. She reared her boys with strong memories of their father but tried to shelter them from the lawsuit. She felt that she had to keep fighting Oxy for Mike's sake. "I don't care if it takes ten years to settle this case," she vowed in the spring of 1981, "I won't give up." Marta undertook such strenuous preparations for her deposition that fall that she lost ten pounds. She studied for it like a final examination, memorizing dates, lying sleepless at night as she tried to recall when Mike had started to get his headaches or when he had gone into the hospital the first time. There were eight lawyers at the deposition, and she was angry because they treated her like a "know-nothing." She was not a woman who exposed her private life willingly or who let her emotions show. Her ideas about bringing up a family were rigidly traditional, and she wanted her boys to grow up normally. She wanted only enough money to replace their father's lost earnings. She did not want a jury to award her "sympathy" money, and nothing irked her more than the thought of people saying, "Oh, look at that poor widow and her two sad little boys."

Her lawyer had cautioned her not even to date after Mike's death, but Marta eventually remarried. There were constant delays in the trial, and she had her own life as well as that of her sons to consider. In the early autumn of 1982, with trial scheduled to begin on November 8 in San Francisco, Marta was determined to take the witness stand, even though she was very obviously pregnant. At home in Manteca, wearing a pink "Round Is Beautiful" T-shirt, surrounded still by pictures of Mike, she spoke of the fury she felt when she turned on the television one day and saw Armand Hammer, chairman of Oxy's parent company, Occidental Petroleum, on the *Merv Griffin Show*. Hammer was chairman of President Reagan's three-member cancer advisory panel, and he had publicly offered $1 million to the scientist who first came up with a cure for cancer. "They were praising him for the money he donated to cancer research," Marta said. "No one bothered to mention that his company is responsible for men dying from cancer."

Marta resented the awesome power the chemical industry exerted over people's lives. She thought the chemical companies had "too much power, too much money, too much clout." She suppressed her anger when she was around the boys, but when she was alone, the obsessiveness of it astonished her. "I get so angry at them," she said, "and I'm a Christian!" Her wrath was focused almost exclusively on Oxy. She was less clear about Dow and Shell's culpability. Oxy was the company that she believed was responsible for Mike's death. Oxy was real—she could see it just by driving a few miles from her home. Marta knew about problems caused by Oxy's sister subsidiary, Hooker, in other parts of the country. And she knew that Oxy was trying to sell the Lathrop plant. It seemed to her that the company was simply trying to pack up and get out of town, leaving the wreckage of its workers' lives behind. Her anger extended to Mike's former Ag Chem supervisor, who lived down the street. She could not stomach the thought of his children playing with her boys and would not allow them inside her home. "I don't want sympathy," she said again. "I just want people to realize the power of these chemical companies."

Marta voted against accepting Oxy's settlement money, but she was in the minority, and she finally relented. "The only way to finance the trial was to take the settlement with Oxy," she said. But she considered the money small recompense for the number of injuries caused the workers. "They put it in the newspaper," she said. "I was embarrassed that people knew that Oxy got off so cheap." At the same time the knowledge that Oxy was out of the game reduced her determination to continue fighting Mike's case in court. "When they said we weren't going to go to court against Oxy, that kind of just took the fight out of it for me," she said. "Oxy was personally vindictive to Michael. Not Dow, not Shell. Oxy was the one I wanted to take to court. It just killed my spirit."

On November 6, the Saturday after her third son had been born, Marta's attorney called her with the proposed settlement from Dow and Shell. She was bedridden with complications from the child's birth and faced a return trip to the hospital. She talked the offer over with Mike's parents, and they encouraged her to accept it in order to avoid the burden of a trial. She called Mike's two brothers and sister, and they agreed. Friends had been telling her that if she won in court, the jury award would be big. But she was worried about the effect a "splashy trial" might have on her boys. Michael, Jr., was smart enough at age eight to read the newspapers. "They understand about the poisoning and what was done at Oxy, but they're so young they don't remember the suffering," she said. Marta was concerned, too, that a large settlement might make it difficult for them to grow up in a normal way. "I want them to grow up and be something," she said. "I don't want them to have everything handed to them on a silver platter. I'm a firm believer in hard work. If they think they're going to come into a lot of money when they're eighteen, it will make them cocky. I saw kids like that in high school, and they had no initiative."

She agonized over the decision during the weekend and finally decided to accept. The settlement was contingent on Marta's never publicly revealing the amount, but newspaper reports put it close to $750,000. "I've been hearing so many ridiculous rumors about what I got," Marta said. "It's fair. It's not spectacular. Just fair. But I wish I could say what it was because people think I'm a millionaire." At Marta's request the judge and the attorneys worked out an arrangement under which the settlement would be paid out over a number of years.

At the end of her long journey Marta was contemplative. "I can't say that I'm especially proud of settling out of court," she said. "It looks kind of like a cop-out unless you're in my shoes."

And then she said that she felt "a sense of quiet" and a great deal of "sadness."

"I feel sad that Michael isn't here to reap the benefits," Marta explained. "We suffered by losing him. But he suffered at their hands every day at work."

On Tuesday morning, December 7, *Arnett* v. *Dow* came to trial in the courtroom of San Francisco Superior Court Judge Daniel Weinstein, a forty-two-year-old ex-federal public defender and a former San Francisco assistant district attorney.

"Day 1—December, 7, 1982," an assistant to the plaintiff's attorneys wrote on his note pad. "David v. Goliath."

The trial lasted for five months, minus a few weeks for holidays and the flu that took out the judge for a few days in January. Two more workers settled out of court, so there were seven plaintiffs. Dr. Charles Hine and the University of California settled halfway through trial, so there were two defendants. The plaintiffs argued that Dow and Shell knew about toxic effects of DBCP on the testicles of laboratory animals in the 1950's but ran such sloppy tests that they never found the dose that caused no effects. They said the defendants failed to warn workers of the hazards of DBCP by passing the information along to Occidental or putting it on their product labels. They accused the two companies of engaging in a conspiracy to suppress the test results by convincing the government of the adequacy of watered-down label warnings, which did not mention reproductive effects.

The defense attorneys balanced the facts of the case against a two-part theme. It was Occidental's shoddy safety programs and the carelessness of the workers themselves that caused their exposure to DBCP, and any resulting fertility injuries were caused not necessarily by the chemical but by a wide variety of influences found in modern life.

Dow and Shell tried to show not only that they had performed extensive tests on DBCP but that the results of the tests were widely available through industry journals. Although the chemical had an effect on the lungs, liver, kidney, and testes at the lowest dose tested, the scientists thought they would see problems in the human target organs before trouble developed with the testes. By signing a contract to buy DBCP from Dow in 1963, Dow said Oxy had been fully warned about DBCP and assumed all responsibility for the workers. Dow said it sent Oxy a formulator's manual containing safety information about DBCP; Oxy said it never got it.

The defense tried to puncture the conspiracy charge by showing that the federal government knew as much about the hazards of DBCP as did the chemical companies when the nematocide was registered in the early 1960's because the 1961 report had been included in the registration data. Not only were joint petitions common industry practice, but they were encouraged by the government to save time and paper work.

Shell had extricated itself from all but the conspiracy claim by settling out of court with the only plaintiff who had worked at Oxy during the two years Shell supplied it with DBCP. Shell attorney Steve Jones said there was no joint effort to conceal but rather a "joint effort by everyone to DISCLOSE information."

Before Dr. Hine left the case having an arrangement under which money was paid to the plaintiffs but no judgment was entered against the scientist, he took the stand in his own defense and repeated, like a saint's daily litany, the progression of tests he'd conducted on DBCP in the 1950's. The lawyers wanted to know why Dr. Hine did not test down to a "no-effect level" and why he thought one part per million was a safe level of exposure when no tests had been done to prove that assumption. Dr. Hine had answered these questions in various ways at previous inquisitions. At the state hearing in San Francisco five years before, he had conceded it was "stupid" not to have tested to a no-effect level. In a newspaper interview he had said, "In all fairness, nobody thought anything but radiation was a sterilizing agent in those days. We just missed the boat." In a magazine interview he said, "I blew it." Here, at the trial, with Duane Miller reading his hearing testimony back to him, Dr. Hine replied, "It was my testimony at that time. I think I was upset and distressed. It is not my opinion today."

Dr. Hine testified that five parts per million had strained the detection capabilities of his laboratory equipment during the 1950's, and he could test no lower. He said the most severe effects occurred at higher dose levels, with significantly less effect at the five parts per million level. By plotting down this "dose response curve" from high to low, he came to a point on the curve where "by judgment, there would be no effect." In consultation with Dow scientists, "it was our best judgment as a group that nothing would occur if it was kept below a part per million, but we didn't know how much below a part per million." Yet, he said, it was "permissible scientific judgment" to make estimates "based on the information you have at hand." By consensus, one part per million became the recommended worker exposure standard, although it was never adopted by the federal government. Dr. Hine conceded, nevertheless, that even now there was no certainty that the one part per million level would provide adequate protection. It "may be too high," he said. " . . . I'm not sure if we know exactly what level would produce an effect in a human being."

Dr. Hine said he had no concern about the effect shown on the testicles at the five parts per million level because he thought it was secondary to the primary effect of the chemical, which was decreased growth and overall poor health. He cited scientific studies dating back to 1918 that showed that if animals didn't grow properly, then the size of their testicles was reduced. He defended Shell's requests to delete health and safety information from a report he wrote on DBCP because the report was prepared for field applicators, who would have found it impractical to wear heavy safety gear. Dr. Hine also reminded the court that he had suggested that the health of people exposed to DBCP should be closely observed. "We didn't know what effects were probable and no way did I think there would be an effect on reproduction," the big scientist said, and tears fell from his eyes.

The plaintiff's attorneys suggested that Dow and Shell had worked together to convince the government that their DBCP workers were healthy and that tougher warning labels were not necessary. The attorneys themselves were startled when Dow's Dr. John Lanham testified that the company had changed its medical treatment cards in 1966 to delete a reference to the link between DBCP and the effect on the testes. The attorneys had never seen the medical cards. They had wangled almost nothing from Dow in the way of internal documents during discovery and privately believed that Dow had shredded anything that might prove legally troublesome or embarrassing (Dow president Paul Oreffice wrote in a 1977 memo to Dow supervisors: " WHEN IN DOUBT, THROW IT AWAY!!!"). Dr. Lanham testified that the cards—which listed medical symptoms associated with the use of all of Dow's chemicals—contained a warning in 1964 about DBCP's possible effect on the testicles. Two years later the warning was removed. He said it was removed because no reproductive problems had surfaced in workers.

The heart of the case was the dispute over whether DBCP had in fact caused the workers' fertility problems. The defendants were boxed into a corner on fertility effects because overwhelming medical evidence linked DBCP with impaired fertility. But each worker had to prove that his fertility problems were directly caused by his exposure to DBCP. The defense tried to dilute their claims by showing that the workers had other medical problems or "social" habits, like marijuana use, which either caused their fertility problems outright or contributed to them. The judge had barred the attorneys from bringing up the "life-style" habits of workers in court unless they could prove they had some direct bearing on their injuries. Marijuana, excessive alcohol, cigarettes, and venereal disease all had been shown to depress male fertility in varying degrees, but full recovery usually occurred shortly after the exposure ended.

In view of the harsh evidence requirements of the law, only one of the seven cases looked to be a sure winner. Richard Perez had started working at Oxy in the summer of 1967 and quit seven years later, when his own physician told him he was sterile. Richard and his wife, Gloria, had a son before he started working at Oxy. During the next ten years they tried without success to have a daughter. Perez had been the first worker to sound the alarm about fertility hazards in Ag Chem. He was still sterile at the time of the trial, and doctors believed his condition was permanent. The rest of the cases were complicated by questions about length of exposure to DBCP and the severity of fertility effects.

Dr. Donald Whorton, the occupational health specialist who had originally tested the Oxy workers, testified that the impaired fertility of five of the seven workers was due to their exposure to DBCP. He was unsure about Frank Arnett and Luther Harrell, who had been exposed to DBCP primarily through their work with stained pallets, not on the Ag Chem line. But he said the role of "life-style" factors, if any, was "overwhelmed" by the effects of DBCP.

The defense's chief fertility expert was Dr. Larry Lipschultz, a Houston urologist who specialized in male reproductive disorders. Dr. Lipschultz said most of the workers' infertility could be partially traced to physical problems like varicose veins in the testicles or obstructions in the tubes that carry sperm. He said three of the seven workers' fertility problems were not due to DBCP at all because they had not been sufficiently exposed to the chemical.

The plaintiffs attacked Dr. Lipschultz's credentials and argued that the workers who he claimed had had little exposure to DBCP had in fact been sprayed with the chemical at times, used rags to wipe up DBCP spills from cans stacked on the pallets, and often spent several hours at a time with their hands covered with gooey chemical substances that accumulated on the pallets. The attorneys also pointed out that no other doctor besides Dr. Lipschultz (and his assistant) had found testicular varicose veins in the workers.

The jury had to rely on the testimony of both the workers and their psychologists to determine whether the fear of cancer and birth defects was real. The workers had originally asked for damages for both the risk and the fear of acquiring cancer and passing on birth defects, but the judge ruled that the science of predicting who would develop cancer was too speculative for consideration in court. He did, however, extract promises from Dow and Shell protecting each worker's right to sue in the future should he contract cancer. The defense attorneys wanted the "fear" claims rejected as well on the ground that exposure to chemicals was not an injury but an "event" experienced by everyone who drank diet sodas or smoked cigarettes. Compensating workers for fears based on such "events" would throw open the floodgates to "insubstantial, transient and fraudulent claims." But the judge did not agree. In a precedential ruling he allowed the Oxy workers to submit evidence to the jury on fear of cancer. Although cancerphobia was a relatively new claim in American product liability litigation, the judge found sufficient "guarantees of genuineness" in the Oxy workers' fertility injuries to permit them to describe how the injury itself, coupled with resulting publicity about DBCP's link to cancer, caused them anxiety. In order to prevail, the workers first had to show that DBCP had physically injured them and that they subsequently developed a fear of cancer based on information from doctors and news reports.

A psychologist who examined the workers at the request of their attorneys testified they were haunted by fears of cancer and "old men's fears of death." He said that every worker expressed fears of cancer and that several had reached the point where every minor illness seemed a major threat. But the defense wondered if the workers' tension might not be partially due to the stresses of their lawsuit and the necessity of having to "visit a San Francisco psychologist."

Finally, the question of DBCP's relationship to birth defects was debated. The issue was central to the case because the workers said they suffered emotional stress as a result of their children's birth defects and feared that future children would be born defective. The workers would also be able to bolster their "fear of cancer" claim if they could prove that DBCP caused a genetic injury to their bodies. Their attorneys argued that DBCP causes genetic changes in everything from bacteria to humans and that such alterations had the potential to trigger cancer, birth defects, and spontaneous abortions. But the research was so speculative, and the scientific consensus so fragmented, that the plaintiffs were allowed very little leeway in what they could present to the jury.

On Wednesday, March 16, 1983, testimony ended. "Take a deep breath," Judge Weinstein told the jurors. That morning, as the jury members walked through the big marble lobby of City Hall on their way to the fourth floor courtroom, the *Chronicle* newsboxes were full of newspapers with headlines shouting about Dow Chemical Company. The acting director of the Environmental Protection Agency was accused of allowing Dow scientists to edit an agency report on the sources and effect of dioxin in the Great Lakes. The judge had to remind the jurors not to read or watch any news related in any way to the trial. It was almost an impossible task because the news was full of reports about Dow and dioxin. The night before testimony ended, ABC's *Nightline* report had been devoted to dioxin, and the network news that evening featured the story of Dow's alleged attempt to suppress information about the hazards of dioxin. Waiting in the empty courtroom at day's end, Chris L'Orange railed against "whatever great cosmic force" had put the company he was defending on a conspiracy charge in a front-page story.

His banter was good-natured, but his concern was genuine. Publicity about Dow and about environmental problems in general had shadowed the trial almost from its beginning. In January and February hardly a day passed without a newspaper or television story about the dioxin contamination of Times Beach, Missouri, or the resignation under fire of Environmental Protection Agency Administrator Anne Gorsuch Burford, who was accused of sacrificing the environment to the interests of industry. In Times Beach a waste hauler had sprayed the streets with dioxin-contaminated oil sludge to help control dust. It was the same waste hauler who had sprayed the Missouri horse stable with contaminated oil in 1971, causing the death of sixty-two horses and illnesses among children who played in the arena. The dioxin wastes were from a defunct hexachlorophene plant that had paid the waste hauler to dispose of them. In December 1982 the EPA found dioxin levels of more than 100 parts per billion in the soil of Times Beach—100 times over the government level considered harmful for long-term contact. After a flood in December and new tests that showed the stubborn dioxin continuing to cling to soil in the one-square-mile city, the EPA decided to buy the town's 800 houses and thirty businesses and close the place down. The federal buyout cost $36.7 million. One of Mrs. Burford's last official acts was to fly to Times Beach to announce it.

The disintegration of the EPA began with the abrupt firing in early March of Rita Lavelle amid charges of political malfeasance and conflict of interest in her administration of the agency's toxic waste cleanup program. The EPA official was accused of cozying up to the chemical industry and of writing a memo that identified the business community as "the principal constituents of this administration." Five congressional subcommittees subsequently launched a wide investigation of alleged wrongdoing at the agency—an investigation that centered on charges of political manipulation in the toxic waste program, possible conflicts of interest and perjury among top agency officials, political "hit lists," and "sweetheart deals" with major industrial polluters. With public opinion polls showing that opposition to President Reagan's environmental policies was building in the country—47 percent of the people surveyed in a *Newsweek* poll were opposed, and a Republican National Committee poll showed that the EPA controversy was "all negative" for Reagan—the clamor for Mrs. Burford's resignation grew louder both inside and outside the White House. In previous polls the majority of Americans had consistently rejected the administration's view that environmental regulations should be eased to promote economic growth. Now, in the wake of Times Beach and other toxic waste threats, the public was registering a new and more pervasive concern in opinion polls: the fear of toxic chemicals. Industry contended that the public fear was fueled by exaggerated media reports, but anxiety only continued to build as new pockets of contamination were discovered. "Hot spots" of dioxin were found near the old Diamond Alkali herbicide plant in Newark, New Jersey; in Missouri, officials suspected 100 more sites were contaminated with dioxin.

In mid-March Mrs. Burford resigned. She was replaced by William D. Ruckelshaus, the nation's first EPA administrator. Once in the director's seat again, Ruckelshaus called for major changes in the way government agencies regulate health, safety, and environmental risk by balancing that risk against the cost of reducing or eliminating it. "We must now assume that life takes place in a minefield of risks from hundreds, perhaps thousands, of substances," the new EPA administrator declared in his first major policy speech. "No more can we tell the public: 'You are home free with an adequate margin of safety.'"

In the late morning of April 8, after eleven days of deliberation, Isaac Williams, the thirty-six-year-old black Social Security Administration office manager who had been elected jury foreman in the DBCP trial, sent word that a verdict had been reached: Dow was convicted of failure to warn about the hazards of the chemical. But the company was acquitted of concealment, and both Dow and Shell were found innocent of conspiracy. The jury was hung seven-to-five on punitive damages. The judge asked the jurors to deliberate one more day on punitives and then to begin making individual damage awards to the workers.

In the late afternoon of April 14 the jury announced a $4.9 million damage award to the Oxy workers. Richard Perez received $2 million for sterility and $25,000 for fear of cancer. Because he was completely sterile, the jury awarded him nothing for fear of birth defects. His wife, Gloria, received $350,000 for loss of consortium and $25,000 for her fear that her husband would get cancer. The jury decided that Perez had been at comparative fault for 5 percent of his own injuries; that reduced the total $2.4 million award by $120,000. "I think everybody felt strongly—even the defense attorneys recognized the fact that this man was obviously sterile and it wasn't likely that he was going to have any kind of recovery," jury foreman Williams said. "I think Perez was probably the most clear-cut. With some on sterility, it was more complicated than others. With Perez, it wasn't."

Steven Sarras was awarded $1 million for sterility, $37,500 for fear of cancer, and $10,000 for fear of birth defects. Sarras had been sterile when the workers were tested in 1977, and his wife was unable to get pregnant. His sperm count eventually improved and a child was born in 1979—with a curved leg. His wife received $100,000 for loss of consortium and $20,000 for fear of birth defects. The total award was reduced 5 percent for comparative fault.

Wesley Jones received $1 million for sterility and $10,000 for fear of cancer. The jury considered him responsible for 10 percent of his injuries because he was a leadman at Oxy and therefore had a greater responsibility to be aware of safety information than the others. The jury had originally given some of the other workers substantially less than it awarded Jones, although all had similar injuries. Rather than lower Jones's damages, it raised the other workers' awards. "Obviously," said a juror who requested anonymity, "some people had some strong feelings about Wesley's character in terms of the kind of individual that he was—you know, turning this man loose with this kind of money. He's not a well person. But the other side of that is that he, like other individuals, had suffered some testicular damage. That's what was being tried, not his character."

Haskell Perry, whose abnormally low sperm count later improved to normal, received $200,000 for sterility, $50,000 for fear of cancer, and $10,000 for fear of birth defects. Ralph Lewis, whose fertility also improved, received $60,000 for sterility. Frank Arnett was awarded nothing for sterility and $10,000 for fear of birth defects. His wife was given $20,000 for fear of birth defects. The jury said all three workers' comparative fault was 5 percent. Luther Harrell was not awarded any damages.

Questions were immediately raised about how Arnett could have a fear of birth defects if the jury did not think he had a sterility injury. "Well, that's kind of strange, isn't it?" the jury foreman conceded. "We didn't feel that Arnett was exposed to DBCP to the extent that it would have had a negative effect on his sperm count. However, the fact that his consciousness was raised concerning the possibility that DBCP was a carcinogen and, in addition to that, could possibly cause birth defects —the fact that he was working at Occidental, had the potential of being exposed to it, would set in his own mind and in his wife's mind a fear that this DBCP had in fact caused the birth defects in their children."

Williams said the jury gave Ralph Lewis money for sterility because he had worked on the DBCP canning line for a time and "the potential for exposure was certainly much greater for him than for Arnett or Harrell." With Harrell, they believed "the exposure wasn't there. His on the stand testimony actually killed him, because there were clear differences in what he said at one time and what he said at another."

Before the workers received their awards, money would be subtracted for the percentage of comparative fault, the amount received from Oxy in the out-of-court settlement, any previous workmen's compensation awards, and the lawyers' fees. The jury never reached a 9–3 majority verdict on punitive damages, and the judge declared a mistrial on that issue.

Dow's Chris L'Orange was "surprised" at the amount of the verdict. "I was surprised because I thought they would come in and say, 'Yeah, there's a fertility injury, and this is the amount of money we ascribe to the fertility injury,' and I had not anticipated it would be as extensive as it was," he said. "I had more concern about them looking at the fear [claims] and saying, 'This will be a lifelong affliction.' Especially in a society today that is as overwhelmingly concerned about cancer as we apparently are. I think everybody was surprised along those lines."

Dow lawyers surveyed the jury about the verdict. "We said, 'Why so much money?' and the answer kept coming back, 'What do you mean, why so much money? We thought we did you a favor,'" explained Dow attorney Don Frayer. "It was very clear that they really thought they had held the amount down and done us a favor because we really weren't the bad guys and they didn't dislike any of our people —some of them said corporations are what made America great—but they said to us, 'You were a little bit pregnant and for reasons we don't understand, the real bad guys, Occidental, were not there, so we let you off easy.'"

Doug Brown was happy about the verdict. "It was a good healthy compensatory award that has a better chance of sticking [in an appeals court]," he said. "And when you get ten times what they originally offered, you've got to be pretty happy." The awards were later adjusted in a settlement conference with some workers getting more and some less.

"I don't think we were guilty," said Dow's Dr. Torkelson. "I think we did everything that was necessary and responsible that we should have done. Anything short of an acquittal to me was a loss.

"We made a mistake on DBCP," the scientist said. "You can't win 'em all. I think everybody is allowed to make a mistake. It's only when you don't learn from your mistakes that you're in trouble. I think our mistake here was that somehow there was a communications goof. I think now we're going to be more careful to make sure that we don't get into the same situation. In terms of what we would do over again, I'm not sure it would be a lot different, given what we knew at that point in time.

"Let's say it's one out of ten thousand, just for odds," he continued. "If you went to the average newspaper and looked at it critically, and you say you're not going to allow one error out of ten thousand words . . ." When asked about the impact of a newspaper error compared to an error in judging a chemical's potential for harming human beings, Dr. Torkelson shot back: "Is it any more important not to destroy somebody's reputation than it is to destroy his testes? Newspapers make the point that they are not destroying life, but that's not true. That's hiding behind something."

"I'm not real happy myself," Haskell Perry said after he had heard the verdict. Perry, who had been Mike Trout's friend in Ag Chem, was living with uncertainty about his own health (he had some nodules growing on his back, which doctors diagnosed as noncancerous) and the kidney problems of his nine-year-old daughter, who was conceived during the time he worked at Oxy. He could not understand why the jury thought his pain less valuable than some of the other Oxy workers' just because he was single when he was infertile and they were married.

More troubling was his concern that Dow had not been sufficiently punished. "There is no way they are ever going to pay enough," he said, "to satisfy what my feelings are inside."

chapter 12

The nematocide DBCP has been banished from the United States since 1979. It was a chemical that rose on the crest of science as progress and that fell with a smash as science was exposed as imperfect. Most farmers have learned how to live without it. Most of the workers have recovered their fertility, although they carry scars of doubt about their future and their children's future. Most of the people who sold it believe that it was a safe chemical that somehow slipped through their elaborate testing and warning systems. "We thought we had a system that would have taken care of that," said Dow vice-president Dr. Etcyl Blair. "Somebody somehow somewhere didn't understand, or the word didn't get to them, or something broke down."

"If we'd known then what we know now, we'd never have made the damn product," said Dow's Dr. Ted Torkelson, whose toxicological tests helped put it on the market. "We did so much more than was normally done. Our recommendations were extremely stringent. But something went wrong with the system that these guys got hurt, and it shouldn't have happened."

Sitting in his tiny cubicle office at Dow headquarters in Midland, a photograph of his five children and three grandchildren displayed on a shelf next to an award for his contributions to industrial toxicology, the fifty-seven-year-old scientist looked shaken when he talked about DBCP. "It hurts," he said.

Dr. Torkelson tested thousands of chemicals in his thirty-year career at Dow. He said he did not know how to describe the tumult and the finger pointing that trailed him after it turned out that one of those thousands of chemicals sterilized men. "Some people have the idea that because you're in industry, you're a cheap, lying bastard," he said. "And yet people in industry came out of the same schools as the clergy, the teachers, the academics, the government people. You've got immorality in industry; you've got immorality in government."

"I think basically Dow has been a very, very honest company. I've been trying to figure out why. I think I know. We were a small town with this big company, and if somebody got hurt, it was your neighbor. The guy I wrote the standard for, or made the recommendation for occupational exposure for, could be my neighbor. And if he wasn't my neighbor, then he was somebody I knew and had to live with afterwards.

"I think science oversold itself, or maybe it was an overzealous press," Dr. Torkelson said. "It goes back to 'Better Things for Better Living Through Chemistry' and that kind of thing. All you had to do was wait a day, and there'd be a new food additive or a new whitener for your clothes or a new floor polish or a new paint or a new weed killer, a new this, a new that—you know, chemistry will take care of us. And finally the pendulum just went the other way, and people started to see bad things about chemistry, and that's all they saw. But I don't think the people here at Dow have lost faith in technology."

During the trial in San Francisco Dr. Torkelson went out to lunch with Dr. Charles Hine, his research collaborator from Shell. "I had quite a talk with Charlie, and our conclusion was, gee, back in '58, '60 when we did that work, we were years ahead of what was being done on other chemicals," he said. "That's what hurts—how we're being questioned on what was done twenty years ago."

Dr. Hine, sixty-eight years old and still teaching at the University of California Medical School in San Francisco, was most tailed by the specter of DBCP. He was the scientist who generally received the blame, fairly or not, for not testing the chemical down to a no-effect level, which was subsequently determined to be 0.01 part per million. He considers himself the father of occupational medicine at the university, where he started an intensive residency program to qualify doctors for certification as occupational physicians. Early in his career he organized a Bay Area poison control center, which dispensed help over the telephone. "I answered that phone all through my childhood," said Hank Hine, his son and ardent defender. But when Dr. Hine was named codirector of a state-funded occupational health residency program at the university several years ago, so much pressure was brought against him by labor groups and by the school's student newspaper that he resigned. "Charles Hine is part of industry-controlled occupational health, and look what happened. If we don't want to see any more DBCPs, we need to train people with a worker's perspective," one local labor leader declared. Hine's opponents charged that he was indelibly tainted by his many industry consultantships; he has consulted for Shell, Dow, Clorox, Du Pont, Kennecott, and ASARCO and for law firms representing tobacco companies. In particular, they criticized an experiment he designed for ASARCO to lower lead levels in refinery workers' blood, which drew government fines and sanctions against the company. "His work raises grave concerns about scientific standards of conduct, particularly for public university employees," the student newspaper *Synapse* editorialized.

"People don't understand the role of physicians in industry," Dr. Hine said. "And I must admit that in the past some physicians in industry were not practitioners of the highest type. But I think it's an unfair accusation that everybody who practices is proindustry. I'm not proindustry; I'm not prounion; I'm pro the profession." Of his newspaper and labor critics, the scientist said, "I'm a natural target. They associate me with the establishment. With industry. And I say, 'Look, I'm for the profession.' "

In the aftermath of DBCP Dr. Hine's consulting business fell off and his popularity as a teacher declined. "Ten years ago there were applauding students, and now there are half-filled classes," said his son, who added that the "pernicious and unpredictable" criticisms of his father have had a draining effect on the whole family. But Dr. Hine will not leave the university. "You don't get chased off the street," he said, "by a hoodlum."

In the Central Valley of California, 7,000 square miles of groundwater from Lodi to Bakersfield are contaminated with DBCP. Studies have shown a link between the chemical in the groundwater and increased incidence of leukemia and stomach cancer among valley residents. In Lathrop the Occidental Chemical Company paid $10 million to settle a state and federal groundwater contamination suit, sold its chemical plant, and moved out of town.

In Modesto, Clyde McBeth, the nematode expert retired now from Shell, still believes in DBCP. "If it did cause sterility, it was because they were exposed to it in the plant and shouldn't have been," he said.

After a time he said, "I think those guys talked themselves out of a job."

Dioxin has been around since the 1930's or, if you believe Dow, since the advent of fire, but it was not until 1983 that it became a household word. An entire city in Missouri was evacuated and bought by the government, a neighborhood around an old herbicide plant in New Jersey was invaded by men in spacesuits, the "town that a test tube built" on the Tittabawassee River in Michigan was turned on end by the suspicions of the country, and the hierarchy of the Environmental Protection Agency was given a push as it tumbled down the hill—all because of a toxic chemical that no one makes on purpose.

In a list of scare words, "dioxin" ranks right up there with "nuclear war," yet even now, after a quarter century of the slavish devotions of the world's best scientists, no one really knows what dioxin does to people or how it does it.

The editor of *Chemical and Engineering News,* Michael Heylin, called it "the dioxin phenomenon—a brew of uncertain science, unanswered and sometimes unanswerable health questions, regulatory dilemmas, intensive press coverage, and legal maneuverings that has bubbled over to besmirch the chemical industry and leave the public confused and scared.

"It involves the credibility of the chemical industry and of the entire regulatory process," Heylin said. "It crystallizes the issue of how industrial workers and the public should be protected from any man-made chemical that may be—but probably isn't—a health hazard and, very critically, how to do it rationally. It is certainly fair to say that dioxin is far less toxic to humans than the public has been led to believe."

Yet the Environmental Protection Agency contended that "the slightest trace of dioxin in the environment may have adverse effects on the health of both human and animal populations."

"There is too much potency in this compound to think that we are not at risk for *something,*" said John Moore of the government's National Institute of Environmental Health Sciences.

University of Illinois toxicologist Dr. Samuel Epstein described dioxin as "a potent multi-system toxic agent producing a panoply of acute and delayed effects, many of which can progress to the chronic. There is now growing and substantial evidence of a clear consistency between the wide range of toxic effects induced by dioxin in experimental animals of various species and those observed in a wide range of exposed human populations, including occupational and Vietnam veterans. This consistency relates to multi-system disease, reproductive toxicity, and carcinogenicity."

And so the dioxin debate continues. There is almost no middle ground. One side contends that dioxin has caused no human deaths, that the chief impact on industrial workers exposed to large amounts of dioxin is chloracne, and that there are no hard data to link dioxin to cancer, birth defects, or reproductive problems in humans. It says dioxin is unquestionably toxic but is species-specific—which is to say, hamsters can tolerate 5,000 times more dioxin than guinea pigs can. It contends that data from industrial accidents show that humans are closer to hamsters than to guinea pigs in their relative resistance to the immediate toxic effects of dioxin.

The other side argues contrarily that dioxin can be linked to birth defects, reproductive problems, and soft-tissue sarcoma, a rare form of cancer that affects muscle, nerve, and fat tissue. It says chloracne is but one of many possible dioxin effects and that its role has been overemphasized to the exclusion of other, more serious disease. The only point on which both sides agree is that there are uncertainties about irreversible, long-term effects of dioxin exposure.

In looking back at the hundreds of factory workers exposed to large doses of dioxin, scientists have been able to draw few conclusions about the future. The 121 men who developed chloracne after the reactor explosion at Monsanto's Nitro, West Virginia, trichlorophenol plant in 1949 have been scrutinized for more than thirty years. A recent follow-up report by a Monsanto epidemiologist and the director of the University of Cincinnati's Institute of Environmental Health says they are not developing cancer or dying in abnormal numbers. But the report has been criticized by Dr. Epstein and others for excluding exposed workers who did not develop chloracne, for using the general population rather than a similar industrial population as a comparison group, and for misinterpreting the cancer data.

Of particular concern were two deaths from soft-tissue sarcoma, including one worker who had never had chloracne. When the two deaths from soft-tissue sarcoma among Monsanto workers were combined with two similar deaths among Dow Midland workers exposed to dioxin in 199 Building in 1964, the results represented a fortyfold excess of the number normally expected to occur in a worker population of that size. Two other deaths from soft-tissue sarcoma at Monsanto are in dispute. One man worked as a truck driver and maintenance man at the Nitro plant and was not engaged in the production of trichlorophenol, although he could have been exposed to the chemical. The other was a thirty-three-year-old worker at Monsanto's Illinois chlorophenol plant whose father also worked at the plant and who also had the rare skin cancer. The son died; the father did not.

Few chemicals leave "fingerprints" at the scene of the crime. Asbestos causes mesothelioma, a rare cancer of the lining of the lungs and abdomen. Vinyl chloride causes angiosarcoma, an uncommon type of liver cancer. But most chemicals cause health effects that are indistinguishable from the common diseases of modern life. Some scientists believe that one of dioxin's "fingerprints" is soft-tissue sarcoma, a group of rare cancers that are found in fewer than 1 percent of all cancer victims. A Swedish study found that paper pulp, forestry, and sawmill workers exposed to 2,4,5-T had from five to six times greater risk of developing the skin cancer than unexposed people had. Dow scientists and others have criticized the methodology of the Swedish studies, including the fact that test subjects were chosen by asking people whether or not they could remember ever being exposed to the herbicides. But closer to home, statistics show Midland County, Michigan —where Dow made dioxin-contaminated products for more than thirty years—had an 800 percent increase in soft-tissue sarcomas from 1970 through 1978. (There was one case in the 1950's, five cases in the 1960's, and eight cases in the 1970's.) Yet studies in Washington State and New Zealand have not turned up the same connection between dioxin and soft-tissue sarcoma.

In Globe, Arizona, Bob McCray, who steam-cleaned herbicide barrels in his auto repair shop in 1968, found a tumor growing on his shoulder six years later. He says it grew from the size of an aspirin to the size of a golf ball. When surgeons removed it, they diagnosed it as a soft-tissue sarcoma.

McCray said his allergies and problems in concentration have left him incapable of holding a steady job. He said his whole family has health problems, including a daughter with cancer. His youngest son, Paul, who was five months old when the helicopter flew over Globe that hot Sunday morning in 1969, has never completely recovered from the host of illnesses that plagued him afterward. "We almost lost him to leukemia," McCray said. "He still has a lot of emotional problems.

"I ended up with cancer. I can't work," McCray said. "You get nervous, anything that happens to you, you think it's from dioxin. The fear is always there. All you can do is live day to day and hope that your sanity holds up.

"Don't let anybody kid you that dioxin is not bad," he said. "They're lying."

Few clues are forthcoming from examinations of the other production workers exposed to dioxin around the world. In the aftermath of the explosion at the BASF plant in West Germany in 1953, four workers died from gastrointestinal cancer and two from lung cancer, five died from cardiovascular diseases, one from cirrhosis of the liver, and one from urogenital disease. All other deaths were accidental or suicidal. In the Netherlands eight workers exposed to dioxin in a 1963 accident have died, six of them apparently from myocardial infarctions. Exposure in a Czechoslovakian plant in the mid-1960's has left many workers with severe liver and neurological damage as well as hypertension and elevated blood levels of lipid and cholesterol. A ten-year follow-up published in 1980 noted that complete recovery was rare, except for anxiety and depression, and that persistence of an impaired metabolism of fats and carbohydrates could be a major cause of premature arteriosclerosis. The British are tight-lipped about the aftermath of the reactor explosion in the Coalite trichlorophenol plant in 1968, but it is known that one worker has died from coronary thrombosis and that some workers continued to have abnormalities in blood chemistry and immune function ten years after the explosion.

In Seveso, Italy—where a dioxin-contaminated cloud was propelled across a town after a trichlorophenol reactor explosion in 1976—recent and disputed studies contend that except for quickly healed chloracne in children, no organs or body functions were impaired and that despite the panic over birth defects and the wave of abortions that followed, no unusual numbers of defects or spontaneous abortions occurred. Analysis of the livers of 144 sheep showed that they died from intestinal disease, not from dioxin exposure.

In 1983, the McKusicks still lived in their low-slung home in Kellner Canyon in Globe. They have never considered moving because, as Bob McKusick sees it, "This is the safest place on earth. They won't come back here to spray because of the 'paranoid' residents." The house is a crazy quilt of animals and people—a whistling cockatiel, a jealous parrot that screams at their granddaughter as she crawls across the floor, three dogs and a cat that come in and out, two turtles whose paralyzed hind legs have been replaced by wheels, and a roomful of animal bones that Charmion McKusick is collating for an anthropology report. A ruddy, congenial man in his early fifties, Bob McKusick has reopened his pottery studio after a four-year break; his wife, Charmion, is teaching anthropology; their daughter and her husband have moved into a trailer home across the road.

The doctor who examined the plaintiffs in the Globe lawsuit at the request of Dow found only minor health problems in the McKusicks —a blurring of eyesight in Bob, greater sensitivity to a pinprick in Charmion's left arm than in her right. But the McKusicks believe they came close to dying more than once after the herbicide spray. They say they are feeling healthy for the first time in years, although, as Charmion said, "We know that we are going to die. We're just putting it off as long as we can. We've taken an aggressive attack against this problem. The people who did not are dead. And I have no sympathy for them." They take massive daily doses of vitamins and a yucca seed extract. They drink sage tea and bottled water. They used to buy two bottles of Anacin, a bottle of Contac, and muscle spasm medicine every week and fall into bed exhausted after dinner. After a flood in 1978, which they theorize brought contaminated water and soil down from the mountain, McKusick "tried to die three times in one year" from heart spasms. His wife had surgery for an intestinal blockage. "For a period of four years," McKusick said, "we could not use our yard because it was contaminated so badly. You couldn't be in it for more than one hour a day. We practically ran to feed the animals." They trucked in new soil for their yard, hosed out the machinery in the pottery shop, and removed the top three feet of soil from the clay pit with a backhoe.

They don't like to talk about the past much anymore. McKusick is tired of taking reporters on guided tours of the mountain. He is tired of hearing skeptical remarks about the deformed goat and duck. ("We can absolutely prove that the goat was born after the spray," he said.) He is convinced that his property and the property of others who had complained about the herbicide spray in 1968 were purposely sprayed again in 1969 on orders of antagonistic Forest Service officers. He is equally convinced that his fight against the herbicide spray was necessary and worthwhile and that the Pinals will never be sprayed again. (His belief that the helicopter will never spray herbicides on Globe again is partially based on psychic perceptions.) But he is distressed to find 2,4,5-T still on sale at the local hardware store.

"We had problems—and still would—convincing people in Globe," he sighed. "A quarter of the people probably believe what happened. Fifty percent think there's a possibility that something happened. Twenty-five percent think we cooked it up trying to make a million dollars."

Willard Shoecraft still lives up the road in Ice House Canyon. Sometime after his wife died, he married the nurse who cared for her in the hospital in Oregon. His radio station has expanded, and stacked up next to piles of KIKO lighters and felt-tip pins in his office closet are copies of *Sue the Bastards*. He, too, seems exhausted with questions about the herbicide spray. "I don't know what causes cancer," he said. "I think the herbicide could act as the trigger. I think that it hurt Billee."

"I believed Bill," declared her daughter-in-law, Mari. "She had proof. And she was an extraordinary person because she was willing to sacrifice everything. The bottom line to her was, don't make it anymore. She said, 'The only thing Dow understands is money, and I want them to pay and pay hard.'"

Despite the fact that it settled out of court with the Globe residents, Dow held firm to its conviction that 2,4,5-T and silvex were safe herbicides and that even though the products represented just a fraction of the company's total sales, to lose the battle on these herbicides would be to lose the battle on all toxic chemicals. So convinced of the rightness of their research and the righteousness of their mission were the Dow scientists that several volunteered to eat 2,4,5-T and silvex. Apart from a metallic taste, no adverse effects were reported.

"There are those in the company who had second thoughts a long time ago about whether we should [defend 2,4,5-T]," said Donald Frayer, Dow's legal claims manager. "I don't mean to suggest there was a lot of controversy about it. But we weren't [of] a monolithic opinion that we ought to keep doing this. I think the prevailing view was that where you have a lot of data, and where you—quote—know that you're right—unquote, and where you have literally hundreds of products in this area . . . if you allow the anecdotes and the fears to prevail, then how can you really sell any of these things? Because it's just a matter of time before anecdotes and fear will attack all of these, and they'll all go down the drain."

Dow didn't settle the Globe case because the company thought it was meritorious but because the massive Vietnam veterans' Agent Orange suit was looming darkly in the distance. Nine thousand plaintiffs were suing Dow and four other herbicide makers on the ground that they knew but failed to inform the government of the hazards of Agent Orange. Since any trial involving dioxin may establish legal precedents about its effect on humans that can be used in future trials, Dow did not want to risk an unfavorable verdict in the Globe case that could carry over to the Agent Orange case. "You can't worry about being stripped of your defense in a very large case in litigating a very small case," Frayer said.

Only one major claim involving dioxin has actually gone to trial in the United States: Forty-seven Illinois railroad workers who sued Monsanto and the Norfolk & Western Railway for health effects they claimed to have suffered in cleaning up a chemical spill reportedly won $7 million from Monsanto in an out-of-court settlement and more than $58 million from the railroad in a 1982 jury trial. The verdict is on appeal. But a Nova Scotia judge ruled in September 1983 that dioxin-contaminated herbicides presented no health hazard to people and denied an injunction sought by landowners to stop the spraying of nearby forests. Justice Merlin Nunn wrote in a 182-page decision:

> I am satisfied that the overwhelming currently accepted view of responsible scientists is that there is little evidence that, for humans, either 2,4-D or 2,4,5-T is mutagenic or carcinogenic and that [dioxin] is not an effective carcinogen, and further, that there are no-effect levels and safe levels for humans and wildlife for each of these substances.
>
> Were I required to do so and perhaps to allay public fears, I will add that the strongest evidence indicates that these substances sprayed in the Nova Scotia environment will not get into or travel through the rivers or streams, nor will they travel via groundwater to any lands of the plaintiffs who are adjacent to or near the sites to be sprayed. Further, if any did, the amount would be so insignificant that there would be no risk. . . .

In the Globe case, Dow's Frayer explained—apart from the fact that the herbicide was apparently applied in a sloppy fashion, which Dow had nothing to do with—"It was an emotional situation. They kept complaining—and I'm sure they were doing it in good faith, I'm sure they were scared of the material, and I'm sure they thought there were effects—but they kept up such a hue and cry that people kept sending investigative teams down and they just never found anything. They canvassed all the doctors in the area to see if they had observed changes . . . and they were pretty consistently reporting no." As for Mrs. Shoecraft's ovarian cancer, Frayer said: "You let one of these things [lawsuits] go on long enough, and it's just like going away on vacation and leaving your windows open. If you're gone for two weeks, you can bet it's going to rain sometime during those two weeks.

"I would suppose if we'd gone out there with a relatively small amount of money and bought our peace, whether we were wrong or not, in 1970 or 1971," he added, "we'd have never known the headache that was going to occur. But you go along, and these things keep piling up: She writes the book, the regulatory actions come along, the Agent Orange case comes along, and for ninety-seven other reasons you've got a difficult problem on your hands."

Once the Globe case had been settled, Dow turned its full attention to the Agent Orange suit and to the defense of its herbicides before the Environmental Protection Agency—two cases that illustrate the sprawl of modern litigation in that Dow had produced 144,000 pages of documents in the EPA case and somewhere beyond 3 million pages of documents in the Agent Orange case. In March 1981 Dow and the EPA began negotiating a closed-door settlement. "Could the EPA, a government agency mandated to protect citizens and the environment, stand up to the political and economic pressure that Dow and its allies would exert?" an Oregon group devoted to fighting herbicides asked rhetorically after the deal had been announced. "The answer is emphatically no. Dow with its tremendous mobilization of legal and scientific advisors put the squeeze on the EPA staff from the outside. On the inside, pressure came from new Reagan appointees dedicated to removing regulations. We think it is safe to say that Dow, more than any other single corporation, has been responsible for the most government money spent on chemical regulation. Their maneuvers to keep 2,4,5-T registered and free of restriction have cost taxpayers literally millions of dollars."

Not to mention the $10 million that Dow estimates it has spent to defend the herbicides and the $3 million it is spending to test the air, water, and soil around its Midland plant for dioxin, to help the state Health Department study soft-tissue sarcoma in Midland County, and to finance an independent scientific review of the health effects of dioxin in the environment. "We believe we have been forthright and honest in responding to the many questions that have been asked of us," Dow president Paul Oreffice declared at a press conference in the summer of 1983. "Sometimes we have been asked questions for which there are no absolute answers. For this we have been accused of equivocation or talking out of both sides of our mouth. When we respond unequivocally, we are accused of arrogance, or self-righteousness. I guess we can accept these accusations. But what we CANNOT and WILL NOT accept is the level of anxiety and concern the publicity surrounding this issue has generated in this country, in this state, and for some in this community."

In interviews at Dow headquarters in Midland in the fall of 1983, executives defended their products and their company, which had increasingly come under attack in the press not only for questions of safety but for questions of attitude. In its dealings with the EPA and with the outside world in general, Dow had—as its president put it—exhibited an arrogant self-righteousness. Oreffice also had speculated that Dow was still paying in many ways for its involvement with napalm in the 1960's, and Ted Doan, who ran Dow during those days, agreed. "You don't have to argue against anybody to say that there are a lot of people who think Dow shouldn't have made that stuff," he said. "We did it for our own good reasons . . . but we've got a raft of people out there that think, Dow did it wrong there, they're probably doing it wrong again.

"Somebody once said if Dow was named Wujinewski, we'd probably be better off because 'Dow' and 'Mao' rhyme—you know, everything worked very nicely for the students," Doan said with a smile. "I think this dioxin thing approaches the napalm thing in being as severe on Dow—not in the same sense, that kind of nonsense of campus recruiting, but in the overall effect and the difficulty of trying to get our message across. Because we never have been able to do it. We can't do it to this day."

Doan and his wife had recently spent some time digging through his grandfather's letters for research on a play about the founder of Dow. "They were quite original—there were no management consultants then, no lawyers, my God, it must have been a paradise!" Doan said. "These were his letters on 'Why profit?' and 'What is the system all about?' The sense of an ethical approach to running a business was the base of the thing. And it's been true ever since. And it's the thing that confounds our people today. Because they still have trouble understanding that the world doesn't have to look at you the way you look at yourself. We think we're the good guys."

Several weeks before, while Doan was having dinner in Washington with Frank Press, an old friend who is president of the National Academy of Sciences, the subject of dioxin came up.

"I said, 'Frank, it's just really overdone. This stuff is not that hazardous,'" Doan recalled.

"And Frank said, 'Well, I don't know, maybe you're being a little cavalier about that.'

"Now there's the representative of private science in this country," Doan said. "And a very bright guy. But that's the scientist's approach. If you don't know an absolute answer, you equivocate. And the scientist has a terrible time coming to our defense, understandably, I think. But in terms of what's going on today, I don't know whether science can afford that luxury. They've got to make risk assessments. They've got to say to themselves, 'Aflatoxin occurs in wheat, and it causes deaths in the United States every year. These things are all around us, and they are exactly the same as toxic chemicals, and they've been here forever.'

"Everything is poisonous," Doan declared. "But you cannot prove that something doesn't do harm. It may be that one molecule of dioxin in the right site at the right time may cause cancer and kill a person. The chances of it being true are almost zero. But you can't prove it isn't true."

Dow's theory—which is both backed and disputed by the scientific community—is that life itself is a risk and that society has progressed to the point where the onslaught of fatty foods, sunlight, radiation, cigarette smoke, alcohol, and stress cause cancer and disease either individually or in combination with toxic chemicals in the environment. According to this theory, risks imposed on people by the chemical or nuclear power industry, for instance, are exaggerated, and the benefits underestimated. In one of its public relations brochures Dow quotes a Harvard University professor who has somehow calculated that living two days in New York City presents the same million-to-one risk of death as living near a polyvinyl chloride chemical plant for twenty years.

But many scientists believe the value of human life is becoming smudged in the conceptual battle of risks and benefits. "The risk of permitting dangerous substances to be uncontrolled is substantially greater than the temporary loss of use of some substance which later proves not to be carcinogenic. If we need demonstration that we have in the past erred in not being sufficiently cautious, consider the long history of substances thought to be safe and later shown to have caused much suffering and death," contended Albert Einstein College of Medicine professor Robin Briehl. "Speaking as a scientist, the central problem with our society is not a lack of sufficient scientific rigor, much as that may unfortunately occur. The real problem is an insufficient consideration of real human needs and values; not the needs of marketing and profit."

Dow believes, nevertheless, that society is slowly coming around to the idea that risks have to be put in perspective or "managed." As one bit of proof they point to the apparent conversion of Dr. Bruce Ames, the biochemistry department chairman at the University of California at Berkeley who testified at the 1977 DBCP hearing in San Francisco that the "modern chemical world" might be responsible for future increases in cancer and mutations in the "human gene pool" and that "much of the modern chemical world we just haven't seen yet in terms of its effect on cancer." Yet in an article in *Science* magazine in the fall of 1983 Dr. Ames concluded that most human cancer was probably caused by a variety of exposures, that there was no good evidence that increased cancer rates in the United States could be blamed on anything but cigarettes, and that one of the biggest risks after cigarettes was the large number of potent carcinogens in our food. "I don't know if I have been 'converted,' but I am much more sympathetic to industry than I used to be," Dr. Ames said. "People damned industry more than they deserved." In *Science* he wrote: ". . . whether or not any recent changes in lifestyle or pollution in industrialized countries will substantially affect future cancer risk, some important determinants of current risks remain to be discovered among long established aspects of our way of life."

"This fear thing is grossly overplayed because as a society we live much longer than our ancestors did and in much better health," said Dow vice-president Dr. Etcyl Blair. "I've often thought it would be a marvelous experiment to take *The New York Times* Sunday edition and subject it to pyrolysis [chemical changes brought about by the action of heat] and all kinds of analysis to analyze what in the world is in that newspaper. You know, if you allow yourself to be engaged in dissecting everything down as far as you can go scientifically, that's part of the problem. With dioxins, we're now down to parts per quadrillion! What does it mean? It's like ghosts or something."

"It seems to me there were two trends developing which have now come to full flower," said Dow chairman of the board Robert Lundeen. "One is the increasing public sensitivity to pollution problems . . . parallel with that and perhaps supporting that increasing sensitivity were these advances in science. And I'm not sure which was the chicken and which was the egg. We find ourselves now where 2,3,7,8-tetrachlorodibenzodioxin is a household word. Most Dow executives couldn't even spell it ten years ago!

"The chemical industry, in some sense, created some of its current problems itself—and I don't want to be misinterpreted on this—because we now find out that things we didn't even know we were producing are in fact there," he continued. "Some of those things are bad actors. Now we know the problem exists, and we'd better do something about it, and we *are* doing something about it. But because of the way information is transmitted instantaneously now, small events, small pieces of information and data out of context with the whole on these complicated issues get magnified in the average citizen's mind out of proportion to the real significance of the event."

Elements of Risk / 303

On October 14, 1983, Dow announced that it was abandoning its fight for 2,4,5-T and silvex. It did not intend to produce the herbicides ever again, and it would drop its efforts to have the chemicals certified as safe by the Environmental Protection Agency.

Dow said the decision was based not on questions of safety but on public anxiety about dioxin.

notes and sources

CHAPTER ONE

Pages 3–6: Early history of the Dow Chemical Company and Midland, Michigan:
Don Whitehead, *The Dow Story: The History of Dow Chemical Company* (New York: McGraw-Hill, 1968).
Murray Campbell and Harrison Hatton, *Herbert H. Dow: Pioneer in Creative Chemistry* (New York: Appleton-Century-Crofts, 1951).
Robert Conot, *American Odyssey* (New York: William Morrow, 1974).
Bruce Carlson, *History Along the Highways of America* (Midland, Mich.: Historical Collection of the Grace A. Dow Memorial Library).
Midland Historic District Report, 1978.
"How Midland, an Abandoned Lumber Camp, Came to Be a Center of Scientific Fabrications," *Michigan Manufacturer,* March 9, 1935.
"Michigan Centennial Historical Sketch/Midland, Michigan," *Detroit Free Press,* August 22, 1937.
Page 5: "'Mr. Dow no doubt has . . .'": Campbell and Hatton, *op. cit.*
Pages 6–7: Early chemical industry history:
Whitehead, *op. cit.*
Thomas Robert Evans, *Salt & Water, Power & People: A Short History of Hooker Electrochemical Company* (Niagara Falls, N.Y.: 1955).
Dan J. Forrestal, *Faith, Hope and $5,000: The Story of Monsanto* (New York: Simon & Schuster, 1977).
Alexander Findlay, *A Hundred Years of Chemistry* (New York: Macmillan, 1937).
John Maxon Stillman, *The Story of Alchemy and Early Chemistry* (New York: Dover, 1960).
Milton Moscowitz, Michael Katz, and Robert Levering, *Everybody's Business: An Almanac* (San Francisco: Harper & Row, 1980).
"'I must tell you . . .'": Findlay, *op. cit.*
Page 7: "'Go out on a starry . . .'": Whitehead, *op. cit.*
"So did the government, which contracted with Dow . . . ": Whitehead, *op. cit.*
"Midland was so crowded . . . ": Campbell and Hatton, *op. cit.*
"One of Herbert Henry Dow's many preoccupations . . . ": Whitehead, *op. cit.*

Page 8: "By the 1930's the chemical industry . . . ": "Chemical Leadership," *Library Digest* (April 27, 1935).

"BIG FIELD AWAITS . . . ": "Big Field Awaits Young Chemists," *Detroit Free Press*, June 27, 1932.

"He once marveled . . . ": Campbell and Hatton, *op. cit.*

Page 9: "He was an avid fruit tree grower . . . ": "Fifty Years in Agricultural Chemicals," *Down to Earth—A Review of Agricultural Chemical Progress* (Midland, Mich.: Dow Chemical Company, Winter 1957).

Pages 9-11: Dow's contributions to Midland:

Whitehead, *op. cit.*

"Midland, Michigan," *The Old AAA Traveler* (date unknown).

Lawrence M. McCracken, "Midland Earns Right to Title of Michigan's Different City," *Detroit Free Press*, April 8, 1938.

"Gift of Mrs. Dow," *ibid.*, September 24, 1938.

"6,000 Christmas Lights," *ibid.*, December 22, 1938.

"Out of Debt," *ibid.*, April 12, 1939.

"Midland Gets New Hospital," *ibid.*, March 3, 1944.

J. Dorsey Callaghan, "Only Midland Could Play This," *ibid.*, December 16, 1952.

J. Dorsey Callaghan, "Industry and Labor Back Cultural Setup," *ibid.*, December 17, 1952.

James S. Pooler, "Midland Dedicates a New House to God," *ibid.*, February 23, 1953.

"Dow Library Is Opened in Midland," *ibid.*, January 25, 1955.

"Dow Offers $200,000 for Jail," *ibid.*, April 14, 1956.

Jerry Sullivan, "Midland Turning Factory Town into Little 'Garden of Eden,'" *Detroit News*, September 21, 1955.

Judd Arnett, "Midland Stands Alone," *Detroit Free Press*, August 31, 1961.

"Accept Dow Funds for Juveniles," *ibid.*, July 17, 1963.

"Dow Foundation to Finance Arts Center in Midland," *ibid.*, December 24, 1966.

Page 10: "'Growth is unlimited . . . '": Sullivan, *op. cit.*

"To Dow's way of thinking . . . ": Campbell and Hatton, *op. cit.*

"'Like that aunt . . . '": Sullivan, *op. cit.*

Page 11: "In the late 1930's . . . ": McCracken, *op. cit.*

"Like his father, Willard Dow . . . ": Whitehead, *op. cit.*

Page 12: "In the 1950's . . . ": "Portrait of a Professional," *Down to Earth* (Summer 1972); "The Dow Medalists," Dow Annual Report (1979); "Over a Patent a Day," *Dow Diamond* (Midland, Mich.: Dow Chemical Company, Fall 1963).

"Like many conservatives, Willard Dow . . . ": Whitehead, *op. cit.*

Pages 12–13: Origins of Dow's toxicology laboratory:

Author's interviews with Dr. Theodore Torkelson and Dr. Wendell Mullison.

D. D. Irish, "Multiple Responsibilities of the Insecticide Chemist," *Advances in Chemistry Series*, No. 1, American Chemical Society (1950).

"Chemical Hawkshaws," *Down to Earth* (Fall 1952).

"Dow Working on Wood That Won't Burn," *Detroit Free Press*, April 8, 1956.

Page 14: "Willard Dow's aggressive leadership . . . ": "Willard Dow," obituary, *Down to Earth* (Summer 1949).

Pages 14–15: Early insecticide history:

Thomas R. Dunlap, *DDT: Scientists, Citizens, and Public Policy* (Princeton, N.J.: Princeton University Press, 1981).

W. W. Fletcher, *The Pest War* (New York: Wiley, 1974).

Page 15: Early chlorophenols development:

Dow Diamond (November 1946, September 1947, June 1949, August 1950, February 1952, June 1952, November 1959).

R. W. Baughman, "Tetrachlorodibenzo-p-dioxins in the Environment," Ph.D. dissertation, Harvard University, 1974.

Page 15: "In the months before Dow first described . . . ": Karl O. Stingily, "A New Industrial Chemical Dermatitis," *Southern Medical Journal* (1940).

"By 1937 Dow's own workers . . . ": Baughman, *op. cit.*, and Milton G. Butler, "Acneform Dermatoses Produced by Ortho (2 Chlorophenyl) Phenol Sodium and Tetra-Chlorophenol Sodium," *Archives of Dermatology and Syphilology* (1937).

Page 16: "Dow has no record . . . ": Author's interview with Garry Hamlin, Dow public affairs.

Pages 16–20: Early herbicide development:

Author's interviews with Dr. Wendell Mullison and Dr. John Davidson.

Dow Diamond (May 1946, January 1947, July 1947, March 1948, May 1948).

Gale E. Peterson, "The Discovery and Development of 2,4-D," *Agricultural History* (July 1967).

Bruce F. Meyers, "Soldier of Orange," *Environmental Affairs* (1979).

Donald Davis, "Herbicides in Peace and War," *Bioscience* (February 1979).

Thomas Whiteside, "Defoliation," *The New Yorker* (February 7, 1970).

Robert Harris and Jeremy Paxman, *A Higher Form of Killing: The Secret Story of Chemical and Biological Warfare* (New York: Hill & Wang, 1982).

Michael Uhl and Tod Ensign, *GI Guinea Pigs* (New York: Playboy Press, 1980).

John Lewallen, *Ecology of Devastation: Indochina* (Baltimore: Penguin, 1971).

Seymour Hersh, *Chemical and Biological Warfare and America's Hidden Arsenal* (Indianapolis: Bobbs-Merrill, 1968).

Page 19: Chemical industry in wartime:

Evans, *op. cit.*

Forrestal, *op. cit.*

Dow Diamond (January 1942, February 1943, March 1944, Exhibit Issue, 1944).

Pages 19–21: Growth of chemical industry:

Evans, *op. cit.*

Dow Diamond (August 1943, September 1944).

"1950 Is Banner for Chemicals," Associated Press, June 22, 1950.

"Big Boom Sighted for Chemicals," United Press, September 23, 1957.

Pages 21–22: DDT history:

"U.S. May Be Bugless Utopia," Associated Press, June 1, 1944.

Allan Shoenfield, " 'Age of Plastics' Due at World War's End," *Detroit News,* July 9, 1944.

Frank Carey, "DDT Air Barrage Protected Yanks," *Detroit Free Press,* November 26, 1944.

Harold Schachern, "Dow Opens Its House of Wonders," *ibid.,* July 28, 1950.

Russell MacFadden, "Plastics—Growing Giant," *ibid.,* July 12, 1960.

Page 22: "Dow announced at war's end . . . ": *Dow Diamond* (October 1945).

Page 24: "'Manufacturers rushed . . .'": Peterson, *op. cit.*

Pages 22–26: Early herbicide development:

Author's interviews with Dr. Wendell Mullison and Dr. John Davidson.

Charles Hamner and H. B. Tukey, "The Herbicidal Action of 2,4 Dichlorophenoxyacetic and 2,4,5 Trichlorophenoxyacetic acid on Bindweed," *Science* (August 18, 1944).

Paul Marth and John Mitchell, "2-4 Dichlorophenoxyacetic Acid as a Differential Herbicide," *The Botanical Gazette* (December 1944).

E. J. Kraus and John Mitchell, "Growth-Regulating Substances as Herbicides," *The Botanical Gazette* (March 1947).

Joe Schwendeman, "Dandelions Don't Grow Under Her Feet," *Golf Journal* (March 1975).

Dow Diamond (May 1946, July 1947, March 1949, June 1950, December 1953, June 1954).

Down to Earth (May 1945, November 1945, Spring 1948, Summer 1949, Summer 1954, Fall 1954, Winter 1954).

Silvex Technical Bulletin No. 1, Dow Chemical Company, January 1954.

What About Kuron?, Dow Chemical Company, May 21, 1957.

CHAPTER TWO

Pages 27–29: Nematode history:

Author's interview with Clyde W. McBeth.

"On the Track of the Invisible Worm," *Shell News* (June 1961).

C. W. McBeth, "Nematodes," *Agrichemical West* (January 1965).

C. W. McBeth, "Preparation and Fumigation of Soil." Unpublished paper.

C. W. McBeth, "D-D and Nemagon Soil Fumigants—Soil Preparation, Application Methods and Diffusion." Unpublished paper.

C. W. McBeth, "All About Nematodes." Unpublished paper.

Ehud Yonay, "The Nematode Chronicles," *New West* (May 1981).

Pages 28–32: Development of early nematocides:

Yonay, *ibid.*

C. W. McBeth and Glenn B. Bergeson, "1,2-Dibromo-3-Chloropropane—A New Nematocide." Unpublished paper.

"The War Against Soil Pests," *Down to Earth* (March 1946).

"New Soil Fumigant Meets Pineapple Growers' Needs," *ibid.* (June 1946).

Pages 32–33: Early toxicological testing:

Author's interview with Dr. Theodore Torkelson.

Testimony of Dr. Julius Johnson, Dow Chemical Company, before the Senate Subcommittee on Reorganization of the Committee on Government Operations, Hearings on Environmental Hazards Coordination, 1963.

Dow Chemical Co. and Union Carbide Chemical Co., "Toxicological Basis of Threshold Limit Values and Pathological Biochemical Criteria," *American Industrial Hygiene,* Vol. 20 (1959).

P. J. Gehring, V. K. Rowe, and Susan B. McCollister, "Toxicology: Cost/Time," *Food and Cosmetics Toxicology,* Vol. 11 (1973).

Page 33: " 'The deadliest sin was to be controversial.' . . .": William Manchester, *The Glory and the Dream* (Boston: Little, Brown, 1974).
" 'It must be crystal clear to every American . . . '": *Dow Diamond* (January 1951).
"Dow employees were sent to school . . . ": *ibid.* (May 1951).
"In speeches Dow president . . . ": *ibid.* (December 1951).
Page 34: "In one of his first public . . . ": *Down to Earth* (Spring 1950).
Pages 34–38: Early testing of DBCP:
Author's interviews with Dr. Charles Hine and Dr. Theodore Torkelson.
Testimony of Dr. Charles Hine, *Arnett et al.* v. *The Dow Chemical Company et al.*, San Francisco Superior Court, December 9–10, 1982.
"Class B Evaluation of 1,2-dibromo-3-chloropropane," UC Confidential Report No. 226, July 15, 1954.
"Chronic Feeding Experiment in Rodents," UC Confidential Report No. 228, November 15, 1954.
"Percutaneous Toxicology and Irritation Studies," UC Confidential Report No. 230, January 7, 1955.
"Acute and Chronic Vapor Exposure of Rodents," UC Confidential Report No. 231, January 12, 1955.
"50 Vapor Exposures and Ancillary Blood Studies," UC Confidential Report No. 278, April 21, 1958.
"Repeated Exposure of Laboratory Animals to 1,2-dibromo-3-chloropropane," Biochemical Research Laboratory, Dow Chemical Company, July 23, 1958.
T. R. Torkelson, S. E. Sadek, V. K. Rowe, J. K. Kodama, H. H. Anderson, G. S. Loquvam, and C. H. Hine, "Toxicologic Investigations of 1,2-dibromo-3-chloropropane," *Toxicology and Applied Pharmacology* (1961).
Pages 37–38: "As for finding an exposure level . . . ": Dow Chemical Co. and Union Carbide Chemical Co., *op. cit.*
Page 41: Shell's internal memoranda on DBCP:
Documents filed with the San Francisco Superior Court, *Arnett et al.* v. *The Dow Chemical Company et al.*
Page 44: "'Monsanto has five hundred . . . '": Author's interview with Dr. Etcyl Blair.

CHAPTER THREE

Page 50: "In September 1954 the Dow . . . ": *Dow Diamond* (September, 1954).
"Dow's promotions came at a time . . . ": *ibid.* (June 1954).
"'The chemical industry rapidly . . . '": *ibid.* (February 1953).
Page 51: "'With safety woven so deeply . . . '": *ibid.* (Summer 1960).
"Dow's annual growth rate . . . ": Whitehead, *op. cit.*
"Dow called its petrochemical stacks . . . ": *ibid.* (August 1952).
"In the summer of 1952 Dow asked . . . ": *ibid.* (June 1952).
"Dow said its scientific research . . . ": *ibid.* (March 1960).
"And as much as Dow fought . . . ": *ibid.*
Page 52: "In 1948, 38 million pounds . . . ": *ibid.* (March 1955).

Elements of Risk / 309

"By 1952 Americans were buying . . . ": *ibid.* (March 1955).
"Six years later, with American . . . ": *ibid.*
"As Dow pointed out . . . ": *ibid.*
"The estimated annual loss . . . ": *ibid.*
"And industry was selling . . . ": *ibid.*
"The productive American farm . . . ": *Dow Diamond* (Summer 1960).
"'Four out of five . . . '": *ibid.*
"'Abandoning the use . . . '": *ibid.*
Page 54: "'No nation in the . . . '": Louis Bromfeld, "Bromfeld on Food Poisons," *Cleveland Plain Dealer,* September 9, 1951.
"'The fabulous Dow . . . '": Schachern, *op. cit.*
"'The regulatory is not the . . .' ": Testimony of Edmund Feichtmeir before the Senate Subcommittee on Reorganization of the Committee on Government Operations, Hearings on Environmental Hazards Coordination, 1963.
Page 55: "The West German workers were . . . ": Drs. E. W. Baader and H. J. Bauer, "Industrial Intoxication Due to Pentachlorophenol," *Industrial Medicine and Surgery,* Vol. 20 (1951); Drs. H. J. Bauer, K. H. Schulz, and U. Spiegelberg, "Occupational Intoxication in the Production of Chlorinated Phenol Compounds" (trans. from German), *Archives of Industrial Pathology and Industrial Hygiene,* Vol. 18 (1961).
"In America, meanwhile, Monsanto . . . ": "Dioxin Leaves Mark on a City Called Nitro," *The New York Times,* August 4, 1983.
Page 56: "But an accident in 1949 . . . ": Drs. Judith Zack and Raymond R. Suskind, "The Mortality Experience of Workers Exposed to Tetrachlorodibenzodioxin in a Trichlorophenol Process Accident," *Journal of Occupational Medicine,* Vol. 22 (January 1980); J. E. Huff, J. A. Moore, R. Saracci, and L. Tomatis, "Long-term Hazards of Polychlorinated Dibenzodioxins and Polychlorinated Dibenzofurans," *Environmental Health Perspectives* (November 1980).
"United States Public Health Service . . . ": "Dioxin Health Studies Disappear," *Chemical Week* (June 29, 1983).
"There were four more . . . ": Huff et al., *op. cit.*; Rebecca L. Rawls, "Dioxin's Human Toxicity Is Most Difficult Problem," *Chemical and Engineering News* (June 6, 1983); "The Health Effects of 'Agent Orange' and Polychlorinated Dioxin Contaminants," Technical Report, American Medical Association, October 1981.
Pages 56–59: Summary of German production accidents, experiments, and isolation of the dioxin contaminant:
J. Kimmig and K. H. Schulz, "Occupational Acne Due to Chlorinated Aromatic Cyclic Ethers" (trans. from German), *Dermatologia* (1957).
K. H. Schulz and J. Kimmig, "Chlorinated Aromatic Cyclic Ethers as the Cause of Chloracne" (trans. from German), *Die Naturwissenshaften* (1957).
Bauer, Schulz, and Spiegelberg, *op. cit.*
Baughman, *op. cit.*
Pages 59–60: Boehringer's communications with Dow:
Author's interview with Garry Hamlin, Dow public affairs.
Unsealed court files, "Agent Orange" consolidated product liability litigation, U.S. District Court for the Eastern District of New York.
Page 60: "A similar explosion the . . . ": Drs. Jacob Bleiberg, Marven Wallen, Roger Brodkin, and Irvin Applebaum, "Industrially Acquired Porphyria," *Archives of Dermatology* (June 1964).

"Five years later . . . ": "Dioxin Health Studies Disappear," *Chemical Week* (June 29, 1983).

"The foundation had launched . . . ": Daniel M. Berman, *Death on the Job* (New York: Monthly Review Press, 1978).

"IHF president Dr. Daniel Braun said . . . ": Author's interview with Dr. Daniel Braun.

"It was predicated on the idea . . . ": Berman, *op. cit.*

"The study itself no longer exists . . . ": "Dioxin Health Studies Disappear," *Chemical Week* (June 29, 1983).

"Said IHF president . . . ": Author's interview with Dr. Daniel Braun.

Pages 61–62: Follow-up German health studies:

Schulz, *op. cit.*

Bauer, Schulz, and Spiegelberg, *op. cit.*

Baughman, *op. cit.*

Page 63: "In 1957, when the Germans . . . ": "Dow Co. to Spend 2 Million to Curb Midland Air Pollution," Detroit *News,* August 7, 1957.

"But if change was slow . . . ": *ibid.*

"'Residents long ago . . . '": *ibid.*

"In the high-growth post-World War II . . . ": Whitehead, *op. cit.*

Pages 63–66: DDT history:

Dunlap, *op. cit.*

Page 65: "Dow executive W. W. Sutherland . . . ": Dunlap, *op. cit.*

Pages 66–72: *Silent Spring* and its impact:

Rachel Carson, *Silent Spring*. (Boston: Houghton Mifflin, 1962).

Frank Graham, Jr., *Since Silent Spring* (Boston: Houghton Mifflin, 1970).

Forrestal, *op. cit.*

Hearings on Environmental Hazards Coordination before the Senate Subcommittee on Reorganization of the Committee on Government Operations, 1963.

John M. Lee, "Silent Spring Is Now Noisy Summer," *The New York Times,* July 22, 1962.

The Silent Spring of Rachel Carson, CBS Reports, April 23, 1963.

"Experts to Report on Bug-Killer," *The New York Times,* May 6, 1963.

"Seek Pesticide Link to Child Ills," United Press International, June 5, 1963.

"Lawmakers Doubt Pesticide Danger," *ibid.,* April 20, 1965.

Author's interviews with Herbert Dow Doan and Dr. Etcyl Blair.

Page 68: "'Miss Carson forfeits . . . '": Walter Sullivan, "Silent Spring," *The New York Times Book Review,* September 23, 1962.

"Miss Carson was called . . . ": Graham, *op. cit.*

Page 73: "Dr. William Darby of the . . . ": Graham, *op. cit.*

"Dow's Dr. Julius Johnson . . . ": "In Defense of Accomplishment," *Dow Diamond* (Summer 1963).

CHAPTER FOUR

Page 74: "In the years following . . . ": Whiteside, *op. cit.*

"Every year an average . . . ": *ibid.*

Pages 74–76: World War II and military research on herbicides:
Whiteside, *op. cit.*
Meyers, *op. cit.*
Victor Yannacone, Jr., W. Keith Kavenagh, and Margie T. Searcy, "Dioxin—Molecule of Death," *Trial* (December 1981).
"Agent Orange Litigation," *Trial* (February 1982).
"Agent Orange and the Pentagon," *Chemical Week* (July 20, 1983).
Lewallen, *op. cit.*
Uhl and Ensign, *op. cit.*
Harris and Paxman, *op. cit.*
Fred A. Wilcox, *Waiting for an Army to Die* (New York: Vintage Books, 1983).
Unsealed court files, "Agent Orange" consolidated product liability litigation, U.S. District Court for the Eastern District of New York.
Page 74: "'Even though wartime . . . '": Lewallen, *op. cit.*
Page 75: "'The capability of destroying . . . '": *ibid.*
"Representatives of companies, including Dow . . . ": "Dioxin—Molecule of Death," *loc. cit.*
"Delmore declared . . . ": *ibid.*
"Albert Hayward, chief of . . . ": *ibid.*
"Dow advised General Delmore . . . ": Unsealed court files, "Agent Orange" consolidated product liability litigation, U.S. District Court for the Eastern District of New York.
"A judge presiding . . . ": *ibid.*
Page 76: "They were also the same . . . ": Carson, *op. cit.*
"In his trip report . . . ": Unsealed court files, "Agent Orange" consolidated product liability litigation, U.S. District Court for the Eastern District of New York.
"An explosion in a herbicide factory . . . ": Thomas Whiteside, *The Pendulum and the Toxic Cloud* (New Haven, Conn.: Yale University Press, 1979); Huff, Moore, Saracci, and Tomatis, *op. cit.*
Page 77: "That same year a U.S. Public . . . ": Unsealed court files, "Agent Orange" consolidated product liability litigation, U.S. District Court for the Eastern District of New York.
"In addition, a report by . . . ": "Agent Orange and the Pentagon," *loc. cit.*
Herbicide use in Vietnam:
Meyers, *op. cit.*
Capt. A. L. Young, USAF, "The Toxicology, Environmental Fate and Human Risk of Herbicide Orange and Its Associated Dioxin," *USAF Occupational and Environmental Health Laboratory Technical Report* (October 1978).
General Accounting Office, "Report by the Comptroller General: Health Effects of Exposure to Herbicide Orange in South Vietnam Should Be Resolved," April 1979.
Ralph Blumenthal, "Files Show Dioxin Makers Knew of Hazards," *The New York Times,* July 6, 1983.
Pages 77–78: Diamond Alkali chloracne outbreak:
Bleiberg, Wallen, Brodkin, and Applebaum, *op. cit.*

"Agent Orange Papers: What Companies Knew," *Chemical Week* (July 13, 1983).

"Toxic Chemicals Cloud Memories of a Life of Work," Philadelphia *Inquirer,* June 8, 1983.

Ralph Blumenthal, "Doctor in Jersey Cited Ailments at Dioxin Plant," *The New York Times,* June 7, 1983.

Rawls, *op. cit.*

"The Health Effects of 'Agent Orange' and Polychlorinated Dioxin Contaminants," *loc. cit.*

Page 78: "In 1962, 4,949 acres . . . ": Whiteside, "Defoliation," *loc. cit.*

"In 1963 herbicide-sprayed . . . ": *ibid.*

"In 1964 it tripled. . . . ": *ibid.*

Pages 78–88: Dow's chloracne problem, its meetings with the herbicide industry, and the aftermath:

Author's interviews with Garry Hamlin, Dow public affairs, and Harold H. Gill, Dow research manager.

Unsealed court files, "Agent Orange" consolidated product liability litigation, U.S. District Court for the Eastern District of New York.

Pages 88–89: Kligman experiments:

Aaron Epstein, "Human Guinea Pigs: Dioxin Tested at Holmesburg," Philadelphia *Inquirer,* January 11, 1981.

Page 90: "A Rand Corporation report . . . ": "Agent Orange and the Pentagon," *loc. cit.*

"The Army Chemical Corps nevertheless . . . ": *ibid.*

"During the spring and early summer . . . ": "Dioxin Trial Papers May Soon Be Made Public," *Chemical Week* (June 8, 1983).

"But Donald Hornig . . . ": *ibid.*

"Yet the State Department . . . ": Harris and Paxman, *op. cit.*

"The government had launched . . . ": Meyers, *op. cit.*; Lewallen, *op. cit;* Harrison Wellford, *Sowing the Wind* (New York: Grossman, 1972).

Page 91: "Yale botanist Arthur W. Galston . . . ": "Herbicides in Vietnam," *The New Republic* (November 25, 1967).

"The Pentagon instead commissioned . . . ": Young, *op. cit.*

"A committee of the National Academy . . . ": Meyers, *op. cit.*

"Fred Tschirley, a plant ecologist . . . ": *ibid.*

"Two independent scientists . . . ": *ibid.*

Pages 91–93: British chloracne outbreak:

George May, "Chloracne from the Accidental Production of Tetrachlorodibenzodioxon," *British Journal of Industrial Medicine,* Vol. 30 (1973).

N. E. Jensen, "Chloracne: Three Cases," Proceedings of the Royal Society of Medicine, Vol. 65 (August 1972).

Page 93: "Meanwhile, back in the United States . . . ": Unsealed court files, "Agent Orange" consolidated product liability litigation, U.S. District Court for the Eastern District of New York.

"Monsanto and Dow . . . ": *ibid.*

"(Thompson Hayward told . . . ": *ibid.*

"In March 1970 . . . ": *ibid.*

CHAPTER FIVE

The story of the herbicide spraying of Globe, Arizona, allegations of subsequent illnesses, and the ensuing conflict is based on author's interviews with Willard Shoecraft, Mari Shoecraft, Ernie Gardner, Mary Lou Gardner, Robert McKusick, Charmion McKusick, Robert McCray, Dr. Keith Barrons, and attorneys Jerry Sullivan and Jack Slobodin; on the depositions of Billee Shoecraft, Willard Shoecraft, Robert Shoecraft, Ernie Gardner, Mary Lou Gardner, Robert McKusick, Charmion McKusick, Robert McCray, William Moehn, Dr. Granville Knight, Dr. Paul Singer, and William Fleischman; on the medical records of the Shoecraft family, the Gardner family, the McKusick family, the McCray family, and the Satama family; on Billee Shoecraft's book, *Sue the Bastards* (Phoenix: Franklin Press, 1971); on newspaper accounts published in the *Arizona Record,* the *Arizona Daily Star,* the Tucson *Daily Citizen,* the *Arizona Republic* and the Phoenix *Gazette;* on Dow Chemical Company's transcript of "Apparent Telephone Conversation Between R. Wurm and J. Hansen on July 18, 1969"; on laboratory reports from Stoner Laboratories in Campbell, California, The Diagnostic Laboratory in Phoenix, GHT Laboratories in Brawley, California, and the University of Arizona Community Pesticide Studies Laboratory; on U.S. Forest Service publicity releases about the spray project; on bulletins and information brochures published by the Dow Chemical Company, including *Silvex Technical Bulletin No. 1* (January 1954), *What About Kuron?* (May 21, 1957), *Information from Dow on Kuron Weed and Brush Killer,* released July 25, 1969; on a summary transcript of the Globe City Council meeting, July 28, 1969; on the Investigation of Spray Project Near Globe, Arizona, February 1970 (the Tschirley task force report); on the Kellner Canyon-Russell Gulch Spray Project Field Evaluation, conducted July 10, 1969; on letters from Tonto National Forest supervisor Robert Courtney to plaintiffs' attorneys; and on various other documents, including affidavits, interrogatories, answers, and trial briefs filed with the U.S. District Court in Phoenix, Arizona, in the case of *Shoecraft et al. v. The Dow Chemical Company.*

Information on the Bionetics report, the Mrak report, and the DuBridge directive against 2,4,5-T is based on the "Evaluation of the Carcinogenic, Teratogenic and Mutagenic Activity of Selected Pesticides and Industrial Chemicals," Bionetics Research Laboratories, Bethesda, Maryland, 1968; on Lewallen's *Ecology of Devastation: Indochina, loc. cit.*; Wellford's *Sowing the Wind, loc. cit.*; Whiteside's *The Pendulum and the Toxic Cloud, loc. cit.*; on "The Teratogenic Evaluation of 2,4,5-T," *Science* (May 3, 1970) and "Herbicides: Order on 2,4,5-T Issued at Unusually High Level," *Science* (November 21, 1969); on *The Report of the Secretary's Commission on Pesticides and Their Relationship to Environmental Health,* U.S. Department of Health, Education and Welfare, December 1969; and on testimony before the Senate Committee on Commerce Subcommittee on Energy, Natural Resources and the Environment in hearings on the Effects of 2,4,5-T on Man and the Environment, April 1970.

CHAPTER SIX

Page 123: "President Richard Nixon forcefully . . . ": "That Nixon-Ford-Iacocca Meeting: Delaying Action?" *Detroit Free Press,* December 1, 1982.

"Then came 1969 and its attendant disasters . . . ": John Quarles, *Cleaning Up*

America (Boston: Houghton Mifflin, 1976); Rice Odell, *Environmental Awakening* (Cambridge: Ballinger, 1980).

Page 124: "By the end of the year the federal . . . ": John Walsh, "Environment: Focus on DDT, the Uninvited Additive," *Science* (November 21, 1979).

"It was the worst bear market . . . ": Quarles, *op. cit.*

"Environmental lawyers and professors . . . ": Malcolm Baldwin and James Page, Jr., *Law and the Environment* (New York: Walker, 1970).

Page 125: "John Quarles, an Interior Department . . . ": Quarles, *op. cit.*

"*The New York Times* called it . . . ": Editorial, *The New York Times*, July 12, 1970.

"Historian Theodore H. White . . . ": Theodore H. White, "Untangling the Environment Jungle," *Life* (June 26, 1970).

Page 126: "'We think pesticides contribute . . . '": Testimony before the Senate Subcommittee on Reorganization of the Committee on Government Operations, 1970.

"The bureaucracy itself had grown . . . ": Richard Kazis and Richard Grossman, *Fear at Work* (New York: Pilgrim Press, 1982).

"Nor were business policy makers . . . ": Kazis and Grossman, *op. cit.*

"The chemical industry was well . . . ": *ibid.;* Mark Green and Norman Waitzman, *Business War on the Law* (Washington, D.C.: The Corporate Accountability Research Group, 1981).

"During World War II . . . ": *Dow Diamond* (April 1953).

"Since the early 1950's Dow had . . . ": *ibid.* (Fall 1965).

Pages 127–128: USDA's ineffective regulation of pesticides:
Wellford, *op. cit.*

Testimony before the House Subcommittee on Reorganization of the Committee on Government Operations in hearings on Deficiencies in Administration of Federal Insecticide, Fungicide and Rodenticide Act, 1969.

General Accounting Office, "Comptroller General's Report to Congress on the Need to Improve Regulatory Enforcement Procedures Involving Pesticides," September 10, 1968, and "Comptroller General's Report to Congress on the Need to Resolve Questions of Safety Involving Certain Registered Uses of Lindane Pesticide Pellets," February 20, 1969.

Page 128: "On the first day of the . . . ": "Earth, Energy and the Environment," *Congressional Quarterly Report* (1977).

"In mid-February Nixon . . . ": "Nixon Starts the Cleanup," *Time* (February 23, 1970); "Text of the President's Message to Congress," *The New York Times,* February 11, 1970.

Page 129: "'For a president . . . '": "This Ecology Craze," *The New Republic* (March 7, 1970).

"From January through March . . . ": "Congress and the Environment: What's Getting Done?" *World Ecology 2000* (March 1970).

"The press marveled . . . ": Irving Howe and Michael Harrington, "The Seventies: Problems and Proposals" (New York: Harper & Row, 1972).

"*The New York Times* noted . . . ": Walter Sullivan, "We Lay Waste the World," *The New York Times,* February 15, 1970.

"By the spring of 1970 . . . ": "Earth Day," *Newsweek* (April 13, 1970).

"Leaders of Environmental Action . . . ": *ibid.*

Pages 130–134: Dow president Herbert Dow Doan and the napalm years:
Author's interview with Herbert Dow Doan.

Robert L. Heilbroner and others, *In the Name of Profit—Profiles in Corporate Irresponsibility* (New York: Doubleday, 1972).

Eric Ludvigsen, "Students Picket at Dow Plant as Protest Against War Role," Detroit *News,* August 8, 1966.

Roger Simpson, "Dow Sees 8% Gain Continuing for 1967," *Detroit Free Press,* May 4, 1967.

"150 Protest Dow's War Production," United Press International, October 19, 1967.

"Dow Campus Hunt Not Hurt by Hot Napalm Argument," *Detroit Free Press,* October 29, 1967.

John Reiter, "Dow Chemical Not Alone as Target of Protestors," *ibid.,* January 29, 1969.

"9 Face Charges in Raid on Dow," United Press International, March 24, 1969.

"350 War Protesters Circle Dow Office in Southfield," Detroit *News,* March 29, 1969.

"Miniskirted Nun in Trouble for Dow Chemical Fracas," *ibid.,* April 12, 1969.

David C. Smith, "How Now Dow's Jones?" *Detroit Free Press,* April 13, 1969.

"100 Pastors Read a Protest Against Napalm," *ibid.,* May 5, 1969.

"The Old Ritual Is Losing Appeal," *ibid.,* May 11, 1969.

"Anti-War Group Vandalizes Firm," Associated Press, November 9, 1969.

"Civic Leader's Son Tied to Dow Vandalism," Detroit *News,* November 20, 1969.

Saul Friedman, "Stolen Dow Files List Gifts to Congressmen," *Detroit Free Press,* February 5, 1970.

"Dow Names C. B. Branch as President," *ibid.,* February 5, 1971.

Hugh McCann, "Dow OKs 3-for-2 Split," *ibid.,* May 6, 1971.

Tom Opre, "Big Business Needs Ethics," *ibid.,* April 12, 1970.

Page 134: "The ground swell had begun . . . ": *Arizona Record,* January 15, 1970.

Page 135: "The government did not determine . . . ": Wellford, *op. cit.*

"'The ban is so full . . . '": Tucson *Daily Citizen,* December 30, 1969.

"Mrs. Shoecraft learned . . . ": Lab report from Stoner Laboratories, Campbell, California, February 1970.

Pages 135–137: Illnesses in Globe residents:
Author's interviews with Robert and Charmion McKusick and Robert McCray; depositions of McKusicks and McCrays; medical reports of Dr. Granville Knight.

Pages 137–141: McCarthy hearings in Globe:
"Probe into Use of Herbicides by Congressman Richard D. McCarthy, D-NY," Globe, Arizona, February 13, 1970.

Page 141: "Dr. Granville F. Knight specialized . . . ": Deposition of Dr. Granville F. Knight, *Shoecraft et al.* v. *The Dow Chemical Company.*

Pages 142–147: The Tschirley task force:
The Investigation of Spray Project Near Globe, Arizona, February 1970.
Shoecraft, *op. cit.*

Page 148: "'There's a good possibility . . . '": "Defoliants, Deformities: What Risk?," *Medical World News* (February 27, 1970).

"Neither side felt comfortable . . . ": Author's interviews with Robert McKusick and Robert McCray.

"According to *Time* . . . ": "Globe's Mystery," *Time* (February 23, 1970).

Page 149: " 'Don't you see?' . . .": Shoecraft, *op. cit.*

Page 150: "'Why are so many . . .'": *ibid.*
"In a report on . . . ": Dr. Matthew Meselson, *Environmental Health Perspectives* (September 1973).
"'I say when in doubt . . .'": "Defoliants, Deformities: What Risk?," *loc. cit.*
"Another scientist . . . ": *ibid.*
Pages 151–152: Duplication of Bionetics tests:
Wellford, *op. cit.*
Dr. Samuel Epstein, "A Family Likeness," *Environment* (July/August 1970).
"Defoliants, Deformities: What Risk?," *loc. cit.*
"Dow Accentuates the Positive," *Down to Earth* (Fall 1970).
Pages 152–161: Senate hearings on 2,4,5-T and Mrs. Shoecraft's trip to Washington:
Testimony before the Senate Commerce Committee Subcommittee on Energy, Natural Resources and the Environment hearings on Effects of 2,4,5-T on Man and the Environment, April 1970.
"Widely Used Herbicide Tied to Birth Defects," *The Washington Post,* April 8, 1970.
Shoecraft, *op. cit.*
Page 158: "Looking back on her work . . . ": Deposition of Jacqueline Verrett, *Shoecraft et al. v. The Dow Chemical Company.* July 1974.
Page 161: "Meetings were held around the country . . . ": "Dow Accentuates the Positive," *loc cit.;* Editorial, *Down to Earth* (Winter 1970).
"University of California . . . ": "Agricultural Chemicals and Range Management," *ibid.,* (Spring 1972).
Page 162: "After the Senate hearing . . . ": Shoecraft, *op. cit.*
"She and her husband had bought . . . ": Deposition of Billee Shoecraft, *Shoecraft et al.* v. *The Dow Chemical Company.*
"She filed a claim . . . ": Shoecraft, *op. cit.*

CHAPTER SEVEN

Mrs. Shoecraft's continuing fight against herbicides is based on author's interviews with Willard Shoecraft, Mari Shoecraft, Ernie Gardner, Mary Lou Gardner, Robert McKusick, Charmion McKusick, Robert McCray, Dr. Charles Hine, and attorneys Jerry Sullivan and Jack Slobodin; on the depositions of Billee Shoecraft, Willard Shoecraft, Robert Shoecraft, Ernie Gardner, Mary Lou Gardner, Robert McKusick, Charmion McKusick, Robert McCray, Dr. Granville Knight, Dr. Paul Singer, Dr. William Bishop, Dr. Matthew Ellenhorn, Dr. William S. Wilson, Dr. E. Paul Wedel, Dr. David James, and Dr. Robert Reynolds; on Mrs. Shoecraft's medical records; on Mrs. Shoecraft's book, *Sue the Bastards, loc. cit.*; on newspaper accounts published in the *Arizona Republic,* the Phoenix *Gazette* and *The New York Times;* on laboratory reports from GHT Laboratories in Brawley, California; and on various other documents, including affidavits, interrogatories, answers, and trial briefs filed with the U.S. District Court in Phoenix, Arizona, in the case of *Shoecraft et al.* v. *The Dow Chemical Company.*
Page 167: "'One hundred pounds of love . . .'": Dr. Frank Egler, Introduction, Shoecraft, *op. cit.*
Page 168–169: The 1970 restrictions on 2,4,5-T and Dow's efforts to keep it on the market:

Elements of Risk / 317

"The Status of 2,4-D, 2,4,5-T, Silvex and MCPA Herbicides," *Down to Earth* (Spring 1971).
Editorial, *ibid.,* (Summer 1972).
"2,4,5-T Chronology," *ibid.,* (Spring 1974).
Wellford, *op. cit.*
Whiteside, *loc. cit.*
Epstein, *op. cit.*
"T on Trial," *American Forests* (February 1979).
Davis, *op. cit.*
"The War on 2,4,5-T," *Environmental Action* (November 5, 1977).
Page 170: "'I would say purely . . . '": Testimony of Air Force Major General Garth Dettinger before House Committee on Veterans Affairs, February 21, 1980.
"In the aftermath . . . ": Professor Ton That Tung, "Clinical Effects of Massive and Continuous Utilization of Defoliants on Civilians," *Vietnamese Studies,* Vol. 29 (1971).
Page 172: "An independent advisory . . . ": "Critics Weigh EPA Herbicide Report, Find It Wanting," *Science* (July 23, 1971); Davis, *op. cit.*
"The fracas was . . . ": "Critics Weigh EPA Herbicide Report, Find It Wanting," *loc. cit.*
"Public-interest lawyer . . . ": *ibid.*
"But Professor Davis . . . ": Davis, *op. cit.*
Pages 172–173: Seesaw battle over 2,4,5-T:
"2,4,5-T Chronology," *Down to Earth* (Spring 1974).
Davis, *op. cit.*
"The War on 2,4,5-T," *loc. cit.*
"The Return of T," *Ncap News* (Fall/Winter 1981–82).
Pages 172–175: New facts about dioxin:
Huff, Moore, Saracci, and Tomatis, *op. cit.*
Rawls, *op. cit.*
"The Health Effects of 'Agent Orange' and Polychlorinated Dioxin Contaminants," *loc. cit.*
Alan Poland and Edward Glover, "Studies on the Mechanism of Toxicity of the Chlorinated Dibenzo-p-dioxins," *Environmental Health Perspectives* (September 1973).
Maurice King, Alan Shefner, and Richard Bates, "Carcinogenesis Bioassay of Chlorinated Dibenzodioxins and Related Chemicals," *ibid.,* (September 1973).
B. N. Gupta, J. G. Vos, J. A. Moore, J. G. Zinkl, and B. C. Bullock, "Pathologic Effects of 2,3,7,8-Tetrachlorodibenzo-p-dioxin in Laboratory Animals," *ibid.*
B. A. Schwetz, J. M. Norris, G. L. Sparschu, V. K. Rowe, P. J. Gehring, J. L. Emerson, and C. G. Gerbig, "Toxicology of Chlorinated Dibenzo-p-dioxins," *ibid.*
Coleman Carter, Renate Kimbrough, John Liddle, Richard Cline, Matthew Zack, and William Barthel, "Tetrachlorodibenzodioxin: An Accidental Poisoning Episode in Horse Arenas," *Science* (May 1971).
"Death of Animals Laid to Chemical," *The New York Times,* August 28, 1974.
R. M. Oliver, "Toxic Effects of 2,3,7,8 Tetrachlorodibenzo 1,4 Dioxin in Laboratory Workers," *British Journal of Industrial Medicine* (February 1975).
Page 175: Cancellation of 2,4,5-T hearings:
Whiteside, *loc. cit.*
"The War on 2,4,5-T," *loc. cit.*
"The Return of T," *loc. cit.*

Page 181: Seveso:
"The Dioxins: Toxic and Still Troublesome," *Environment* (January/February 1981).
Whiteside, *loc. cit.*
Pages 184–186: Herbicide debate:
Agent Orange: Vietnam's Deadly Fog, WBBM-TV News, Chicago, March 25, 1978.
Politics of Poison, KRON-TV News, San Francisco, 1979.
"Dioxin Summary," *Comments from CAST* (August 25, 1978).
The Phenoxy Herbicides, Council for Agricultural Science and Technology, Report No. 77, August 1978.
"Scientific Dispute Resolution Conference on 2,4,5-T," sponsored by the American Farm Bureau Federation, June 1979.
Marian Burros, "The CAST Controversy," *The Washington Post,* March 8, 1979.
Robin Marantz Henig, "CAST-Industry Tie Raises Credibility Concerns," *Bioscience* (January 1979).
"Dow Finds Support, Doubt for Dioxin Ideas," *Chemical and Engineering News* (February 12, 1979).
Pages 190–193: EPA RPAR against 2,4,5-T and Dow's response:
"2,4,5-T Pesticide Being Reviewed By EPA," EPA press release, April 12, 1978.
"EPA Takes Emergency Action to Halt Herbicide Spraying," EPA press release, March 1, 1979.
"2,4,5-T Update," *Down to Earth* (Fall 1978).
"Dow Comment on Emergency Suspension of 2,4,5-T and Silvex," March 1, 1979.
"A Scientific Critique of the EPA Alsea II Study and Report," Environmental Health Sciences Center/Oregon State University, October 25, 1979.
"Herbicide Ban Upheld by Judge," *Detroit Free Press,* April 13, 1979.
"Fallout from Agent Orange Dogs a Herbicide," *Business Week* (March 24, 1980).
"Hazards for Export," *Newsday* special reprint, 1979.
Robert Matthews, "The Herbicide Controversy," speech to the Washington State Forestry Conference, November 2, 1979.
"Statement by Dr. Samuel Epstein on Behalf of the Interfaith Center on Corporate Responsibility: 2,4,5-T and the Dow Track Record," May 9, 1980.
"Report from the Dow Chemical Annual Stockholders Meeting May 9, 1980 in Midland, Michigan," Interfaith Center on Corporate Responsibility.
Politics of Poison, loc. cit.
"T on Trial," *loc. cit.*
Wilcox, *op. cit.*

CHAPTER EIGHT

The story of Mike Trout, the manufacture of DBCP at the Occidental Chemical Company, and the subsequent infertility of the chemical workers is based on author's interviews with Marta Trout, Robert Trout, Betty Trout, Haskell Perry, Farnham Soto, Dr. Charles Hine, Dr. M. Donald Whorton, Dr. Theodore Torkelson, Dr. Etcyl Blair, and John Mendes; on the depositions of Dr. Hine, Dr. Whorton, Dr. Torkelson, Dr. Ira H. Monosson, Jack Hodges, and Clifford Enos; on newspaper and magazine accounts published in the Stockton *Record,* the San Francisco *Chronicle* and *Examiner,* the *Los Angeles Times,* the Associated Press, and *New West;* on Dr. Whorton and Eric Skjei, *Of Mice and Molecules: Technology and Human Survival* (New York: Dial, 1983); on records and exhibits filed in the case of *Arnett et al.* v. *The Dow Chemical Company et al.,* including medical reports, interrogatories, answers, trial briefs, motions, Dow, Shell, Occidental, and Hooker internal memorandums and correspondence, USDA correspondence with Dow and Shell, workmen's compensation claims, formulator's manuals, material safety data sheets, product labels, technical advisory bulletins, work rule sheets, employment and safety rules, industrial hygiene surveys, annual production and sales reports, U.S. Department of Labor Guidelines for Control of Occupational Exposure to DBCP; on court testimony in the trial of *Arnett et al.* v. *The Dow Chemical Company et al.,* including Dr. Whorton, Dr. Hine, Wesley Jones, Dr. Stephen Raffle, Dr. Maurice Beaulieu, Dr. Donald Lunde; and on scientific papers, including "The Male Factor in Fertility and Infertility," *Journal of Urology* (1951); "Toxicologic Investigations of 1,2,-Dibromo-3-Chloropropane," *Toxicology and Applied Pharmacology* (1961); "Hygienic Characteristics of the Nematocide Nemagon in Reaction to Water Pollution Control," *Hygiene and Sanitation* (1971); "Induction of Stomach Cancer in Rats and Mice by Halogenated Aliphatic Fumigants," *Journal of the National Cancer Institute* (December 1973); "Semen Analysis: Evidence for Changing Parameters of Male Fertility Potential," *Fertility and Sterility* (1974); "Carcinogenicity of Ethylene Dibromide and 1,2-Dibromo-3-Chloropropane after Oral Administration in Rats/Mice," *Toxicology and Applied Pharmacology* (1975); "Infertility in Male Pesticide Workers," *Lancet* (December 17, 1977); and "Health Hazard Evaluation of the Occidental Chemical Company," Report for NIOSH, October 1977.

Pages 203–205: The EPA and OSHA:

Quarles, *op. cit.*

Berman, *op. cit.*

Robert Boyle and the Environmental Defense Fund, *Malignant Neglect* (New York: Vintage Books, 1980).

Kazis and Grossman, *op. cit.*

Green and Waitzman, *op. cit.*

CHAPTER NINE

The story of the aftermath of the DBCP sterility scandal is based on author's interviews with Marta Trout, Dr. Etcyl Blair, and Peter Weiner; on newspaper and magazine accounts published in the *Arkansas Gazette,* San Francisco *Chronicle* and *Examiner, Los Angeles Times, The Wall Street Journal, Detroit Free Press, The New York Times,* Associated Press, and *New West;* on records and exhibits filed in the case of *Arnett et al.* v. *The Dow Chemical Company et al.,* including medical reports, interrogatories, answers, and Dow, Shell, Occidental, and Hooker internal memorandums and correspondence; on court testimony in the trial of *Arnett et al.* v. *The Dow Chemical Company, et al.,* including Wesley Jones, Dr. Stephen Raffle, Dr. Maurice Beaulieu, Dr. Donald Lunde; on testimony in the October 1977 California Department of Industrial Relations hearings on DBCP; on "Dibromochloropropane (DBCP): Suspension Order and Notice of Intent to Cancel," *Federal Register,* November 9, 1979; on the NIOSH DBCP Registry, and on scientific papers, including "Effect of Pesticides on Testicular Function," *Journal of Urology* (March 1978); "Testicular Function

Among Shell Denver Plant Employees," a Report to the Shell Oil Company by Dr. M. Donald Whorton and Dr. Thomas Milby, November 1978; "Testicular Function in DBCP Exposed Pesticide Workers," *Journal of Occupational Medicine* (1979); "Chronology of Studies Regarding Toxicity of 1,2-Dibromo-3-Chloropropane," *Annals of the New York Academy of Sciences* (1979); "Epidemiological Assessment of Occupationally Related, Chemically Induced Sperm Count Suppression," *Journal of Occupational Medicine* (1980); "Dibromochloropropane and Its Effect on Testicular Function in Man," *Journal of Urology* (1980); and "A Method for Monitoring the Fertility of Workers," *Journal of Occupational Medicine* (March 1981).

Page 235: "'There is a very strong . . .'": "Notes on Dibromo-Chloropropane for September Board of Directors Meeting," September 26, 1977.

CHAPTER TEN

The continuing story of Mike Trout and DBCP is based on author's interviews with Marta Trout, Robert Trout, Betty Trout, Haskell Perry, Farnham Soto, Dr. Charles Hine, Dr. M. Donald Whorton, Mel Rice, and Harold Eisenberg; on "The Nematode Chronicles," *New West* (May 1981) and "Trading Off at Hooker Chemical," *APF Reporter* (April 1981); on testimony and records in the EPA suspension hearing on DBCP, October 1979; on "Dibromochloropropane (DBCP): Suspension Order and Notice of Intent to Cancel," *loc. cit.*; on records and exhibits filed in the case of *Arnett et al.* v. *The Dow Chemical Company et al.*, including medical reports, interrogatories, answers, trial briefs, motions, Dow, Shell, Occidental, and Hooker internal memorandums and correspondence; on workmen's compensation claims; on records filed in the case of *The United States and People of the State of California* v. *Occidental Petroleum Corporation et al.*; on Occidental Chemical interoffice communications obtained by the House Subcommittee on Oversight and Investigations of the Committee on Interstate and Foreign Commerce; and on scientific papers, including "Mutagenicity, Carcinogenicity and Reproductive Effects of Dibromochloropropane," Environmental Health Associates, Inc., October 1982.

Pages 249–256: Business backlash against regulation and passage of TOSCA: Author's interviews with Robert Lundeen and Dr. Etcyl Blair.

"Earth, Energy and the Environment," *Congressional Quarterly Report* (1977).

William Tucker, "Of Mites and Men," *Harper's* (August 1978).

"Dow Expects 10% Rise in Profits," Detroit *News*, March, 10, 1977.

"Dow Cuts CMU Aid After Talk by Actress," United Press International, October 29, 1977.

"CMU Will Tell Dow How Funds Are Used," Associated Press, January 8, 1978.

Alex Taylor, "Dow Says It's Choking on Red Tape," *Detroit Free Press*, April 23, 1978.

"Dow Unseated as Profit Leader in Chemicals," *ibid.*, May 21, 1978.

Walter B. Smith, "Dow Chief Takes Swing at Government," Detroit *News*, September 19, 1978.

Allan Sloan, "U.S. Driving Us Bananas—Dow," *Detroit Free Press*, September 19, 1978.

"Dow President Isn't Shy in Defense of Big Business," *ibid.*, October 29, 1978.

Elements of Risk / 321

"Dow Chemical Needs Another Saran Wrap," *ibid.*, December 11, 1978.

Murray Weidenbaum, "The Costs of Government Regulation of Business," prepared for the Subcommittee on Economic Growth and Stabilization, Joint Economic Committee, 95th Congress, April 10, 1978.

Susan R. Pollack, "Dow Veils Threat on College Gifts," Detroit *News*, January 7, 1979.

"Dow Calls U.S. Rules 'Hidden Tax' on Products," *ibid.*, February 6, 1979.

"Dow and Apartheid," *The Point Is (A Summary of Public Issues Important to the Dow Chemical Company)*, March 18, 1983.

Lili Francklyn, "Debunking Madison Avenue," *Environmental Action* (May 1981).

E. H. Blair and F. D. Hoerger, "Toxic Substances Legislation and an Incompatibility with Innovation and Science," *Environmental Policy and Law* (June 1976).

"Lobby Against Cancer Policy," *Congressional Quarterly* (February 23, 1980).

Elizabeth Whelan, "The Politics of Cancer," *Policy Review* (Fall 1979).

"What Cancer Epidemic?" *The Point Is (A Summary of Issues Important to the Dow Chemical Company)*, October 26, 1979.

"Cancer—Debunking the Myth Makers," *ibid.*, September 17, 1982.

Philip M. Boffey, "Cancer Experts Lean Toward Steady Vigilance, but Less Alarm on Environment," *The New York Times*, March 2, 1982.

Glenn Frankel, "The TOSCA Tragedy," *The Washington Monthly* (July/August 1979).

Testimony before the Senate Committee on Commerce, Science and Transportation, April 1977.

Quarles, *op. cit.*

"EPA Is Ignoring Pesticide Risks, Report Charges," Associated Press, January 3, 1977.

"Nation's Pesticide Law Snarled," *The Washington Post*, October 2, 1977.

Green and Waitzman, *op. cit.*

Dr. Samuel Epstein, *The Politics of Cancer* (San Francisco: Sierra Club Books, 1978).

Mary Douglas and Aaron Wildavsky, *Risk and Culture* (Berkeley: University of California Press, 1982).

Pages 273–276: The EPA under President Reagan:

"Anne and Bob Go to Washington," *Denver Magazine* (August 1981).

"Move Over, Jim Watt, Anne Gorsuch Is The Latest Target of Environmentalists," *National Journal* (October 24, 1981).

"The Rhinestone Cowboys," *The Living Wilderness* (Fall 1981).

"EPA in Disarray," *The Washington Monthly* (December 1981).

Eliot Marshall, "Revisions in Cancer Policy," *Science* (April 1, 1983).

Michael Wines, "Scandals at EPA May Have Done In Reagan's Move to Ease Cancer Controls," *National Journal* (June 18, 1983).

"Critics Take Fresh Aim at the Toxics Control Act," *Chemical Week* (June 29, 1983).

Kenneth R. Noble, "Fulfilling a Promise on Deregulation," *The New York Times*, August 29, 1983.

"The Information Content of Premanufacture Notices," background paper of the Office of Technology Assessment, April 1983.

Lou Cannon, *Reagan* (New York: Putnam, 1982).

CHAPTER ELEVEN

The story of the DBCP trial in San Francisco is based on author's interviews with Marta Trout, Haskell Perry, Dr. Charles Hine, Dr. Theodore Torkelson, Duane Miller, Douglas Brown, Frederick Duda, H. Christian L'Orange, Burt Ballanfant, Isaac Williams, and Don Frayer; on the trial testimony of Dr. Charles Hine, Art Schober, Dr. Kenneth Lyons Jones, Dr. Frederick Hecht, Dr. John Lanham, Dr. Larry Lipschultz, Dr. M. Donald Whorton, Dr. George Bach-Y-Rita, Wesley Jones, and Gloria Perez; on newspaper and magazine accounts published in the Fresno *Bee,* Stockton *Record,* San Francisco *Chronicle* and *Examiner;* and on briefs, records, and exhibits filed in the case of *Arnett et al.* v. *The Dow Chemical Company et al.*

Page 284: "(Dow president Paul Oreffice wrote...": "Records Retention," September 15, 1977.

Pages 286–288: The Dow report and the disintegration of the Reagan EPA:

Wayne Biddle, "10 Years Later, Missourians Find Soil Tainted by Dioxin," *The New York Times,* November 10, 1982.

Eleanor Randolph, "Dioxin Scare Leaves Town Bewildered," *Los Angeles Times,* December 31, 1982.

Philip Shabecoff, "Forecast for EPA Was Stormy from the Start," *The New York Times,* February 20, 1983.

Philip Shabecoff, "Dismissed Official Faults EPA Chief," *ibid.,* February 24, 1983.

Robert Reinhold, "Uproar at EPA Blends Politics, Pollution and Public Health," *ibid.,* February 27, 1983.

"Cleaning Up the Mess," *Newsweek* (March 7, 1983).

Stuart Taylor, Jr., "EPA Inquiries Center on Four Issues," and Philip Shabecoff, "Seven Days of Decision: Why Head of EPA Quit," *The New York Times,* March 13, 1983.

Leslie Maitland, "EPA Aides Recall Pressure to Alter Dow Dioxin Study," *ibid.,* March 19, 1983.

Paul Magnusson, "Scientists Say EPA Officials Used Threats," *Detroit Free Press,* March 19, 1983.

Philip Shabecoff, "Ruckelshaus Gives Pledge to Enforce Environmental Laws," *The New York Times,* May 5, 1983.

CHAPTER TWELVE

The DBCP epilogue is based on author's interviews with Dr. Theodore Torkelson, Dr. Charles Hine, Hank Hine, and Clyde McBeth; on articles published in the University of California at San Francisco newspaper, *Synapse;* and on *Water—A Clear and Present Danger, ABC-News Closeup,* August 15, 1983.

The dioxin epilogue is based on author's interviews with Robert McCray, Robert McKusick, Charmion McKusick, Willard Shoecraft, Mari Shoecraft, Garry Hamlin, Don Frayer, Herbert Dow Doan, Dr. Etcyl Blair, Robert Lundeen, and Dr. Bruce Ames; on "Dioxin," a *Chemical and Engineering News* special issue (June 1983); on "Agent Orange—Dioxin," a technical report published by the American Medical Association, October 1981; on "The Return of T," *loc. cit.;* on "Jury Awards $58 Million to 47 Railroad Workers Exposed to Dioxin," *The New York Times,* August

27, 1982; "Dioxin's Peril to Humans: Proof Is Elusive," *ibid.*, January 23, 1983; on a letter from Robin Briehl, *Chemical and Engineering News* (February 14, 1983); on "Concern Growing over Unclear Threat of Dioxin," *The New York Times*, February 15, 1983; on Dr. Samuel Epstein's testimony on Agent Orange before the House Veterans Affairs Subcommittee in April 1983; on "Dioxin, How Great a Threat?," *Newsweek* (July 11, 1983); on "Do Phenoxy Herbicides Cause Cancer in Man?," *Lancet* (May 8, 1982); on "Herbicides, Occupation and Cancer," *ibid.* (June 26, 1982); on "The Fate of 2,4,5-T Following Oral Administration to Man," *Toxicology and Applied Pharmacology* (1973); on "Fate of Silvex Following Oral Administration to Humans," *Journal of Toxicology and Environmental Health* (1977); on "The Mortality Experience of Workers Exposed to Tetrachlorodibenzodioxin in a Trichlorophenol Process Accident," *Journal of Occupational Medicine* (January 1980); on "A Mortality Analysis of Employees Engaged in the Manufacture of 2,4,5-T," *ibid.*; on "Mortality Experience of Employees Exposed to 2,3,7,8-tetrachlorodibenzo-p-dioxin," *ibid.* (August 1980); on the September 15, 1983, decision of Nova Scotia Supreme Court Justice D. Merlin Nunn; on remarks by Paul Oreffice, president and chief executive officer, Dow Chemical Company, at June 1, 1983, press conference; on "Nothing Ventured, Nothing Gained," *The Point Is (A Summary of Public Issues Important to the Dow Chemical Company)*, October 31, 1980; on "Dietary Carcinogens and Anticarcinogens," *Science* (September 23, 1983); and on "Dow Halts Fight to Sell Herbicide," *The New York Times*, October 15, 1983.

index

Abrams, Creighton, 170
Acetylcholinesterase, 200
Adams, E. M., 67
Agee, William, 249
Agent Orange, 77–79, 104, 116, 119, 127
   chloracne and, 87, 93
   composition of, 76, 114
   spraying of, 77, 83
      last, 170
   Vietnam veterans and, 184, 278
      class action suit on behalf of, 190, 299, 300
   warnings about, 90–91, 116
Agent Orange: Vietnam's Deadly Fog, 184
Agricultural chemicals, 14–49
   American farm and, 52
   "dermatitis" from, 15–16, 54–62
   exaggerated claims for, 66
   herbicides, *see* Herbicides
   insecticides, 21–22
   nematocides, 27–49
      see also DBCP
   rise in use of, 52
   wood preservatives, 15–16
   see also specific chemicals
Ahnfeldt, A. L., 21
Albert, Dr. Roy E., 262, 264
Allan, James, 185, 188
Allied Chemical Company, 6, 11
American Association for the Advancement of Science (AAAS), 91, 119
American Conference of Government and Industrial Hygienists, 48
American Cyanamid Company, 6, 11, 68, 185
American Enterprise Institute, 250, 255
American Industrial Health Council, 252

American Journal of Digestive Diseases, 53–54
American Medical Association, 53, 64
Ames, Dr. Bruce, 241–42, 265, 302–03
AMVAC Chemical, 248–49, 261, 264
Anderson, Dr. Earl, 31
Aniline, 88
Archives of Dermatology and Syphilogy, 16, 78
Arizona Daily Star, 106, 111
Arizona Fish and Game Department, 101, 102
Arizona Record, 101, 104
Arizona Republic, 194–95
Arkansas Gazette, 232
Armco Inc., 203, 204
Arnett, Dorothy, 256, 289
Arnett, Frank, 228, 256–57, 285
   damage award received by, 289
Arnett v. Dow, 281–86
   verdict, 288–90
ASARCO, 292, 293
Asbestos, 185, 295
Ash, Roy, 124
Ash Council, 124–25
Associated Press, 21, 110–11, 185, 222–23, 243
Astrocytomas, 209–10
Atropine, 200, 201, 212
Auchter, Thorne, 274
Azoospermia, DBCP-induced, 212–48
   in animals, 202, 205
   Cal-OSHA and, 222, 225, 226, 227
   chemical industry and, 234–35
   CDIR hearings and, 237–44
   EPA hearings and, 263–66
   FSH and, 219, 225
   impotence and, 228
   jokes about, 232

Azoospermia (cont'd)
  kepone and, 216, 217
  lawsuit over, see DBCP, lawsuit over
  as male contraceptive, 230–31
  Occidental's plans to deal with, 226–27
  older or sterile workers and, 231
  recovery from, 226, 246–47, 256, 291
  scandal over, 222–25, 230
  sperm count and, 219, 221, 225, 228, 235, 243–47, 264
  testing for, 218–22, 225, 227, 228, 233
  union meeting about, 227–28
  Whorton and, 219–21, 225–28
  Zavon and, 220, 223–26

Badischer Anilin & Soda-Fabrik (BASF), 56, 57, 62, 92–93
Baeder, Donald, 231, 270–71
Barnes, Earle B., 191
Barrons, Keith, 105
Baughman, Robert, 16
Bayley, Dr. Ned, 154–56, 163
Beech-Nut Packing Company, 64
Benson, Ezra Taft, 52, 72
Berry, Lowell, 196–97
Best Fertilizers, 39
  early years of, 196–97
  pollution from, 197
  purchase of, by Occidental, 48, 197
Beutel, A. P., 126
Biological Abstracts, 42
Biological warfare, 18–19, 21
  in Vietnam, 74–79, 87, 90–91, 93–94
  see also Agent Orange
Bionetics Research Laboratories, 90
  pesticide study of, 116–20, 150, 151
    Congressional hearings and, 153, 155, 157
Bioscience, 185
Birth defects, 241–44, 247, 256–57
  EPA hearings and, 265–66
  fear of, 288, 289
Bishop, Dr. William, 148, 180
Biskind, Dr. Morton, 53–54
Bittner, E. Ross, 134
Blair, Dr. Etcyl, 43–44, 169, 191, 252, 254, 303
  DBCP and, 229, 232, 233
    CDIR hearings on, 239–40
    in retrospect, 291
  Silent Spring and, 71–72
Blum, Arlene, 241, 243, 265
Blum, Barbara, 191

Boehringer Sohn, C. H.:
  manufacturing control of, 2, 4, 5-T and, 59–61, 83–85
  outbreak of chloracne at, 56–57
Botanical Gazette, The, 23
Braden, Tom, 119
Branch, C. B. "Ben," 133, 134
Braun, Dr. Daniel, 60
Bricker, Ted, 214, 218, 222
Briehl, Robin, 302
Brine, 4–5
Brinkley, David, 129, 249
Brinkley, Parke, 126
Bromine, 4–5
Brown, Douglas, 278, 290
Buck, Bill B., 101
Buckley, M. S., 87
Bumb, Dr. Robert, 186
Burbank, Luther, 9
Burford, Anne Gorsuch, 273–76, 286–87
Business Advisory Council, 127
Business and Defense Services Administration, 126
Business Roundtable, 204, 251, 255
Business Week, 272
Butler, Dr. Milton G., 16
Byerly, Dr. T. C., 154, 156, 157

Caesar, Sid, 50
California Department of Food and Agriculture, 229
California Department of Health, 229
California Department of Industrial Relations
  hearings on DBCP, 237–44
California Occupational Safety and Health Agency (Cal-OSHA), 222, 225, 226, 227, 272–73
Cancer, 62, 71, 249
  cranberry scare of 1959 and, 72
  DBCP and, 205–06, 213–14, 227–29, 232–44, 272–73, 278
    EPA hearings on, 262–63, 266
    risk assessment model and, 263
  dioxin and, 295–96
  early linking of chemicals to, 52–53
  EPA and, 274, 276
  in retrospect, 295–303
  lawsuit over, see DBCP, lawsuit over
  mutations and, 241–43
  OSHA's ranking of chemicals causing, 252–53, 276
  2,4,5-T and, 190

Cancerphobia, 285–86
  damages awarded due to, 288, 289
Carbon disulfide, 29
Carson, Rachel, 63, 121
  death of, 71
  Silent Spring, 66–73, 76
Carter, Jimmy, 253, 255
Cartwright, Sucherman, Slobodin & Fowler, 188, 193
Castle & Cooke, 262
CBS Reports, 69
Center for Disease Control, 174
Center for Study of Responsive Law, 153
Central Michigan University, 250
Central Valley, California, 293
Chandler, E. L., 85
Chemical Abstracts, 42, 59
Chemical and Engineering News, 186, 294
Chemical industry:
  agricultural chemicals, *see* Agricultural chemicals
  American farm and, 52
  Ames and, 242
  cold war years and, 33
  DBCP and, 234–35
    CDIR hearing on, 237–44
    defense of, 247–48
    EPA hearing on, 261–66
    lawsuit over, *see* DBCP, lawsuit over
  dioxin and, 185–86, 293–304
  EPA and, 203–04, 261–66
  establishment of, 6
  labels on products, 13–14, 54
  in the 1930's, 8
  organic chemistry and, 6–8
  post-World War II growth in, 20–21
  public relations and, 50–52, 72, 73
  regulation of, 51–52, 65, 69–73, 126–30
    backlash against, 248–56, 273–76
  respectability of, 50–51
  safety and, 51, 54
  science and, 51–52
  Silent Spring and, 66–73
  toxicological studies by, 13–14, 47
    see also Toxicological studies
  workers in, *see* Workers in the chemical industry
  World War II and, 18–21
Chemical Manufacturers Association, 250, 252
Chemical Progress Week, 50
Chemical warfare, 20, 21, 22, 29, 35, 199
Chemical Week, 67, 233

Chirurgical Observations of Cancer of the Scrotum of Chimney Sweeps (Pott), 53
Chloracne, *see* Workers in the chemical industry, chloracne among
"Chloracne-Dow Experience," 87–88
Chlorine, 4–5, 15, 20
  dioxin and, 57–58
Chlorophenols, 15–19
  as herbicides, 16–19
  reaction to, 16, 54–55
  as wood preservatives, 15–16
Chloropicrin, 20, 29
Chromosomes, 265–66
Cleveland Plain Dealer, 54
Coalite and Chemical Products Ltd., 91–93
Coal tar chemicals, 7
Coca, Imogene, 50
Cook, Rex, 214–15, 218, 219–20, 227
Costle, Douglas, 244, 253
  banning of DBCP, 266
Council for Agricultural Science and Technology (CAST), 185
Council of Environmental Quality (CEQ), 128
Connolly, Cyril, 73
Conrad, Laurence A., 4
Courtney, Robert, 107, 110–11, 114
Cranberry scare of 1959, 72
Crown Zellerbach, 192
Curtice, Harlow, 33
Cuyahoga River, 124, 203

Darby, Dr. William, 73
Darwin, Charles, 17
Davidson, John, 25–26
Davis, Donald, 172
Davis, Fanny Fern, 23
Day, Boysie E., 161
DBCP (1,2-dibromo-3-chloropropane), 31–49, 198–248, 256–304
  application of, 248
  azoospermia and, *see* Azoospermia, DBCP-induced
  ban on, 234, 266, 291
  birth defects and, 241–44, 247, 256–57
    EPA hearings on, 265–66
  black market in, 248
  cancer and, 205–06, 213–14, 227–29, 232–44, 272–73, 278
    EPA hearings on, 262–63, 266
    risk assessment model and, 263
  CDIR hearings on, 237–44

Elements of Risk / 327

DBCP (cont'd)
  chemical industry and, 234–35
  defense of, 247–48
  EPA and, 206, 213–14, 234, 244, 247–48, 258–59
    hearings on, 261–66
  in retrospect, 291–93
  lawsuit over, 244–46, 271, 277–90
    Arnett v. Dow, 281–86, 288–90
    birth defects and, 277, 286
    Brown and, 278–79
    cancer and, 277–81, 285–86
    damages awarded, 288–89
    defense team, 278
    factors influencing, 277–78
    Hine and, 244, 277, 282, 283–84
    infertility and, 278, 279, 282–85
    interrogatories for, 245–46
    settlement of, 277–82
    Shell and, 268–69
    Trout and, 245–46, 267, 273, 277, 279–81
    verdict, 288–90
  as male contraceptive, 230–31
  peach industry and, 202, 231, 261–62
  pineapple industry and, 262, 266
  produced by Dow and Shell, 31–49
    cancer and, 213–14, 232–44
    CDIR hearings on, 237–44
    concern over, 37–38, 40–41, 198
    demonstrations of, 38–39
    effectiveness of, 31, 48
    EPA and, 213–14
    exposure guidelines for, 48, 216–17
    infertility and, 220–21, 228–29, 232–37, 264
    information sheets on, 201–02, 232–33
    in retrospect, 291–93
    labeling of, 36, 41, 43, 45–48
    manufacture of, 41, 44, 47, 48, 198
    mechanism of action, 31–32
    patents governing, 39, 228
    price war over, 48
    safety precautions when handling, 42, 205, 232
    St. Clair and, 235–36
    shutdown of production, 229, 233–34
    summary report on, 44–45
    toxicological testing of, 32–49
    USDA and, 45–48, 198
    "warning odor," 42, 233
  produced by Occidental, 198–248, 256
    cancer and, 205–06, 213–14, 227–29, 236–37
    company management and, 215–16, 234–35, 236–37
    data sheets on, 201–02, 213
    dumping of wastes, 259–61, 270
    as a formulator, 198–99
    groundwater pollution by, 196, 259–61, 293
    health hazards of, 200–01, 205–06
    illnesses of workers, 208–13
    infertility and, see Azoospermia, DBCP-induced
    measuring of, 216–17
    organophosphates and, 199–201
    safety rules for, 199–201, 206
    shutdown of production, 222, 227, 231
    working conditions, 199–202, 214
D-D (dichloropropane-dichloropropene), 29–30, 31
DDT (dichlorodiphenyltrichloroethane), 36, 135, 199–200
  contamination of coho salmon by, 117, 124
  controversy surrounding, 64
  described, 21–22
  effects of long-term, low-level exposure to, 63–66, 69
  phasing out of, 119, 121, 124
  sales of, 52, 63, 64
  Silent Spring and, 66–72
  spraying with, 65–66
Death to Weeds, 24
Defense Production Act, 87
Defoliation, see Vietnam War, defoliation during
Del Monte, 262
Delmore, Fred J., 75
Detrick, Fort, 19, 74, 75
Detroit Economic Club, 249
Detroit Free Press, 8, 21, 22, 54
Dettinger, Garth, 170
Diamond Alkali, 60, 120
  dioxin and, 75–78, 83, 85, 86, 287
Dibenzofuran, 57
Dioxin, 57–62
  aerial spraying in Arizona and, 143, 144, 147
    Congressional hearings on, 153–59
    see also Globe, Arizona, Forest Service spraying of
  chemical structure of, 58, 80–81
  chloracne and, 76–94, 173
  debate over, 294–95
  effects of exposure to, 60–62, 87, 90, 150–52, 173–75, 185, 188, 189, 192–93
  EPA and, 190, 294, 300

328 / Cathy Trost

Dioxin (*cont'd*)
  "fingerprint" of, 295–96
  in retrospect, 293–304
  lawsuits over, 299
  manufacturing control of, 59–61
  mystery of, 294
  potency of, 58, 76, 81, 173, 189, 193
  proindustry view of, 185–86
  ranking of contamination by, 88
  research on, 58–59
  soft-tissue sarcoma and, 295–96
  Times Beach and, 286–87
  trichlorophenol and, 58–62, 80–94, 174, 181
  2,4,5-T and, 120, 130, 150–63, 172–75, 185, 189
    suspension of use of, 158–63, 169–75
Djerassi, Carl, 230–31
DNA, 241, 247, 276
  EPA hearings and, 265–66
Doan, Herbert "Ted" Dow, 71, 301–02
  background of, 130–31
  environment and, 130
  as president of Dow, 131, 132–34, 274
Doan, Leland I. (Lee), 33–34, 130
Dow, Alden, 9, 10
Dow, Grace Anna Ball, 4, 10
Dow, Herbert Henry, 3–11
  death of, 8
  described, 3–4
  experimentation by, 5–9
  lawsuits against, 5
  marriage of, 4
  public-spirited contributions of, 9–11
  success in life of, 5
Dow, Willard, 8
  death of, 15, 33–34
  described, 14
  politics and, 12
  research and, 11
  toxicology lab and, 12–13
Dow Chemical Company:
  agricultural chemicals and, *see* Agricultural chemicals
  bunker mentality of, 63
  business ethics and, 43–44, 133, 134
  DBCP and, *see* DBCP, produced by Dow and Shell
  dioxin and, 75–94, 150–52, 157, 159–62, 185–86, 189–95, 293–304
  dumping of wastes, 5, 130, 134
  environmental issues and, 130, 133–34, 251
  financial situation of, 255
  founding of, 4
  government regulation and, 12, 51–52, 126–30, 249–56, 261
  growth of, 6, 51
  Industrial Health Board, 240
  industry self-regulation and, 83–87
  in-house toxicology lab, 12–14, 26, 32, 33, 36–49, 130, 178
  in retrospect, 300–04
  lobbyists for, 126–27, 131, 133, 274–75
  Midland, Michigan, and, 3–5, 7, 9–11, 63, 186
  napalm and, 117, 127, 131–33, 301
  nematocides and, 27, 30–49
  OSHA and, 252–53
  research, 5–9, 11–12, 25, 51, 70–71
  safety of products, 13–14, 43–44, 51, 70–71, 79–89
  scope of, 54
  silvex spraying in Arizona, *see* Globe, Arizona, Forest Service spraying of
  Shoecraft lawsuit against, 121, 162, 167–95
    settlement of, 193–95
  stockholders of, 191–93
  Times Beach and, 286–87
  TOSCA and, 253–54
  2,4,5-T and, 59–60, 79–93, 103–05, 117–20, 130, 150–52, 157–62, 169–75, 185, 189–95
  Vietnam War and, 75–79, 87, 93, 127
    class action suit by veterans, 190
  World War I and, 7
  World War II and, 20, 22
Dowfume W-10, 30
Dowicides, 15, 84, 87
Down to Earth, 22, 169
Dow's Reaction to DBCP, 236
Drayton, William, 255
Drum, Fort, 74
DuBridge, Dr. Lee, 117–18, 135, 153, 162
Dumping of chemical wastes, 5, 130, 134, 259–61, 270, 275
Dunn, C. L., 85
Du Pont, 5, 11, 13, 19

"Earth Day," 129, 148
Eastman Kodak, 13
EDB (ethylene dibromide), 30–31
  cancer and, 205
Edge of the Sea, The (Carson), 63
Edgewood Arsenal, 75, 76
Edson, Robert, 259–61
Egler, Frank, 121, 167

Endrin, 135
Enos, Clifford, 215
Environment, 63, 123
 backlash against issues of, 249–56
 jobs versus, 204, 249
 responsive and responsible business community and, 132
 Silent Spring and, 66–73
 Times Beach and, 287
 war against abuses of, 123–30, 133–34
Environmental Action, 129, 148, 254
Environmental Protection Agency (EPA):
 creation of, 124–25
 DBCP and, 206, 213–14, 234, 244, 247–48, 258–59
  hearings on, 261–66
 dioxin and, 190, 294, 300
 early actions of, 203–04
 forces acting upon, 125–26, 205
 Gorsuch and, 273–76, 286–87
 Oreffice and, 251
 Rebuttable Presumption Against Registration (RPAR) and, 190, 213, 247–48
 regulation of pesticides and, 126, 127–30
 Ruckelshaus and, 172–73, 203, 287–88
 Times Beach and, 286–87
 TOSCA, 253, 276
 2,4,5-T and, 172–73, 175, 190–91, 195, 248, 304
 weakening of, 205
Epstein, Dr. Samuel, 117, 158, 192–93, 252, 253, 294
Excellence (Gardner), 133

Factlines, 272
Fannin, Paul, 104
Federal bureaucracy, 126
Federal Insecticide, Fungicide and Rodenticide Act (FIFRA), 35, 127, 205, 253–54
Federal Register, 47
Feichtmeir, Edmund, 40
Finch, Robert H., 117, 158
Finkbine, Sherri, 73
Fitzhugh, Edwin, 137–38
Fitzpatrick, Tom, 194–95
Fleischman, William, 101, 115
Follicle-stimulating hormone (FSH), 219, 225
Fonda, Jane, 249–50
Food and Drug Administration (FDA), 43, 45, 65, 83
 DBCP and, 234

 regulation of pesticides and, 69, 128, 135, 153
Food and Cosmetic Act, 43
 Delaney Amendment to, 72
 Miller Amendment to, 65
Ford, Gerald R., 253
Ford, Henry, II, 123
Frawley, Dr. John, 70
 dioxin and, 86–87
Frayer, Donald, 290, 298–300
Freeman, Barry, 107, 121
Fumazone, 39, 202
 competition with Nemagon, 48
 manufacture of, 48
 medical file of, 48–49
 registration of, 47–48, 206

Galston, Dr. Arthur W., 91, 138, 140
Galvin, James, 226
Gardner, Ernie, 97
Gardner, John, 133
Gardner, Mary Lou, 165–66, 188, 194
 Billee's cancer and, 179–84, 187
Gehring, Dr. Perry, 240, 242
General Accounting Office, 127, 128
Genetics, 241–43, 265–66, 276, 286
Genghis Khan, 27
GHT Laboratories, 104, 180
Gibbon, Edward, 33
Givaudan, 59
Globe, Arizona, Forest Service spraying of, 94–122, 134–95
 "A Record of Abnormal Occurrences . . . ," 98–99
 Bionetics report and, 116–20
 described, 95–97
 herbicides used for, 103–04
 illness as a result of, 97, 109–10, 113, 115, 116, 135, 136–37, 140, 141, 144–46
 in retrospect, 297–300
 lawsuit over, *see* Shoecraft, Billee, lawsuit of
 McCarthy hearings and, 137–42
 mock funeral and, 110–11
 plant and animal life and, 97–103, 135–36, 139–52, 185, 189
 reasons for, 96
 Shoecraft and, *see* Shoecraft, Billee
 USDA task force and, 142–48
Goldwater, Barry, 104
Gordon Toxicology Conference, 40
Goring, Cleve, 191
Gorsuch, Anne McGill, 273–76, 286–87

330 / Cathy Trost

Governor of Arizona's task force, 101–03
Graham, Frank, Jr., 66
Great White Hope, The, 120
Green, Hetty, 4
Groth, Dr. David, 81
Growth-regulating compounds, 17–19
Guthrie, David, 258, 270

Hamlin, Garry, 59, 79, 86
Hammer, Armand, 197, 272, 280
Hamner, Charles, 18, 22
Hansen, Jim, 105–06
Hardin, Clifford M., 156, 158, 162
Hargraves, Dr. Malcolm, 53
Harper's magazine, 249
Harrell, Luther, 285, 289
Hart, Philip, 153, 155, 159, 161, 162
Harwood, Gerald, 262–66
Haskell Laboratory of Industrial Toxicology, 53
Hawaii, 262, 266
Hayward, Albert, 75
Heacock, James, 229
Heilbroner, Robert L., 132
Heinrichs, Jay, 192
Herbicides, 16–19
  agricultural use of, 22–26
  growth in use of, 52, 74
  military use of, 18–19, 74–79, 87, 90–91, 93–94
  restrictions on use of, 135, 138, 153
  skin disorders caused by, *see* Workers in the chemical industry, chloracne among
  UN resolution banning wartime use of, 117
  U.S. Forest Service use of, *see* Globe, Arizona, Forest Service spraying of
Hercules Inc., 70
  dioxin and, 83, 85, 86
Hexachlorophene, 174
Heylin, Michael, 294
Hickel, Walter J., 125, 158
Higginson, John, 252–53
Hine, Dr. Charles:
  DBCP and, 35–47
    aftermath of, 292–93
    CDIR hearings on, 237–40
    Jones and, 217–18
    lawsuit and, 244, 277, 282, 283–84
  DDT and, 70
  Shoecraft and, 177–79, 217
Hitchner, Lea S., 65
Hodges, Jack, 215, 222

Hoffmann, Dr. Friedrich, 76
Holder, Dr. Ben, 79–80, 84
Hooker, Elon Huntington, 6, 197
Hooker Chemical Company, 6
  chemical warfare products, 20
  DBCP-infertility connection, 226, 247–48, 258
  Love Canal and, 259, 271–72
  purchase of, by Occidental, 197
  60 Minutes and, 270–71
  working conditions at, 5
Horner, Jack, 48
Hornig, Donald, 90
Houghton Mifflin, 66, 67–68
House Select Committee to Investigate the Use of Chemicals in Food Products, 65
Hueper, Dr. Wilhelm, 52–53
Huxley, Thomas, 33
Hyman Company, Julius, 38
Hypospermatogenesis, 245

"If it didn't kill you immediately, then it was okay," 13–14, 52, 140
  DDT and, 65
  as principle of toxicology, 53–54
Industrial Medicine and Surgery, 55
Infertility, *see* Azoospermia, DBCP-induced
Informed Citizens Union (ICU), 108
Insecticides, 21–22
  DDT, *see* DDT
Insull, Samuel, 126
International Association of Cancer Victims, 176
Interstate Commerce Commission, 34–35
Irish, Dr. Don, 12–14

Jackson, Henry "Scoop," 104, 108
Jefferson, Thomas, 19
Johns-Manville, 274
Johnson, Dr. Julius, 70–71, 73, 117
  2,4,5-T suspension and, 151–52, 159–61
Johnson, Lyndon B., 117
Jones, Deborah, 211–13, 243–44
Jones, Franklin D., 24
Jones, Kathleen, 211
Jones, Lesley, 212
Jones, Steve, 282
Jones, Wesley, 211–16
  damage award received by, 288–89
  examination of, 217–18
  illness of, 212–13, 217

Elements of Risk / 331

Jones (cont'd)
  infertility of, 212–16
  suicide attempt of, 212–13
  wife's pregnancy and, 243–44
Jones & Laughlin Steel Company, 203
Journal of the National Cancer Institute, 205

Kansel weed killer, Scott's, 154
Kapp, Dr. Robert, 265–66
Kate Smith Hour, 50
Kepone, 216, 217
Kerr-McGee Chemical Corporation, 279
Kessler, Dr. Alexander, 230
KIKO, 104–05, 165, 298
Kimmig, Dr. Josef, 57
Klauder, Joseph V., 60
Kligman, Dr. Albert, 88–89
Knight, Dr. Granville, 141–42, 175, 176, 179, 180, 184, 187
Kraus, Dr. E. J., 18–19, 22, 23, 25
Kuron, 25–26
Kusnetz, Howard, 240

Laetrile, 180, 181, 183
La Follette, Johnson, Schroeter & De Hass, 179
Lanham, Dr. John, 284
Lathrop, California, 39, 196–97
  polluted water in, 196, 259–61, 270, 278
Lavelle, Rita, 275, 287
Leary, Dr. John, 45
Lewis, Ralph, 289
Lewis, Richard, 143–44, 146
Life magazine, 125
Lindane, 128, 135
Lindley, James, 220, 222, 223, 226–27, 231, 236, 247
Lipschultz, Dr. Larry, 285
L'Orange, H. Christian, 278, 279, 286, 289
Love Canal, 259, 271–72
Lundeen, Robert, 251, 254–55, 303
Lykken, Louis, 40, 44–45

McBeth, Clyde W., 29, 31, 38, 48, 293
McCarthy, Richard D., hearings held by, 137–42, 153
McCray, Bob, 97, 112–13, 142, 148, 184, 187, 188, 194, 296
McCray, Paul, 97, 113, 136, 145, 296
McCray, Rosalie, 97, 113, 136, 187

McKusick, Bob, 96–111, 116–21, 141, 179, 185, 187, 194
  described, 100, 297–98
  illnesses of, 136–37, 139, 297
  9-Point Proposal of, 100, 104
  threats by, 100, 148
  USDA task force and, 146–47
McKusick, Charmion, 97–99, 101, 116, 135–36, 187
  illnesses of, 137, 297
McKusick, Kathleen, 136, 194
McKusick, Randy, 136
McKusick, Stephanie, 136
McLean, Louis A., 67
McLeod, Dr. John, 219
McNamara, Robert S., 74, 90
Magnesium, 20, 127
Male contraceptives, DBCP-induced infertility and, 230–31
Manchester, William, 33
Mankiewicz, Frank, 119
Manufacturing Chemists Association, 50
Marijuana, 192, 284
Markert, Clement L., 118–19
Martin, Dr. Paul, 139–40
Matthews, Robert, 192
Maui Land & Pineapple, 262
Mayo Clinic, 53
Medic, 50
Mendes, John, 196
Merck, George W., 19
Merszei, Zoltan, 236, 272
Merv Griffin Show, 280
Meselson, Dr. Matthew, 116–17, 150, 155
Miami Inspiration Hospital, 109, 115
Michigan Department of Public Health, 79, 83, 160
Michigan Manufacturer, 5
Michigan State University, 250
Midland Chemical Company, 4
Midland, Michigan, 3–11
  air pollution controls in, 63
  described, 9, 11, 63
  Dow Chemical and, 3–5, 7, 9–11, 63, 186
  Green Rush and, 3
Midwest Research Institute, 91
Miller, Duane, 277, 283
Miller, Rocky, 110
Mitchell, John, 18, 22, 23
Mize, Stanley, 224–25
Moehn, William, 96, 100, 101, 105, 115
Monosson, Dr. Ira, 273
Monsanto, 6, 44, 185

Monsanto (cont'd)
  dioxin and, 83–84, 86, 155, 299
  synthetic rubber and, 20
  2,4,5-T accident at, 55–56, 60, 75, 93
Montrose Chemical Corporation, 67
Moondust and Other Poems (Shoecraft), 166
Moore, John, 294
Mountain States Legal Foundation, 256, 273
Mount Sinai Hospital, 185
Mrak, Emil, 117
Mrak Commission, 117, 119, 120, 138, 158, 159
Müller, Paul, 21, 52
Mullison, Dr. Wendell, 13, 18, 19, 23–26
Mustard gas, 7, 127
Mutations, 241–44, 247, 256–57
  EPA hearings and, 265–66

Nader, Ralph, 116, 123, 153, 250
Napalm, 117, 127, 278
  furor over manufacture of, 131–33, 301
National Academy of Sciences, 18, 91
National Agricultural Chemicals Association, 65, 66, 126
National Cancer Institute, 53
  stomach gavage testing method, 205, 213, 236–37, 240–41
National Enquirer, 184
National Environmental Policy Act (NEPA), 128
National Health Federation, 138
National Institute for Occupational Safety and Health (NIOSH), 203, 225
National Institute of Environmental Health Sciences, 151, 158, 294
National Naval Medical Center, 35
National Peach Council, 231, 261
National Research Council, 18
Needham, John Turberville, 27–28
Nelson, Gaylord, 129
Nemagon, 38, 41
  cancer and, 213–14
  competition with Fumazone, 48
  labeling of, 45–48
  registration of, 47–48
  workers manufacturing, 47, 48
Nematodes, 27–31
  advertisements concerning, 29–30
  damage done by, 27–28, 202
  DBCP and, *see* DBCP
  D-D and, 29–30, 31
  described, 27
  early attempts at control of, 28–29
  EDB and, 30–31
  life cycle of, 28
  root-knot, 28
  types of, 28
Newark Beth Israel Hospital, 77–78
New Republic, The, 91, 129
Newsweek, 287
New Yorker, The, 66, 152
New York Times, The, 66–67, 68, 125, 129, 272, 303
New York State Agricultural Experiment Station, 22–23
Nightline, 286
1969 as "Year of Ecology," 123–25
Nixon, Richard, 72, 117, 162
  environment and, 123–25, 128–29, 134, 175
  Armco and, 204
Nobel Prize, 52
Norfolk & Western Railway, 299
North Central State Weed Control Conference, 24–25
Northwestern Bell Telephone Company, 17
Nunn, Merlin, 299

Occidental Chemical Company, 6, 196–244
  DBCP and, *see* DBCP, produced by Occidental
  dumping of wastes by, 259–61
  purchase of Best and Hooker, 48, 197
Occupational Safety and Health Agency (OSHA):
  purpose of, 203
  ranking carcinogens, 252–53
  safety standard for DBCP, 234
  weakening of, 204, 205, 274
Occupational Tumors and Allied Diseases (Hueper), 52–53
Oil, Chemical and Atomic Workers International Union, 204, 214, 218, 220, 221, 227–28
O'Keefe, Dr. Patrick, 187
Oligospermia, 219, 221, 225, 264
  recovery from, 246, 247
  see also Azoospermia, DBCP-induced
Oppenheimer, Robert, 33
Oreffice, Paul, 249–51, 252, 255, 284, 300–01
  described, 250–51
  Reagan administration and, 274–75
Organic chemistry, 6–8
Organophosphates, 199–201
  effects of exposure to, 200–01
  fertility and, 215–16

Elements of Risk / 333

Organophosphates (*cont'd*)
 mechanism of action, 200
 poisoning by, 212, 214
 safety rules for, 199–201
Other Side of Love Canal, The, 272

Packard, David, 159
Parathion, 199–200
 effects of, 200
Pentachlorophenol, 15
 reaction to, 55, 57
Perez, Gloria, 215, 284, 288
Perez, Richard, 214, 215, 216, 284
 damage award received by, 288
Perkin Medal, 8
Perry, Haskell, 199, 201, 206, 210–11, 267
 damage award received by, 289, 290
Pesticides:
 carcinogenicity, mutagenicity, teratogenicity testing of, 116–20, 138, 150–58
 dumping of wastes, 259–61
 early warnings about dangers of, 53–54
 Knight and, 141
 regulation of, 69, 127–30, 138–39, 158–63, 205, 253–54
 Silent Spring and, 66–73
 technological risk versus human safety and, 172–73
 see also specific types of pesticides
Petrochemicals, 8
Phenol, 7, 13, 15
Phenoxyacetic acids, 18–19, 22–26
Philip, Prince, 71
Phillips, Robert K., 231
Phoenix Gazette, 111, 137–38
Picciano, Dr. Dante, 265, 266
Picric acid, 7, 20, 127
Pierovich, John, 139
Pineapple Research Institute, 29, 31
Plant parasite nematodes, *see* Nematodes
Politics of Cancer, The (Epstein), 252
Polychlorinated dibenzo-p-dioxins, 57–58
Porphyria cutanea tarda, 78
Pott, Percivall, 53
Press, Frank, 301–02

Quarles, John, 125
Queeny, John F., 6

Rand Corporation, 90

Rappaport, Dr. Stephen, 216–17, 222
Reagan, Ronald, 251, 256, 280
 antiregulatory philosophy of, 273–76
 EPA controversy and, 287
Reynolds, Robert, 171, 180–81
Ribicoff, Abraham, 68–71
Rigterink, Raymond "Ray," 12
Risks, who decides about?, 254
Roos, Reverend Robert E., 191–92
Roosevelt, Franklin D., 12
Root-knot nematodes, 28
Rose, Dr. Lawrence, 273
Rowe, Dr. Verald K., 40–41, 229
 dioxin and, 82–86
Rubber, synthetic, 20
Ruckelshaus, William, 173, 203, 205, 287–88
Rudolf II, Holy Roman Emperor, 6

Safety, 51
 chemical industry on, 51, 54
 early questions about, 52–54
 in retrospect, 297–304
 Miller Amendment and, 65
 "no-effect level" and, 37
 toxicological studies and, *see* Toxicological studies
Sagebrush Rebels, 256, 273
St. Clair, Jack, 235–36
Salem Memorial Hospital, 183–84
Salt River Valley Water Users Association, 102, 105
San Francisco Radiation Defense Laboratory, 35
San Joaquin Valley, California, 196–98, 207, 278
Santa Barbara oil spill, 123–24
Saran Wrap, 10, 132
 promotion of, 50
Sarras, Steve and Sandra, 257
 damage award received by, 288
Satamas family, 97, 187, 194
Schoellkopf Aniline & Chemical Works, 6
Schroeter, Rudolf H., 179, 182–83, 189, 193
Schulz, Dr. Karl, 57–59, 61–62, 76
Science, 22–23, 119, 303
Science Advisory Committee:
 of President Johnson, 90
 of President Kennedy, 68, 69, 70
Scientists, 51–52, 292
 herbicides and, 149–52
 on safety of chemicals, 54, 301–02
Sea Around Us, The (Carson), 63

Selikoff, Dr. Irving, 185
Sevareid, Eric, 69–70
Shell Chemical Company, 6, 204
   DBCP and, see DBCP, produced by Dow and Shell
   nematocides and, 27, 29–48
   on safety of chemicals, 54
   St. Clair and, 235–36
Shoecraft, Billee, 95–96, 99–122, 134–84, 193–95
   background of, 164–65
   Congressional hearings on 2,4,5-T and, 152–59, 162–63
   death of, 183–84, 187
   described, 99, 104, 112, 115, 166–67, 184
   doctors and, 141, 147–48, 175–80
   on femininity, 167
   illnesses of, 109–10, 115, 135, 139, 143, 147–48, 170, 175–84
      cancer, 179–84, 300
   lawsuit of, 121, 162, 167–95
      book about, 167–68, 171
      depositions during, 171–72, 187
      Dr. Hine and, 177–79
      lawyers and, 177, 179, 180, 186–87
      settlement of, 193–95
   at McCarthy hearing, 139
   marriage of, 165–66
   mock funeral, 110–11
   9-Point Proposal of, 100, 104
   notoriety of, 141, 170
   scientists and, 149–50
   speaking engagements of, 169–70, 176
   tactics of, 112–15, 120–21
Shoecraft, Mari, 165, 166–67, 177, 180, 181, 182, 187, 298
Shoecraft, Robert, 165, 166, 175–76
Shoecraft, Willard, 95, 104–05, 115, 120, 162, 168, 170
   described, 164–66, 298
   Billee's cancer and, 179–84
Silent Spring (Carson), 66–73, 76, 90
Silicosis, 60
Silvex, 25–26, 298, 304
   label of, 113, 114, 189
      Shoecraft on, 171
   manufacture of, 55, 58, 82, 174
   reaction to, 60
   U.S. Forest Service use of, see Globe, Arizona, Forest Service spraying of
Simmons, J. S., 21
Since Silent Spring (Graham), 66
Singer, Dr. Paul, 179–80

60 Minutes, 270–71
Skinner, Dr. F. I., 139
Slobodin, Jack, 188–89, 193–94
Society for the Prevention of Environmental Collapse, 138
Society of American Foresters, 161
Soft-tissue sarcoma, 295–96
Sorge, Georg, 57, 59
Soto, Farnham, 200, 267, 269
Southern Medical Journal, 15
Spear, Dr. Robert, 216–17, 222
Stauffer Chemical Company, 67
Steamship Inspection Service, 126
Steiger, Sam, 104
Steinfeld, Dr. Jesse, 158–59
Steinke, Ruth, 146
Sterility, see Azoospermia, DBCP-induced
Stimson, Henry L., 18
Stingily, Dr. Karl O., 15–16
Stockton *Record*, 222
Sue the Bastards (Shoecraft), 167–68, 177
Sullivan, Jerry, 186–88, 193–94
Sullivan, Walter, 68
Summit Labs, 218–20
Superfund, 274
Sutherland, W. W., 65
Synapse, 293

TEPP, 200
Tetrachlorodibenzo-p-dioxin, 57–58, 80–81, 93
Thalidomide, 72–73, 90, 140, 152
Thief in the Soil, 39
This Week, 50
Thompson Hayward Chemical Company, 81–82, 87, 93
Time magazine, 68, 123, 148–49
Times Beach, Missouri, 286–87
Today show, 50
Ton That Tung, 170
Torkelson, Dr. Theodore, 12–14
   DBCP and, 32, 36–38, 40–43, 49, 217, 229
   in retrospect, 291–92
   lawsuit over, 290
Toxicological studies, 40
   DBCP, 32–49
      concern over, 37–38, 40–41
      published report, 41–43
      workers manufacturing, 47
   Dow labs, 12–14, 25, 32, 33, 36–49
   early principle of, 53–54
   LD50, 26, 43
   in the 1950s, 43–44

Toxicological studies (cont'd)
"no-effect level," 37
post-*Silent Spring,* 73
Shell, 34–48
University of California Medical School, 34, 35, 47
Toxicology and Applied Pharmacology, 40, 41
Toxic Substances Control Act (TOSCA), 253–54
ineffectiveness of, 275, 276
Train, Russell, 175, 205
Trichlorophenol, *see* 2,4,5-trichlorophenol
Trout, Betty and Robert, 202, 207–10, 246, 267–71
Trout, Marta, 207–10, 221, 272–73
lawsuit against Oxy and, 245–46, 273, 279–81
Mike's illness and, 257–58, 266–70
pregnancy of, 246–47, 256
Trout, Matthew, 256, 268, 272, 279
Trout, Michael Lee, Jr., 207, 208, 210, 269, 279, 281
Trout, Mike, 202, 204–11, 215, 245–47
brain tumor of, 208–11, 213, 257–58, 266–70
death of, 269–71, 272–73
described, 206–07, 221, 245, 267
infertility of, 218, 221, 222, 244–47, 256
lawsuit of, 245–46, 267, 273, 277, 279–81
Tschirley, Dr. Fred, 91, 142
Tucker, William, 249
Tukey, H. B., 22
20/20, 185
2,3,7,8-tetrachlorodibenzo-p-dioxin, 58, 81, 85, 92, 303
2,4-dichlorophenoxyacetic acid (2,4-D):
agricultural use of, 22–26
combined with 2,4,5-T, 25, 74–78, 103–04, 114
commercial sale of, 24
laborsaving value of, 52
military use of, 74–78, 118
spraying of, in Arizona, 103–04, 113, 135, 139
synthesis of, 17–18
testing of, 18, 19, 22–25, 116
2,4,5-trichlorophenol:
dioxin and, 58–62, 80–94, 174, 181
manufacturing control of, 59–61, 79, 82–94
workers' reaction to, 55–62, 76, 80, 160
2,4,5-trichlorophenoxyacetic acid (2,4,5-T), 32–33, 54, 79–93
agricultural use of, 22, 25

combined with 2,4-D, 25, 74–78, 103–04, 114
Congressional hearings on, 152–59
curtailment in use of, 118–20, 124, 135, 141
dioxin and, 120, 130, 150–63, 172–75, 185, 189
effects of exposure to, 90, 116, 119, 138, 140, 150–52
in retrospect, 298–300, 304
manufacture of, 55–56, 58–60, 79, 82–87
military use of, 19, 74–79, 114, 118, 119
spraying of, in Arizona, 103–05, 116–20, 134–63, 168–95
suspension of use of, 158–63, 169–75, 190–93, 195, 248, 304
testing of, 18, 19, 25, 135

Underground Battlefield, The, 39
Union Carbide, 11, 185
EPA and, 203, 204
United Nations, 117
U.S. Army Chemical Corps, 90, 126
U.S. Department of Agriculture (USDA), 43
Arizona herbicide spraying and, 101, 103, 112, 115–16, 170
Congressional hearings on, 154–58
suspension of 2,4,5-T and, 158, 161, 162, 163
task force investigating, 142–48
DBCP and, 45–48, 198, 261
DDT and, 65, 66, 69
dioxin and, 82–83, 87
nematodes and, 28, 29
regulation of pesticide and, 126, 127–28, 153
U.S. Department of Commerce, 204
U.S. Department of Defense, 74, 75, 77, 83, 90, 91, 118, 159
U.S. Department of Justice, 204, 275
U.S. Department of State, 90
U.S. Forest Service, *see* Globe, Arizona, Forest Service spraying of
U.S. Golf Association, 23
U.S. Joint Chiefs of Staff, 19, 90
U.S. Public Health Service, 56, 128
dioxin and, 77, 79–84
University of Arizona, 101, 104
University of California:
lawsuit against, 244, 277, 283
Medical School, 34, 35, 47, 292
University of Chicago, 18
University of Hamburg, 57, 59, 61
University of Utah Medical Center, 215–16

University of Wisconsin, 185, 187–88
Urea, 6–7

Van Reeven, J. P., 235
Velsicol Chemical Corporation, 67, 264
Verity, C. William, Jr., 204
Verrett, Dr. Jacqueline, 150–51, 157–58, 187
Vervais, Gregory, 220, 227
Vial, Don, 237–39, 243
Vietnam War, 74–79, 87, 90–91, 93–94, 127, 278
 anti-war movement, 123
 defoliation during, 74–79, 87, 90, 104, 114, 116–20, 142, 153, 162, 191
 extent of, 77, 78, 83
 halting of, 118, 137, 170
 see also Agent Orange
 napalm and, 131–33, 192
Vinyl chloride, 273, 295

Wallace, Mike, 270
Wall Street Journal, The, 132
War Research Service, 18, 19
Warskow, William, 102–03
Watt, James, 255–56, 273, 275
Weedone, 24
Weeds, herbicides and, *see* Herbicides
Weidenbaum, Murray, 249, 255
Weiner, Peter Hart, 238–39
Weinstein, Daniel, 282, 285, 286
White, Etta and William, 164
White, Theodore H., 125
Whiteside, Thomas, 152
White-Stevens, Robert H., 68, 69
Whorton, Dr. Donald, 219–21, 225–28, 244, 256, 258
 DBCP lawsuit and, 284–85
 EPA hearings and, 263–64

Trout's pregnancy and, 245–47
Wilkenfeld, Jerry, 258, 271
Williams, Isaac, 288, 289
Wireworm, 30
Wöhler, Friedrich, 6–7
Workers in the chemical industry, 35
 chloracne among, 75–94, 185–86
 in Czechoslovakia, 173, 296
 debate over, 294–97
 in England, 91–93, 174–75, 296
 in France, 89–90
 in Italy, 181, 186, 296–97
 in Michigan, 16, 55, 79–89, 160, 189
 in Mississippi, 15–16, 54–55
 in the Netherlands, 76–77, 296
 in New Jersey, 60, 77–78
 in West Germany, 54–62, 92–93, 296
 in West Virginia, 54, 55–56, 93, 295
 DBCP and, 35, 41, 44, 47–49
 "need to be realistic" and, 38
 "no-effect level" and, 37
 single exposure effects on, 13–14
 toxicological studies of, 34, 47
World Health Organization, 157, 230, 231
World War I, 7
World War II, 18–22, 52–53, 74
Wurm, Ross, 105–08

Yannacone, Victor, 121, 190
Your Show of Shows, 50

Zavon, Dr. Mitchell:
 DBCP and, 40, 45, 46, 220, 247
 cancer and, 236–37
 CDIR hearings on, 240
 infertility and, 220, 223–26, 236, 237
 as a male contraceptive, 231
 DDT and, 70

Elements of Risk / 337